Mechanical Engineering in Biomedical Applications

Scrivener Publishing
100 Cummings Center, Suite 541J
Beverly, MA 01915-6106

Publishers at Scrivener
Martin Scrivener (martin@scrivenerpublishing.com)
Phillip Carmical (pcarmical@scrivenerpublishing.com)

Mechanical Engineering in Biomedical Applications

Bio-3D Printing, Biofluid Mechanics, Implant Design, Biomaterials, Computational Biomechanics, Tissue Mechanics

Edited by

Jay Prakash Srivastava, Dražan Kozak, Vinayak Ranjan, Pankaj Kumar, Ranjan Kumar
and
Shubham Tayal

Scrivener Publishing

WILEY

This edition first published 2024 by John Wiley & Sons, Inc., 111 River Street, Hoboken, NJ 07030, USA and Scrivener Publishing LLC, 100 Cummings Center, Suite 541J, Beverly, MA 01915, USA

For more information about Scrivener publications please visit www.scrivenerpublishing.com.

Wiley Global Headquarters
111 River Street, Hoboken, NJ 07030, USA

For details of our global editorial offices, customer services, and more information about Wiley products visit us at www.wiley.com.

Library of Congress Cataloging-in-Publication Data

ISBN 978-1-394-17452-2

Cover image: Pixabay.Com
Cover design by Russell Richardson

Set in size of 11pt and Minion Pro by Manila Typesetting Company, Makati, Philippines

Printed in the USA

10 9 8 7 6 5 4 3 2 1

Contents

Preface

This book explores the latest research and developments related to the interdisciplinary field of biomedical and mechanical engineering. The book collects contributions from leading international scholars and practitioners, who offer their insights and perspectives on the research, key technologies, and mechanical engineering techniques used in biomedical applications.

The book is divided into several parts that cover different aspects of mechanical engineering in biomedical research. The first part focuses on the role of additive manufacturing technologies, rehabilitation in healthcare applications, and artificial recreation of human organs. The part also covers the advances, risks, and challenges of bio 3D printing. The second part presents insight into biomaterials, including their properties, applications, and fabrication techniques. The part also covers the use of powder metallurgy methodology and techniques of biopolymer and bio-ceramic coatings on prosthetic implants. The third part covers biofluid mechanics, including the mechanics of fluid flow within our body, the mechanical aspects of human synovial fluids, and the design of medical devices for fluid flow applications. The part also covers the use of computational modelling to study the blockage of carotid arteries. The final part elaborates on soft robotic manipulation for use in medical sciences.

We would like to thank all the contributors to this book for their invaluable insights and expertise. Their contributions have made this book a truly interdisciplinary and international exploration of mechanical engineering in biomedical applications. We would also like to acknowledge the support of the publisher and the editorial team who have worked tirelessly to bring this book to fruition. We hope that this will be a useful resource for mechanical engineering students and researchers who are interested in biomedical applications, as well as medical professionals who want to understand the principles behind the design and development of medical devices and related technologies.

Finally, we offer our gratitude to our families and loved ones for their support and encouragement throughout the editorial process.

The Editors
November 2023

Acknowledgments

We would like to express our gratitude to all the contributors to this book, who have generously shared their expertise and insights on the topic of "Mechanical Engineering in Biomedical Applications: Bio-3D Printing, Biofluid Mechanics, Implant Design, Biomaterials, Computational Biomechanics, Tissue Mechanics". Their contributions have made this book a comprehensive and insightful exploration of the challenges and opportunities in the domain.

We would also like to thank the publisher and the editorial team for their support and guidance throughout the editing process. Their dedication and hard work have been instrumental in bringing this book to fruition.

We are grateful to our families and loved ones for their support and encouragement throughout the editing process. Their patience and understanding have been invaluable.

The Editors

Part I
ADDITIVE MANUFACTURING

1

The Role of Additive Manufacturing Technologies for Rehabilitation in Healthcare and Medical Applications

Vidyapati Kumar[1], Ankita Mistri[2*] and Abhishek Mohata[3]

[1]*Department of Mechanical Engineering, Indian Institute of Technology, Kharagpur, West Bengal, India*
[2]*Department of Mechanical Engineering, Indian Institute of Technology, Dhanbad, Jharkhand, India*
[3]*Department of Mechanical Engineering, Jalpaiguri Government Engineering College, West Bengal, India*

Abstract
Additive manufacturing (AM) is a rapidly evolving technology that is being utilized to produce medical components across a wide range of sectors. Since each patient is unique in terms of health and dental care, AM has enormous potential for personalized and customized therapy. In addition to its bespoke design and reduced manufacturing time and cost, it also offers a broad and extended opportunity for the biological mimicry of desired complicated states of physiological devices. Scaffolds with a fitted outer shape and a porous internal structure may be created via additive manufacturing, which is critical for repairing vast segmental bone lesions. The scaffold-building process includes scaffold design, additive manufacturing, and post-treatments. This research attempts to provide a more systematic assessment of the use of various AM processes and their need in the current era with their classifications in the healthcare and medical field utilizing different biomaterials, as well as the scope of future research for advancing the medical field using various AM processes.

***Keywords*:** Biomaterials, additive manufacturing, implants, scaffolds, prosthetics

**Corresponding author*: ankitamistri93@gmail.com

Jay Prakash Srivastava, Dražan Kozak, Vinayak Ranjan, Pankaj Kumar, Ranjan Kumar and Shubham Tayal (eds.) Mechanical Engineering in Biomedical Applications: Bio-3D Printing, Biofluid Mechanics, Implant Design, Biomaterials, Computational Biomechanics, Tissue Mechanics, (3–22) © 2024 Scrivener Publishing LLC

1.1 Introduction

In this contemporary period of the Fourth Industrial Revolution, the marketplace must constantly improve itself via continual innovation and customization, in which the innovation and deployment of unconventional manufacturing processes play a vital role. Additive manufacturing (AM) has already been shown to play a significant part in the unconventional manufacturing process. Additive manufacturing, also known as 3D printing, rapid prototyping, or free-form fabrication, is an operation in which materials are joined by layer-by-layer deposition to produce a three-dimensional product from a computer-aided design (CAD) model thereby eliminating the need for various machining processes. AM technologies, which use metal powders to create intricate and sophisticated 3D geometry, are a burgeoning industry. AM technologies have been highlighted as a critical manufacturing priority and are considered one of the essential components of Industry 4.0, having the potential to amend the global manufacturing industry. Currently, it has been noted that the demand for AM surged fast after the start of COVID-19 in many industries where the economic downturn suffered as a consequence of the efforts implemented to limit the pandemic. Additive manufacturing is already enabling a design and industrial revolution in various industries, including aviation, power, automobile, healthcare, tooling, and consumer products.

The typical manufacturing procedure for printing tissues using the AM technique has been depicted in Figure 1.1. The first phase in this process is data collection, in which the data about the organ or portion of the body

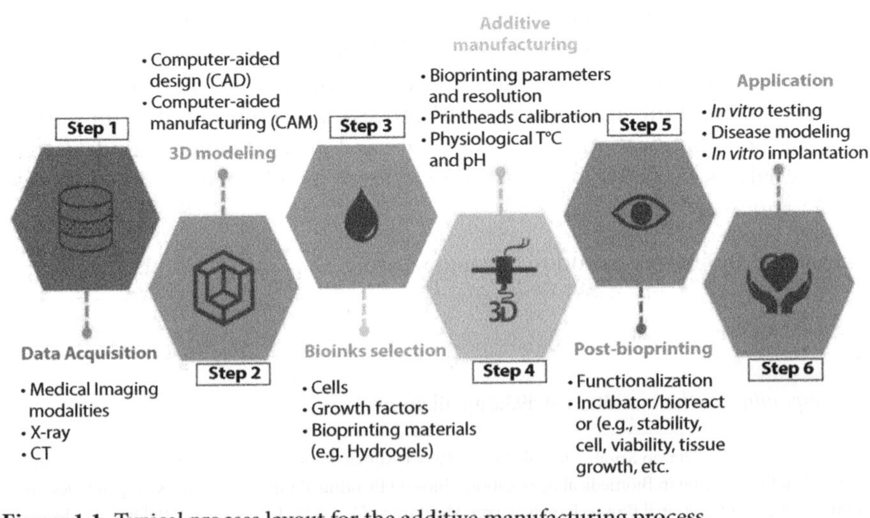

Figure 1.1 Typical process layout for the additive manufacturing process.

that needs to be printed are obtained using medical imaging methods such as X-ray and computed tomography (CT) scan. Once the data are obtained, a 3D model of the component is created using computer-assisted design (CAD) and computer-aided manufacturing (CAM) processes. The bioinks used for 3D bioprinting are then chosen, and the bioprinting settings and resolutions are calibrated. Once the material has been 3D bioprinted, its functionality is tested in an incubator or bioreactor to confirm stability, cell viability, tissue development, and so on. Following these procedures, the printed component is now ready for *in vitro* testing, disease modeling, and *in vivo* implantation in the human body, saving a significant amount of life in healthcare systems.

From this perspective, it is important to grasp the importance of additive manufacturing technologies like 3D bioprinting in healthcare and biological sciences. In the healthcare and medical fields, the general practice is to do preclinical research using animal-based models, notably rodents [1]. From a regulatory standpoint, it is critical and obligatory to conduct this experiment using animal-based models in order to bring forth the most current therapeutic breakthroughs from preclinical to clinical trials. These problems have not been highlighted for many decades, but a recent surge in opprobrium based on humanitarian and scientific considerations has shown that around 80% of putative treatments perish in clinical trials while having effectiveness and safety in preclinical investigations [2]. Potential underlying causes include inadequate characterization of relevant animal models [3, 4], a lack of acceptable experimental quality in *in vivo* investigations, and significant interspecies-related differences to humans in areas such as anatomy, (patho)physiology, and immunology. Many disorders that develop in humans are not ubiquitous in nature in animals; thus, an artificial disease induction is necessary to simulate the diseases, and even if the animals display identical diseases, the underpinning pathophysiology condition is entirely unknown. The mouse-based animal model for SARS coronavirus infection has recently shown severe encephalitis, which is not seen in human pathogenesis. Hence, as seen in Figure 1.2 [5], the translational utility of preclinical animal models has been called into doubt owing to a lack of concordance and reproducibility. Figure 1.2 displays the high expense of drug development as well as their relatively high failure rates in the clinical phase, raising concerns about the current applicability of research methodologies. To address the aforementioned limitations, additive manufacturing technologies play a critical role in which a 3D organ or human-based models may be generated in order to undertake fundamental and preclinical research. As a result, this AM technology has attracted

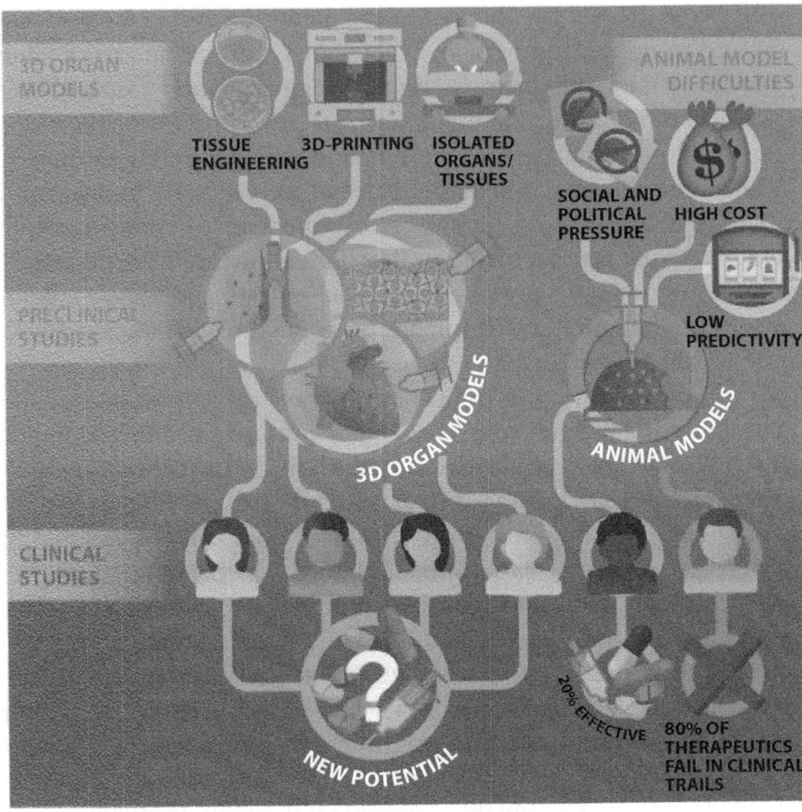

Figure 1.2 Applicability of AM technologies for the production of 3D organ models [5].

conscientious scrutiny due to its usefulness in producing a human-based model as a replacement for animal models for scientific purposes.

1.2 Classification of the Additive Manufacturing Process

Several 3D printing technologies have been created to bioengineer three-dimensional tissue or organ frameworks for biomedical purposes, as demonstrated in Figure 1.3. The most extensively utilized forms of 3D bioprinting processes are extrusion-based bioprinting, laser-assisted bioprinting, and laser-based stereolithography. The effectiveness of each printing technique is heavily reliant on biomaterial choices and functions.

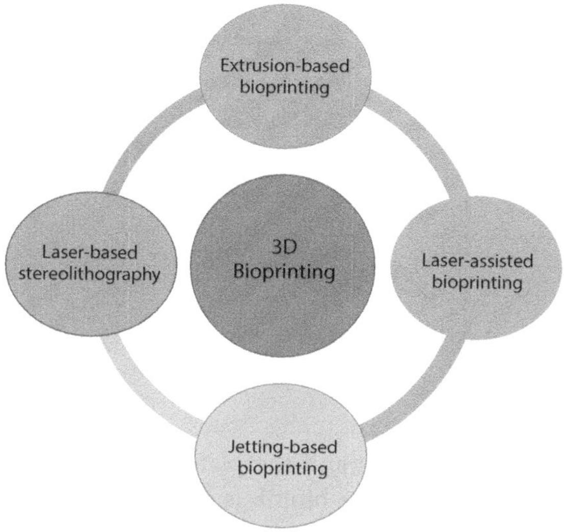

Figure 1.3 Types of 3D bioprinting.

1.2.1 Jetting-Based Bioprinting

Jetting-based bioprinting is the earliest printing method, in which bioinks are used to print. These bioinks may be either natural or manufactured substances that help in cell adherence, propagation, and replication. Bioink is pushed with force via a nozzle in this approach, resulting in a spray of droplets. These printers may have a single or several print heads. A chamber and a nozzle are both included in each print head. The surface tension of the fluid keeps the bioinks near the nozzle opening. In three ways, pressure pulses are injected into the print head chamber. As seen in Figure 1.4, it is provided via piezoelectric inkjet, thermal inkjet, or electrostatic bioprinting. The actuator in the piezoelectric inkjet generates pressure pulses to deposit the bioinks; however, certain print heads need back pressure to complement the pressure pulses to make droplets of bioinks. When a voltage pulse is supplied to a thermal inkjet printer's thermal actuator, it locally warms the bioink solution. Figure 1.4 [6] shows that local heating produces a vapor bubble. This bubble rapidly expands and shrinks, generating a force burst inside the fluid compartment and driving the bioink droplet to defy interfacial tension and accumulate on the scaffold. Thermal inkjet printers may discharge biological materials such as proteins and mammalian cells, among other things. Bioink droplets are created in electrostatic bioprinters by increasing the capacity of the fluid compartment with the aid of a bioink fluid attached to the plate. After that, the pressure plate

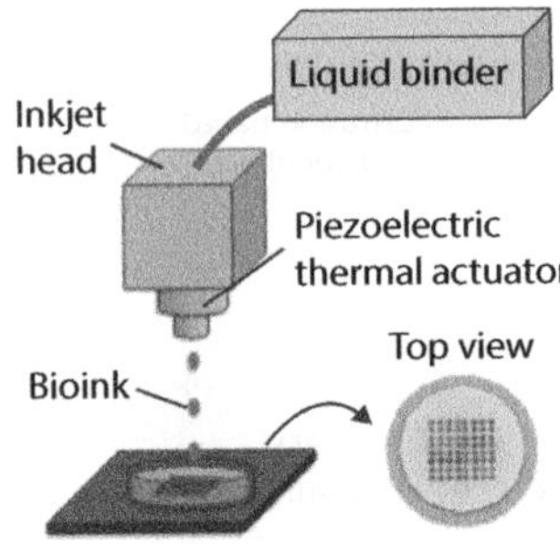

Figure 1.4 Jetting-based 3D bioprinting [6].

deflects between the electrode and the plate once the voltage is applied. Finally, as the voltage drops, the bioink is evacuated as the pressure plate re-establishes its position, and printing happens.

1.2.2 Extrusion-Based Bioprinting

Extrusion-based bioprinting is based on the notion of applying extrusion pressure to the bioink, which is very beneficial for tissue regeneration and repair. The bioink contained in this process is largely deposited using pneumatic pressure, a mechanical pressure in the form of a screw or piston, and lastly, the substrate is extruded out, as illustrated in Figure 1.5 [6]. The

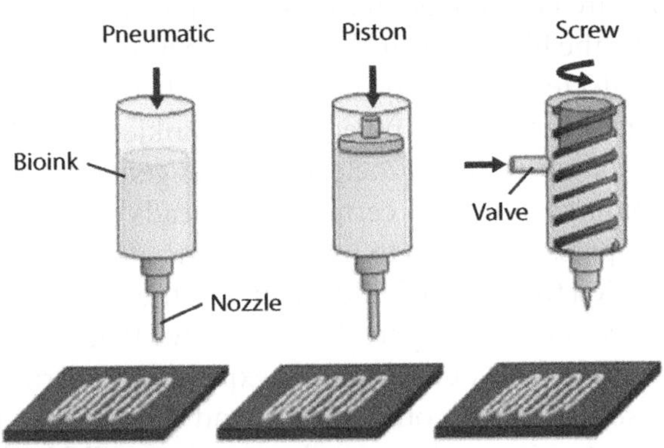

Figure 1.5 Extrusion-based 3D bioprinting [6].

robotic stage controller governs and controls the whole extrusion process of the bioprinter. The head can move in three directions, namely, x, y, and z, and the bioink may be dispensed directly onto the substrate underneath it. These printers are capable of dispensing high cell density bioinks, unique hydrogels with a wide range of viscosities, and biodegradable thermoplastics like polycaprolactone. When compared with inkjet printers, extrusion bioprinting reduces the possibility of bioink clogging. The main disadvantage of extrusion is that we must guarantee that the shear force is not so great that it impairs cell viability.

1.2.3 Laser-Assisted Bioprinting

Laser-assisted bioprinting is another popular method for bioprinting live cells onto a substrate. This printing is made feasible by using a high-intensity light source or light with a long wavelength. A laser pulse, focusing lens, donor slide, energy absorption layer, donor substrate, and collector slide are the main components of a laser bioprinter, as depicted in Figure 1.6. The focusing lens in Figure 1.6 concentrates the high-intensity light, after which the bioink is concentrated on the collection slide and the printed output is formed. Unlike inkjet printers, laser printers have no nozzles and may therefore deposit large densities of bioinks without clogging, as demonstrated in Figure 1.6 [6].

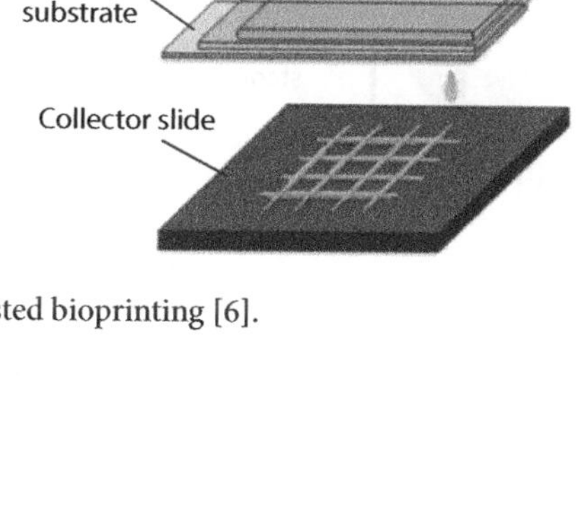

Figure 1.6 Laser-assisted bioprinting [6].

1.2.4 Laser-Based Stereolithography

It is a free-form process for depositing light to cross-linked polymer materials, as demonstrated in Figure 1.7 [6]. Most stereolithography techniques use UV light, directed onto the photocurable resin's surface. When the resin has dried, the platform travels higher and is ready to deposit a new coat of resin. This procedure is repeated till the product is finished. Despite its benefits, stereolithography has an extended processing time.

1.3 AM Materials for Medical Applications

Biomaterials have been employed in regulated pharmaceutical delivery techniques, sutures and adhesives, cardiac bypass, rehabilitative and orthopedic devices, ocular devices including corneas and corrective lenses, and dentistry. Several experiments have been carried out in the field of biomedical engineering, such as neurosurgical bone grinding, in which multicriteria decision-making and optimization tools have been successfully employed to fine-tune the parametric setting, which aids in the 3D bioprinting process [7]. Several components, including titanium, have been utilized to generate therapeutic implants. Bioceramics, polymers, metals, and composites are just a handful of the materials accessible. Table 1.1 outlines the current biological applications for biomaterials.

Different kinds of reinforcing materials may be used to improve the qualities of 3D printed polymer composite products as depicted in Figure 1.8. Nylon was employed as the matrix material, whereas carbon, glass, and Kevlar were used as reinforcing materials to achieve a significant

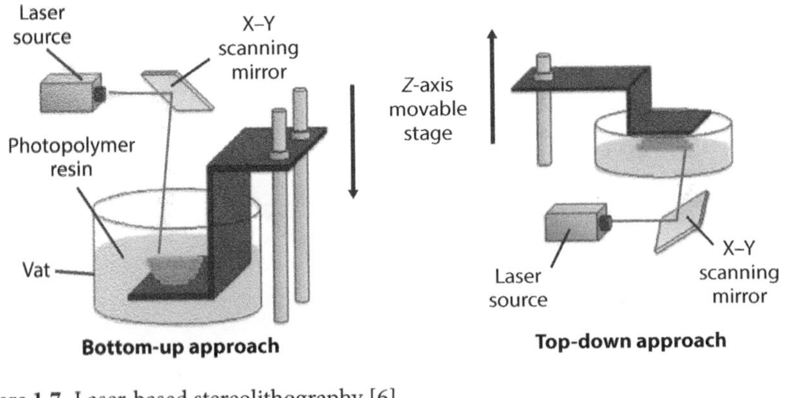

Figure 1.7 Laser-based stereolithography [6].

Table 1.1 3D bioprinting applications of biomaterials.

References	Biomaterial	Applications
[8]	Silicones, hydrogels	Advances in ophthalmology such as intraocular lenses (IOL), contact lenses
[9–11]	Polymeric component, metallic material, ceramic material, PU, PE, PBE, PP, SS	Vascular prosthesis, mechanical gates, stent, cardiovascular pumping systems, blood bags, and catheters are all examples of products that may be used in the field of cardiology.
[12–15]	PCL, silk, collagen	Nerve regeneration scaffolds are used in both the central and peripheral nervous systems.
[12, 16–20]	PLLA, PDLA	Medication transmission, vascular bypass, skeletal fasteners, stabilization pegs, and dermal filler for facial atrophy are all used in the treatment of facial atrophy.
[20]	PLGA	Pharmaceutical delivery
[12, 21]	PCL	Everlasting implantation, medication release, maxillocranial facial implant
[22–24]	PCL-gelatin, PCL-chitosan, PCL-collagen	Rejuvenation of tissues
[25, 26]	PPD	Intrinsic rupture fixation, medical implants in the form of films, foaming agents, and molded scaffold are all possible applications.
[27, 28]	LDPE, HDPE	Rhinoplasty surgeries, comprehensive hip arthroplasty, and osteolysis therapy using polymer-ceramic composites

(*Continued*)

Table 1.1 3D bioprinting applications of biomaterials. (*Continued*)

References	Biomaterial	Applications
[29, 30]	PMMA	Dentistry implantation for rehabilitation and esthetics, orbital medicinal implantation, rhinoplasty, cranioplasty, bone cement in hip joint restoration
[31, 32]	PDMS	Implantable electrical equipment and sensors, clinical implants, esophageal analogs, catheters, shunts, blood pumps, and pacemakers are all included inside this enclosure.
[32, 33]	PA such as nylon and nylon composites	Sutures, denture manufacture, bone regeneration scaffold material, and nanofillers
[32, 34]	CNT and composite material	Metallic sealants for musculoskeletal implantable devices to increase the permeability of the substrate, minimize metallic ionization, and enhance hydroxyapatite production

Note: polyurethane (PU), polyesters (PE), polybutesters (PBE), polypropylene (PP), stainless steel (SS), polycaprolactone (PCL), poly(l-lactic acid) (PLLA), poly(lactic acid), poly(lactic-co-glycolic acid) (PLGA), polycaprolactone (PCL), Poly(l-lactic acid) (PLLA), poly(d-lactic acid) (PDLA), poly-para-dioxanone (PPD), low-density polyethylene (LDPE), high-density polyethylene (HDPE), poly(methyl methacrylate) (PMMA), polydimethylsiloxane (PDMS), polyamides (PA), carbon nanotubes (CNT), and composites.

increase in tensile strength and fracture resistance [35]. To increase the tensile and thermal characteristics of the polymer used in biomanufacturing, acrylonitrile-butadiene-styrene (ABS) as a matrix material and carbon fiber as a reinforcing material were blended [36, 37]. In another work, ABS was combined with graphene to increase the thermal characteristics of polymeric biomaterials [38]. In the instance of Epoxy resin [39], fiberglass was utilized as a reinforcing material to increase tensile strength and fatigue life. When epoxy acrylate [40] is combined with titanium dioxide, it has been discovered

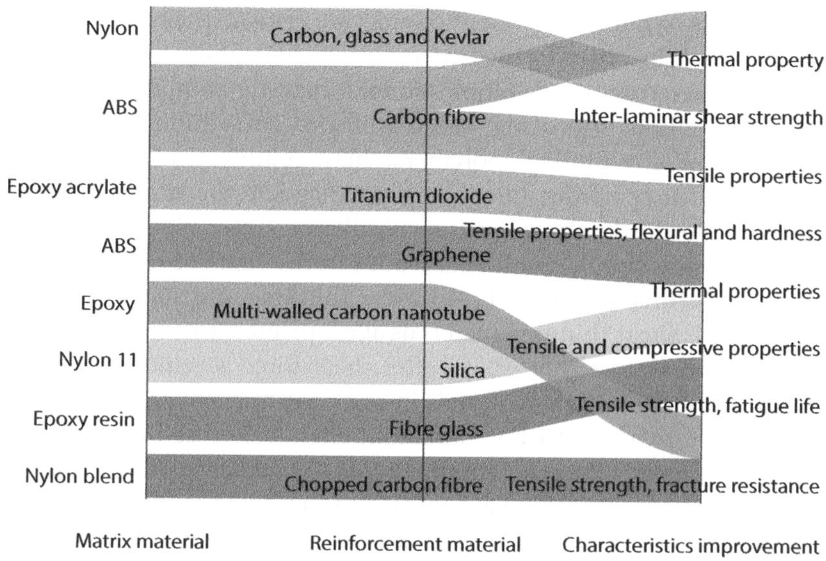

Figure 1.8 Characteristic improvement in the AM material with various matrix and reinforcement materials.

that the polymer improves in terms of tensile and flexural strength and hardness. To achieve an increase in tensile strength and fatigue life, a nylon mix with fiberglass as a reinforcing material is used. Figure 1.8 depicts a list of all matrix materials, reinforcing materials, and their accompanying characteristic enhancements. As a result, it can be shown that polymeric biomaterials, when appropriately combined with matrix and reinforcing material, greatly increase material characteristics, which is highly important during the bioprinting process of various components as needed.

Bioinks are fluid compositions that comprise three or four matrix constituents and are supplied into a bioprinter before being accumulated on the scaffold. Such scaffolds allow cells to connect, survive, bioaccumulate, and replicate after printing [41]. This cell proliferation phenomenon is important in tissue regeneration because it assists in the rehabilitation of dead body tissues. These multicomponent bioinks are utilized to produce various tissue constructions. The reason for employing multicomponent bioinks is because natural polymers such as gelatin and collagen are extensively utilized bioinks that aid in cell adhesion and migration. However, they have limited mechanical capabilities; to compensate for these limitations, various additives and biomaterials are combined to make a multicomponent biomaterial with enhanced qualities. Rheological characteristics of multicomponent bioinks are critical; they must be accurately managed in order to provide optimum printability and structural stability of the structure. Multiple bioinks may be combined in multicomponent bioinks to provide varied stiffness. These bioinks, which are constructed of cross-linked polymers, may solidify immediately after bioprinting. If they harden too soon, they may jam the printer nozzle. If it solidifies too slowly, the structure will collapse. These bioink multimaterial combinations should not be hazardous in the short or long term [42]. Bioinks are further subdivided into shear thinning and shear thickening materials [43]. Shear thinning materials allow printers to avoid using shear force, and they restore their shape after shear force is removed. Gelation occurs in shear thickening materials in response to chemical and physical stimuli. When exposed to UV radiation, chemical cross-linking occurs. The same biomaterial bioink cannot be used to print all tissues and organs because it may not meet the mechanical and functional needs of tissues or organs such as multiplication and dissemination. Some biomaterials, such as polyethylene glycol (PEG), have variable molecular weight and may join printed tissue, but they do not proliferate. Gelatin and fibrin, for example, are examples of natural biomaterials with poor mechanical characteristics. As a result, hybrid bioinks are created by combining more than one bioink, which is responsible for improving component quality as well as providing

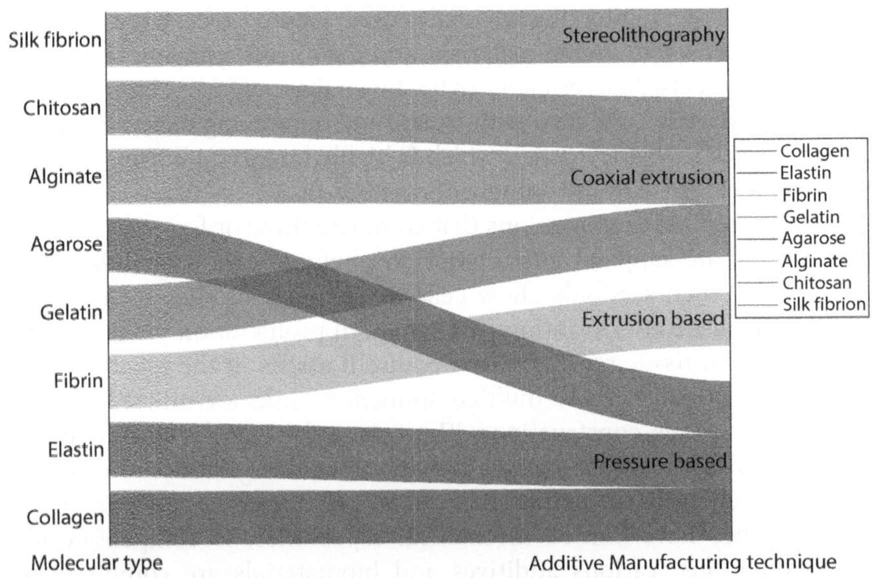

Figure 1.9 The use of several additive manufacturing processes and a range of molecular bioink types.

an environment conducive to cell growth, and these hybrid bioinks have acquired widespread acceptance. Figure 1.9 depicts some of the molecular varieties of bioinks that may be mixed to form a hybrid bioink that can be used in various additive manufacturing processes.

1.4 Biomedical and Healthcare Applications of AM

As illustrated in Figure 1.10, additive manufacturing technologies are widely used in the medical and healthcare fields such as the fabrication of medical

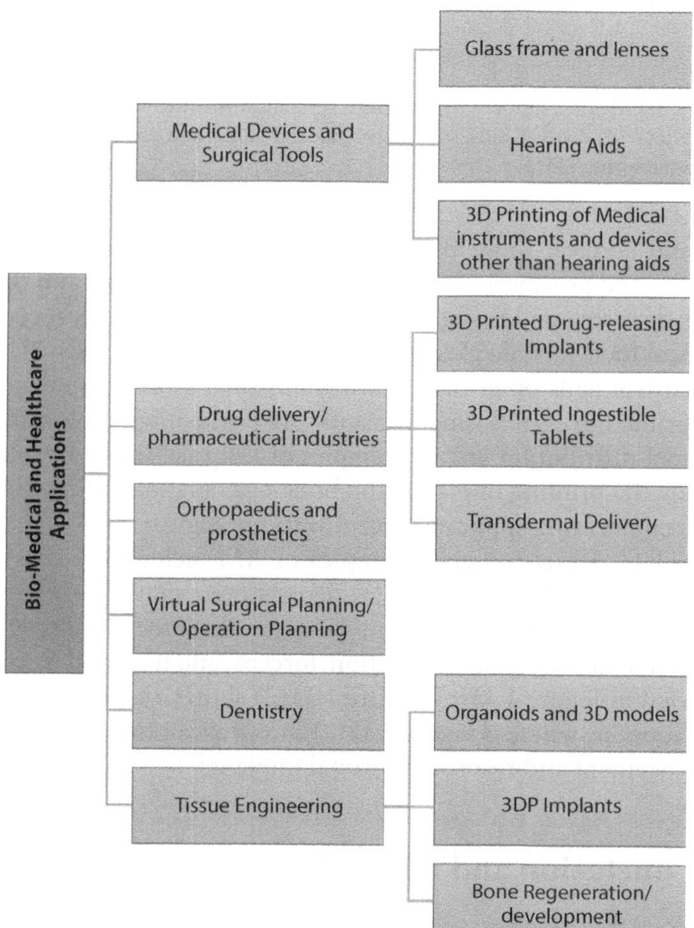

Figure 1.10 Applications of the additive manufacturing in biomedical and healthcare sectors.

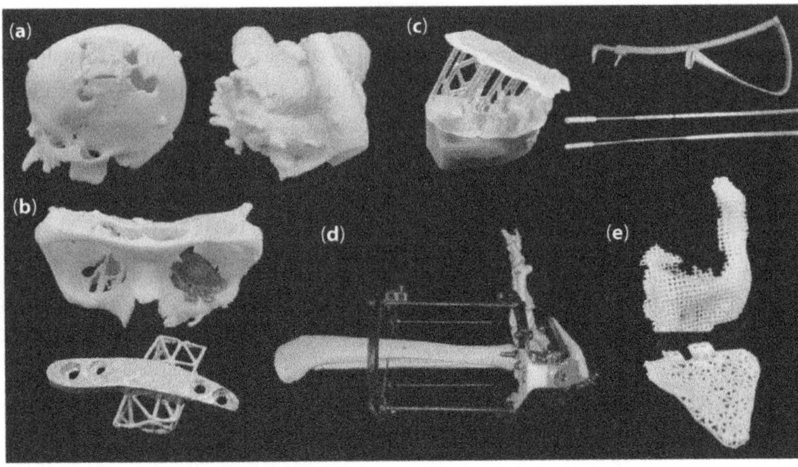

Figure 1.11 (a) Medical models; (b) implants; (c) tools, instruments, and parts for medical devices; (d) medical aids, supportive guides, splints, and prostheses; and (e) biomanufacturing [45].

devices and surgical tools, drug delivery and pharmaceutical industries, orthopedics and prosthesis devices, virtual surgical planning and operation planning, dentistry, and tissue engineering applications. Medical equipment and surgical tools include glass frames and lenses, hearing aids, 3D printed medical instruments other than hearing aids such as stents, and so on [41, 44]. 3D printed drug-releasing implants, 3D printed ingestible tablets, and transdermal distribution are all examples of drug delivery. Organoids and 3D models, 3D printing implants, and bone regeneration are case studies of the tissue engineering applications of AM technology.

Figure 1.11 depicts several examples of AM technology uses in each of the categories [45]. Figure 1.11(a) depicts preoperative skull and heart models. The craniomaxillofacial implants are depicted in Figure 1.11(b), and the dental drilling guide, reduction, forceps, and nose and throat swabs are depicted in Figure 1.11(c). Figure 1.11(d) shows a tailored and mobile external support, whereas Figure 1.11(e) shows a framework for zygomatic bone replacement and resorbable orbital implants.

1.5 Conclusion and Future Outlook

Based on the aforementioned descriptive study, it can be stated that additive manufacturing technologies proved to be extremely successful and efficient in making body tissues, organs, 3D models, and scaffolds using 3D bioprinting procedures. These approaches offer enormous potential for developing implants such as porous scaffolds, cellular structures, and body organs, and by using these methods, diverse tissues and body organs may be repaired, which is highly important in medical areas. Bioinks are employed in this technique to create this complex structural framework

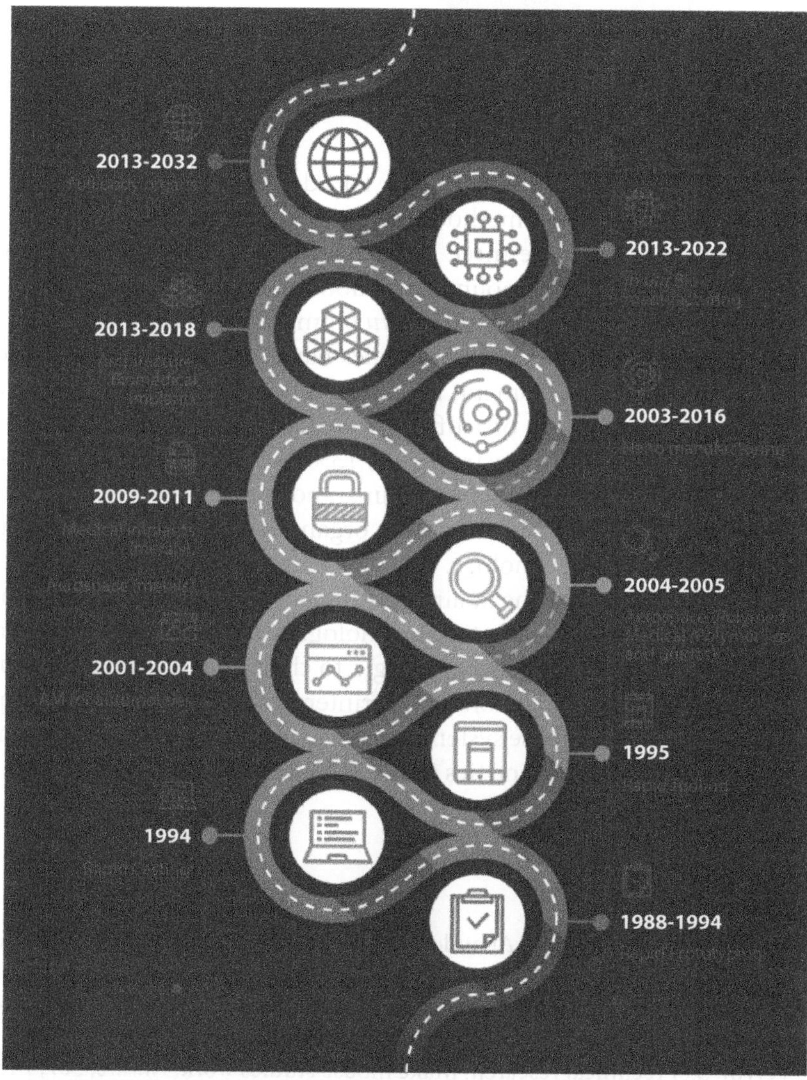

Figure 1.12 Chronology of additive manufacturing scientific and technical advancements, illustrating previous, current, and futuristic possibilities.

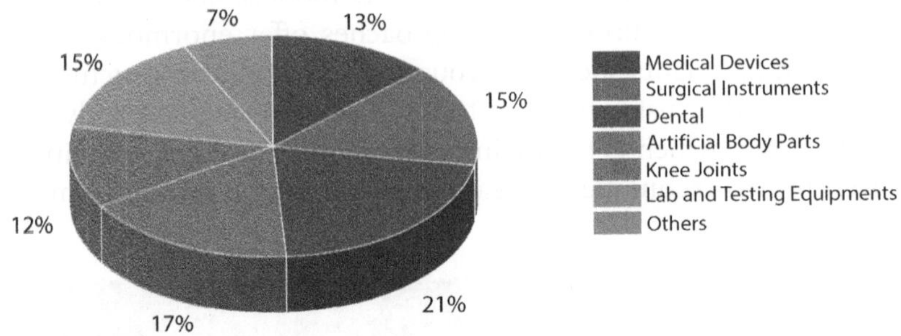

Figure 1.13 Utilization of AM technologies in healthcare fields.

for cell growth and proliferation. Figure 1.12 depicts the evolution of several components using AM methods over a three-decade period. Some of the advances made in these three decades of study include rapid prototyping, rapid casting, rapid tooling, automotives, medical jigs and guides, implants, nanomanufacturing, and *in situ* biomanufacturing.

Figure 1.13 shows that AM technologies have the potential to produce around 21% of dental equipment, including implants and surgical guides for dental applications; 17% of artificial body parts; 15% of lab and testing equipment; 15% of surgical instruments; 13% of medical devices; 12% of knee joints; and the remaining 7% of human body parts [46].

Future potential includes full-body organ development and manufacture utilizing AM technology. Furthermore, various environmental elements such as pressure, heat, air moisture, and radiation may be combined by employing 4D bioprinting technologies, so that the product generated using AM technologies is synergic to the biological human body. Miniaturization and sterilization of bioprinters and accompanying equipment, as well as making these systems user-friendly for end-user clinicians, will be secondary design priorities in the future.

References

1. Osuchowski, M.F. *et al.*, Abandon the mouse research ship? Not just yet! *Shock*, 41, 6, 463–475, 2014.
2. Pound, P. and Bracken, M.B., Is animal research sufficiently evidence based to be a cornerstone of biomedical research? *BMJ*, 348, 1–3, 2014.
3. Perrin, S., Preclinical research: Make mouse studies work. *Nature*, 507, 7493, 423–425, 2014.

4. Couzin-Frankel, J., When mice mislead. *Sci. (New York, N.Y.)*, 342, 6161, 925–923, 2013.
5. Weinhart, M., Hocke, A., Hippenstiel, S., Kurreck, J., Hedtrich, S., 3D organ models—Revolution in pharmacological research? *Pharmacol. Res.*, 139, 446–451, 2019.
6. Huh, J.T., Yoo, J.J., Atala, A., Lee, S.J., Three-dimensional bioprinting for tissue engineering, in: *Principles of Tissue Engineering*, pp. 1391–1415, 2020.
7. Tappa, K. and Jammalamadaka, U., Novel biomaterials used in medical 3D printing techniques. *J. Funct. Biomater.*, 9, 1, 1–16, 2018.
8. Davis, J.R., Overview of biomaterials and their use in medical devices, in: *Handbook of Materials for Medical Devices*, ASM International, pp. 1–12, Novelty, OH, USA, 2003.
9. Zdrahala, R.J. and Zdrahala, I.J., Biomedical applications of polyurethanes: A review of past promises, present realities, and a vibrant future. *J. Biomater. Appl.*, 14, 1, 67–90, 1999.
10. Zhu, Y., Gao, C., He, T., Shen, J., Endothelium regeneration on luminal surface of polyurethane vascular scaffold modified with diamine and covalently grafted with gelatin. *Biomaterials*, 25, 3, 423–430, 2004.
11. Yuan, Y., Ai, F., Zang, X., Zhuang, W., Shen, J., Lin, S., "Polyurethane vascular catheter surface grafted with zwitterionic sulfobetaine monomer activated by ozone. *Colloids Surf. B Biointerfaces*, 35, 1, 1–5, 2004.
12. Stratton, S., Shelke, N.B., Hoshino, K., Rudraiah, S., Kumbar, S.G., Bioactive polymeric scaffolds for tissue engineering. *Bioact. Mater.*, 1, 2, 93–108, 2016.
13. Horner, P.J. and Gage, F.H., Regenerating the damaged central nervous system. *Nature*, 407, 6807, 963–970, 2000.
14. Scheib, J. and Höke, A., Advances in peripheral nerve regeneration. *Nat. Rev. Neurol.*, 9, 12, 668–676, 2013.
15. Cao, J. *et al.*, Induction of rat facial nerve regeneration by functional collagen scaffolds. *Biomaterials*, 34, 4, 1302–1310, Jan. 2013.
16. Liu, Z., Liu, X., Ramakrishna, S., Surface engineering of biomaterials in orthopedic and dental implants: Strategies to improve osteointegration, bacteriostatic and bactericidal activities. *Biotechnol. J.*, 16, 7, 1–23, 2021.
17. Karst, D. and Yang, Y., Molecular modeling study of the resistance of PLA to hydrolysis based on the blending of PLLA and PDLA. *Polym. (Guildf)*, 47, 13, 4845–4850, 2006.
18. Kadambi, P., Luniya, P., Dhatrak, P., Current advancements in polymer/polymer matrix composites for dental implants: A systematic review. *Mater. Today Proc.*, 46, 740–745, 2021.
19. Jin, S. *et al.*, Recent advances in PLGA-based biomaterials for bone tissue regeneration. *Acta Biomater.*, 127, 56–79, 2021.
20. Tan, F., Zhu, Y., Ma, Z., Al-Rubeai, M., Recent advances in the implant-based drug delivery in otorhinolaryngology. *Acta Biomater.*, 108, 46–55, 2020.
21. Jędzierowska, M., Binkowski, M., Koprowski, R., Wróbel, Z., Evaluation of the effect of a PCL/NaNoSiO2 implant on bone tissue regeneration using

x-ray micro-computed tomography, in: *Advances in Intelligent Systems and Computing*, vol. 1186, pp. 107–117, 2021.
22. Hernandez, J.L. *et al.*, Effect of tissue microenvironment on fibrous capsule formation to biomaterial-coated implants. *Biomaterials*, 273, 1–22, 2021.
23. Dong, Q. *et al.*, 3D-printed Mg-incorporated PCL-based scaffolds: A promising approach for bone healing. *Mater. Sci. Eng. C*, 129, 1–14, 2021.
24. Hu, B., Guo, Y., Li, H., Liu, X., Fu, Y., Ding, F., Recent advances in chitosan-based layer-by-layer biomaterials and their biomedical applications. *Carbohydr. Polym.*, 271, 1–17, 2021.
25. Pérez-Merino, P., Dorronsoro, C., Llorente, L., Durán, S., Jiménez-Alfaro, I., Marcos, S., *In vivo* chromatic aberration in eyes implanted with intraocular lenses. *Investig. Ophthalmol. Vis. Sci.*, 54, 4, 2654–2661, 2013.
26. Terrada, C. *et al.*, Cataract surgery with primary intraocular lens implantation in children with uveitis: Long-term outcomes. *J. Cataract Refract. Surg.*, 37, 11, 1977–1983, 2011.
27. Teo, A.J.T., Mishra, A., Park, I., Kim, Y.J., Park, W.T., Yoon, Y.J., Polymeric biomaterials for medical implants and devices. *ACS Biomater. Sci. Eng.*, 2, 4, 454–472, 2016.
28. Zhou, J., Huang, X., Zheng, D., Li, H., Herrler, T., Li, Q., Oriental nose elongation using an L-shaped polyethylene sheet implant for combined septal spreading and extension. *Aesthetic Plast. Surg.*, 38, 2, 295–302.
29. Peltola, M.J., Vallittu, P.K., Vuorinen, V., Aho, A.A.J., Puntala, A., Aitasalo, K.M.J., Novel composite implant in craniofacial bone reconstruction. *Eur. Arch. Oto-Rhino-L.*, 269, 2, 623–628, 2012.
30. Gao, C., Peng, S., Feng, P., Shuai, C., Bone biomaterials and interactions with stem cells. *Bone Res.*, 5, 1–33, 2017.
31. Tayyaba, S., Ashraf, M.W., Ahmad, Z., Wang, N., Afzal, M.J., Afzulpurkar, N., Article fabrication and analysis of polydimethylsiloxane (PDMS) microchannels for biomedical application. *Processes*, 9, 1, 1–31, 2021.
32. Kumar, V., Babbar, A., Sharma, A., Kumar, R., Tyagi, A., Polymer 3D bioprinting for bionics and tissue engineering applications, in: *Additive Manufacturing of Polymers for Tissue Engineering*, pp. 17–39, CRC Press, Boca Raton, 2022.
33. Kubyshkina, G., Zupančič, B., Štukelj, M., Grošelj, D., Marion, L., Emri, I., Sterilization effect on structure, thermal and time-dependent properties of polyamides, in: *Conference Proceedings of the Society for Experimental Mechanics Series*, vol. 3, pp. 11–19, 2011.
34. Li, X., Liu, X., Huang, J., Fan, Y., Cui, F.Z., Biomedical investigation of CNT based coatings. *Surf. Coat. Technol.*, 206, 4, 759–766, 2011.
35. O'Connor, H.J. and Dowling, D.P., Low-pressure additive manufacturing of continuous fiber-reinforced polymer composites. *Polym. Compos.*, 40, 11, 4329–4339, 2019.
36. Tekinalp, H.L. *et al.*, Highly oriented carbon fiber-polymer composites via additive manufacturing. *Compos. Sci. Technol.*, 105, 144–150, 2014.

37. Compton, B.G., Post, B.K., Duty, C.E., Love, L., Kunc, V., Thermal analysis of additive manufacturing of large-scale thermoplastic polymer composites. *Addit. Manuf.*, 17, 77–86, 2017.
38. Wei, X. *et al.*, 3D printable graphene composite. *Sci. Rep.*, 5, 1–7, 2015.
39. Agarwal, K., Kuchipudi, S.K., Girard, B., Houser, M., Mechanical properties of fiber reinforced polymer composites: A comparative study of conventional and additive manufacturing methods. *J. Compos. Mater.*, 52, 23, 3173–3181, 2018.
40. Xu, Y., Qi, S., Xu, Y., Development of new TiO 2 nanoparticles modified photosensitive resin for rapid prototyping. *Sci. Adv. Mater.*, 12, 2, 244–248, 2019.
41. Babbar, A., Tian, Y., Kumar, V., Sharma, A., 3D bioprinting in biomedical applications, in: *Additive Manufacturing of Polymers for Tissue Engineering*, pp. 1–16, CRC Press, Boca Raton, 2022.
42. Zhao, Y., Li, Y., Mao, S., Sun, W., Yao, R., The influence of printing parameters on cell survival rate and printability in microextrusion-based 3D cell printing technology. *Biofabrication*, 7, 4, 1–11, 2015.
43. Guzzi, E.A. *et al.*, Hierarchical biomaterials via photopatterning-enhanced direct ink writing. *Biofabrication*, 13, 4, 1–13, 2021f.
44. Kumar, V., Prakash, C., Babbar, A., Choudhary, S., Sharma, A., Uppal, A.S., *Additive Manufacturing in Biomedical Engineering*, pp. 143–164, CRC Press, Boca Raton, 2022.
45. Salmi, M., Additive manufacturing processes in medical applications. *Materials*, 14, 1, 1–16, 2021.
46. Culmone, C., Smit, G., Breedveld, P., Additive manufacturing of medical instruments: A state-of-the-art review. *Addit. Manuf.*, 27, 461–473, 2019.

2

Artificial Recreation of Human Organs by Additive Manufacturing

Neetesh Soni* **and Paola Leo**

Engineering Innovation Department, University of Salento, Via per Arnesano, Lecce, Italy

Abstract

Additive manufacturing is a technology that allows the creation of complex, customized parts and devices through the layering of materials. In the medical field, additive manufacturing is used in a variety of applications, including the creation of prosthetic limbs, surgical guides, and implantable devices. Prosthetic limbs and other assistive devices can be created using additive manufacturing techniques to ensure a perfect fit for the patient and to reduce costs. Customized surgical guides can also be created to assist in complex surgeries, such as cranial or spinal procedures, by providing a precise template for the surgeon to follow. Additive manufacturing can also create implantable devices, such as dental implants and spinal cages. Those devices can be tailor-made for the individual patient, decreasing the complication rate of surgery. In addition, researchers are exploring using 3D printing technology to create bioprinted tissue and organ models for medical research and drug testing.

Additive manufacturing is, therefore, a promising technology that has the potential to revolutionize the medical field. So, in this chapter, the authors will analyze the current applications of additive manufacturing in the biomedical field and the future perspective.

Keywords**:** Artificial recreation, bioimplantation, additive manufacturing, artificial intelligence programming, cellular tissue mechanics

**Corresponding author*: neetesh.soni@unisalento.it

Jay Prakash Srivastava, Dražan Kozak, Vinayak Ranjan, Pankaj Kumar, Ranjan Kumar and Shubham Tayal (eds.) Mechanical Engineering in Biomedical Applications: Bio-3D Printing, Biofluid Mechanics, Implant Design, Biomaterials, Computational Biomechanics, Tissue Mechanics, (23–42) © 2024 Scrivener Publishing LLC

2.1 Introduction

The additive manufacturing (AM) process, also known as 3D printing, is a technology that can create artificial human organs. This is done using a 3D printer to build up layers of materials, such as cells, to create a functional organ. Creating an artificial organ using 3D printing is known as bioprinting. The technology is still in the early stages of development and has not yet been used to create functional human organs for transplantation [1–3]. However, researchers have been able to use 3D printing to create small parts of organs, such as blood vessels and heart tissue. Artificial recreation of human organs (ARHO) combines cells, materials, and biochemical factors to create functional structures that mimic natural organs [4–7]. This field aims to develop replacement organs for damaged or diseased and to create new organs for people who are born without them or suffer from organ failure. This is done by growing cells in a lab and then using them to build the organ's structure. ARHO by AM is a relatively new technology that has the potential to revolutionize the field of medicine, along with other materials, such as hydrogels or collagen, to create functional human organs. The first step in bioprinting is to create a digital model of the organ to be printed. This can be done using imaging techniques such as computed tomography scan and magnetic resonance imaging (CT scans or MRIs). Once the digital model is created, the 3D printer can use this information to build the organ layer by layer [8–10]. One of the significant challenges in bioprinting is ensuring that the cells stay alive and continue to function correctly once they have been printed. Investigators have found that using hydrogels or collagen as a support material can help to keep the cells alive and healthy during the printing process. Another challenge is printing the complex architecture of human organs [11]. Researchers have been able to print small parts of organs, such as blood vessels and heart tissue, but printing a fully functional organ is still a work in progress and still in the early stages of development [12–15].

Despite these challenges, bioprinting can potentially change how we think about organ transplants. If investigators can develop a way to bioprint functional human organs, it could end the shortage of organs for transplantation. Additionally, bioprinting could be used to create personalized organs for patients with specific genetic conditions, and researchers are also exploring the potential for bioprinting to help in drug discovery and testing.

2.2 Role of Additive Manufacturing for Human Organs

In AM, the role of 3D printing has the potential to revolutionize the field of organ transplantation by creating functional, anatomically correct replacement organs using a patient's cell. Investigators are currently exploring bioprinting for various organs, including the heart, liver, and kidneys [16–18]. However, the technology needs much more development and research, and many challenges need to be overcome before it can be used to create functional replacement organs for human patients. These challenges include developing methods for printing large, complex organs with multiple types of cells and keeping the cells alive and functional during the printing process. Despite these challenges, the potential benefits of bioprinting, including reduced wait times for organ transplants and the elimination of the need for immunosuppressive drugs, make it a promising area of research.

Several factors affect the success of using 3D printing for human organs, including the availability of suitable materials, the ability to create complex structures and vasculature of the organ, and the ability to coax cells to grow and differentiate into the appropriate tissue types. The essential human organs that generally require transplant or replacement are

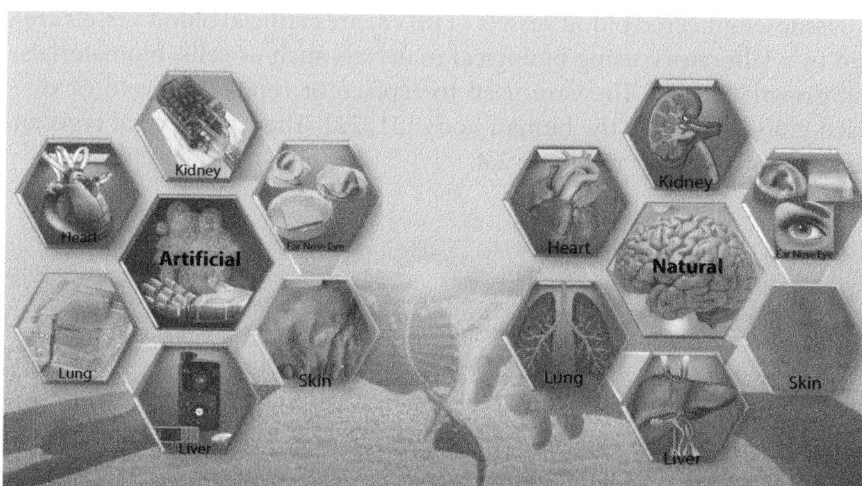

Figure 2.1 Artificial recreation organs vs. natural human organs.

represented in Figure 2.1 as natural organs of the human body. Its applications extend beyond these examples, as a variety of damaged organs can also be replaced. Such as artificial functional skin, lungs, kidney, pancreas, intestines, corneas, bone marrow, ears, limbs, and heart tissue recovering these also can be replaced by AM with the help of the Medical industries.

2.3 Role of Artificial Recreation

Artificial recreation of human cells and organs refers to using technology to create simulations or digital models of the structure and function of human cells and organs. It can include things like 3D models, computer simulations, and bioprinting. The classification of artificial recreation of human cells and organs can be based on the type of technology used, such as 3D printing [18], tissue engineering [19], and organ-on-a-chip [20]. The types of artificial recreation of human cells and organs are represented in Figure 2.2 below.

Medical science, additive manufacturing, and other relative backgrounds have created several applications for the artificial functional recreation of cells and organs for the human body, such as medical research, replacement of damaged organs, medical training, and drug development, as shown in Figure 2.3 below.

Tissue-engineered blood vessels (TEBVs) are artificial blood vessels created in a laboratory using biological materials such as cells, biomaterials, and growth factors. They are used to replace or repair damaged or diseased blood vessels in the human body [21, 22]. There are several types of TEBVs, shown below in Figure 2.4.

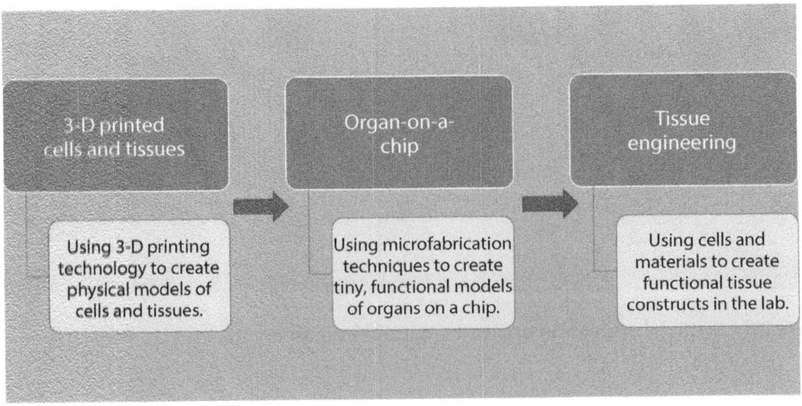

Figure 2.2 Technology relation for artificial recreation of the function of the human body.

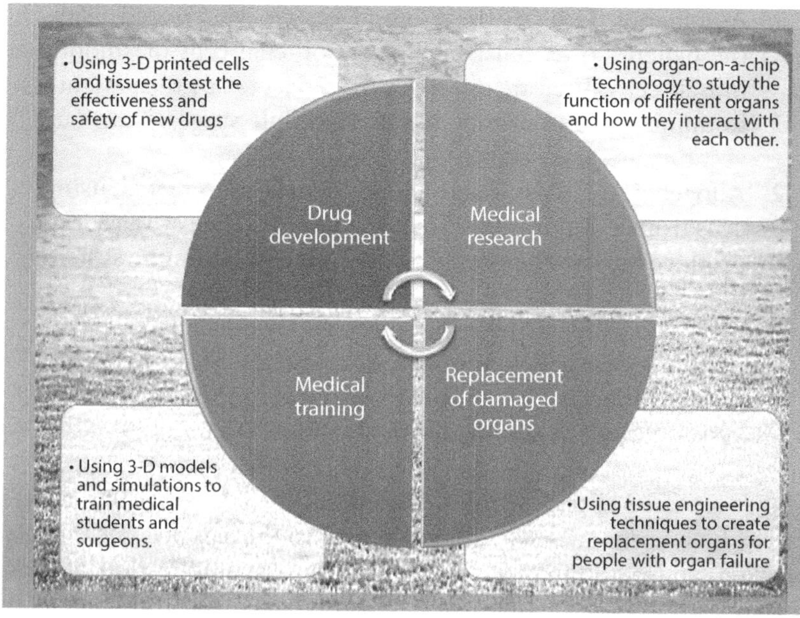

Figure 2.3 Details about artificial functional recreation for biomedical applications.

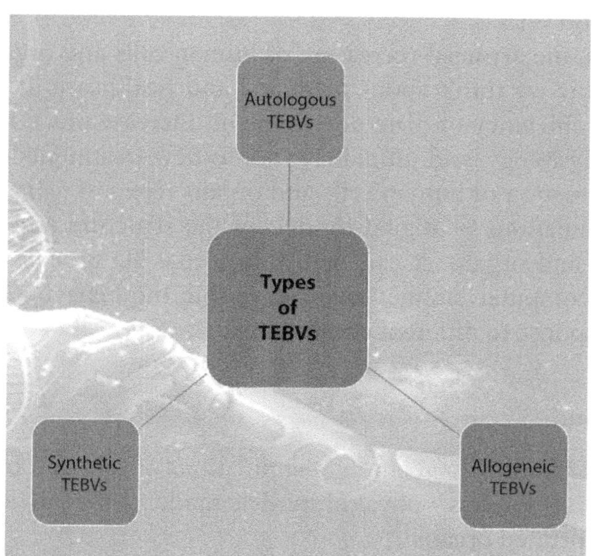

Figure 2.4 Types of TEBVs.

1. Autologous TEBVs: These blood vessels are created using the patient's own cells, such as endothelial cells or smooth muscle cells. They are less likely to be rejected by the patient's immune system but may require multiple surgeries to harvest the necessary cells [23].
2. Allogeneic TEBVs: These blood vessels are created using cells from a donor rather than the patient. They are easier to produce but may be rejected by the patient's immune system [24].
3. Synthetic TEBVs: These blood vessels are created using synthetic materials such as polymers or metals. They are easy to produce but may not mimic the natural properties of blood vessels as well as autologous or allogeneic TEBVs [25].

TEBVs are used to treat various conditions, including peripheral artery disease, which occurs when the blood vessels in the legs become narrowed or blocked, and hemodialysis access, a common complication of kidney failure. Creating a TEBV involves harvesting cells from the patient or a donor and then culturing the cells in a laboratory. The cells are then seeded onto a scaffold, which is a structure that supports the growth of the blood vessel. The scaffold can be made of natural or synthetic materials. Once the cells have grown into a blood vessel, they are implanted into the patient. The process can take several weeks or months, depending on the complexity of the blood vessel.

In general, the artificial recreation of human cells and organs can revolutionize how we think about medicine and healthcare. It can help to improve the efficiency of drug development, increase our understanding of human physiology, and ultimately lead to new treatments for diseases. Artificial recreation of human cells and organs refers to using technology to create simulations or digital models of the structure and function of human cells and organs. It can hold things like 3D models of cells and organs and computer simulations that mimic the behavior of cells and organs in response to different conditions.

Classification:

1. *In silico* models—computer simulations of cells and organs
2. *In vitro* models—physical models made in the lab, such as 3D-printed organs
3. *In vivo* models—living organisms used to study cells and organs, such as animals used in research

Types:

1. Organoids—using 3D cell cultures that mimic the structure and function of organs
2. Tissue engineering—using cells, scaffolds, and other materials to create artificial tissue
3. Organ printing—using 3D printing technology to create physical models of organs

Applications:

1. Drug development—testing the safety and effectiveness of new drugs on artificial models of cells and organs before they are tested on humans
2. Medical education—using artificial models of cells and organs to teach medical students about the structure and function of these systems
3. Tissue replacement—using artificial cells and tissues to replace damaged or diseased cells and organs in the human body

It is essential to analyze the function, reaction, and effects of the operated human organs; after replacing the damaged organs, much research is still needed before they can be widely used in clinical practice. Many investigators and researchers are currently exploring essential areas such as materials, decellularized organs, living cells and biocompatible polymers, and self-healing and shape memory (SHSM) properties inside the organs to create functional bioartificial organs [26].

2.3.1 Decellularized Organ Regeneration

Decellularized organ regeneration is a medical technology that involves removing all living cells from an organ or tissue, leaving behind only the extracellular matrix, which is then used as a scaffold to support the growth of new cells. This process is used to create replacement organs and tissues for transplantation. There are two main types of decellularization techniques: detergent-based and enzymatic. Detergent-based methods involve using chemicals to dissolve the cells, while enzymatic methods use enzymes to break down the cell membranes [27, 28]. Decellularized organ regeneration has many potential applications in the field of medicine, including the following:

1. Creating replacement organs for transplantation, such as hearts, lungs, and livers, and creating replacement tissues, such as blood vessels and skin
2. Generating organoids which are miniature versions of organs that can be used for research and drug development
3. Creating personalized medicine, where a patient's cells can generate replacement organs or tissues less likely to be rejected by the patient's immune system

It is important to note that decellularization is still a developing technology, and not all of the above applications have yet been fully developed or implemented in clinical practice.

2.3.2 3D Bioprinting of Organs and Cells

3D bioprinting is a process that uses 3D printing technology to create structures made of living cells. The process involves using a 3D printer to deposit layers of cells and other materials such as hydrogels to create a 3D structure. In organ printing, cells are taken from a patient and then printed in a specific pattern to form the structure of an organ. The cells are then grown and nurtured in a laboratory until they form a functional organ that can be transplanted into the patient. There are two main types of 3D bioprinting: extrusion-based and laser-based. Extrusion-based bioprinting uses a nozzle to deposit cells, while laser-based bioprinting uses a laser to deposit cells selectively. Developing replacement organs for transplant, including the likes of hearts, lungs, and livers, crafting alternative tissues such as blood vessels and skin, producing miniature organoids for research and drug development purposes, and advancing personalized medicine through the generation of patient-specific organs or tissues with reduced immune system rejection risk all fall under the ambit of 3D bioprinting. It's worth emphasizing that 3D bioprinting remains an evolving technology, and not all of the aforementioned applications have been completely realized or integrated into clinical practice [29].

2.3.3 Self-Healing and Shape Memory for Artificial Organs

Self-healing and shape memory (SHSM) additives can be added to manufactured organs to improve their performance and durability. Self-healing (SH) additives are materials that can repair small cracks or damage to the organ on their own without the need for external intervention, for example, polymethyl methacrylate (PMMA); similarly, many biomaterials are coming nowadays. These materials can be based on different technologies, such as microcapsules filled with healing agents or chemical reactions that cause the materials to repair themselves. SM additives can return to their original shape after being stretched, compressed, or deformed. They can be added to manufactured organs to improve the ability of the organ to withstand mechanical stress and return to its original shape. Combining SHSM additives into manufactured organs can improve their durability and performance and, thus, their ability to function and integrate within the human body. But it is important to note that these technologies are still under research and development and have not yet been fully implemented in clinical practice.

2.4 Role of Additive Manufacturing in Orthopedics

Orthopedic structures refer to medical machine-made body parts and devices used to support or replace bones and joints, such as implants and prostheses. In orthopedics, AM is used to create customized and patient-specific implants, prostheses, and other structures that can improve patient fit, function, and outcomes. Additionally, AM can produce complex geometries, reduce the weight by lattice structure, and improve the overall performance of the implant [30]. One of the most common applications of AM in orthopedics is the production of spinal implants. These devices can be customized to fit the unique curvature and shape of each patient's spine, improving the fit and function of the implant. Joint replacements, such as hips, knees, and ankles, can also be made using AM. These devices can be tailored to match the size and shape of each patient's joint, improving the range of motion and reducing the risk of complications. AM is also used to produce surgical guides and templates, which assist in the alignment and placement of implants during surgery [31]. These guides can be customized to match each patient's unique anatomy, improving the accuracy and precision of surgery.

Furthermore, some recent research is being carried out for the creation of bones and cartilage with the use of AM techniques. AM in orthopedics can improve the fit and function of medical devices, reduce the risk of complications, and improve patient outcomes. It has the potential to revolutionize the field of orthopedics by providing customized and patient-specific solutions for a wide range of conditions and injuries.

Orthopedic AM can be made in most body parts and joints per the specific human body and size. The variation only depends on the materials

Table 2.1 Types of orthopedic structures using additive manufacturing.

Functions	Details of the functions concerning AM
Implants	AM can be used to create customized joint replacements, such as hip or knee implants, that are designed to fit each patient's specific anatomy. It can improve the fit of the implant and improve patient outcomes. Additionally, AM can be used to produce complex geometries that are not possible with traditional manufacturing methods.
Prostheses	Orthopedic prostheses, such as those used to replace a lost limb, can also be created using AM. Using 3D scanning and modeling techniques, these prostheses can be designed to match the remaining limb and provide a better function for the patient.
Scaffolds	Scaffolds are structures that can be used to support bone growth and repair. AM can be used to create scaffolds tailored to each patient's specific needs, such as size, shape, and porosity.
Surgical guides	Surgical guides aid in implant placement or other devices during surgery. AM can be used to create surgical guides that match the specific anatomy of the patient, which can help reduce the time required for surgery and improve the accuracy of the procedure.

and composition, which are relatively beneficial to the human body. There are several technologies and processes which can produce a wide range of orthopedic structures as shown in Table 2.1.

Overall, orthopedic AM allows for the creation of highly customized devices. These patient-specific medical devices can improve the fit and function of orthopedic structures, as well as the outcomes for patients' reactions and after-replacement effects. Figure 2.5 shows the complete process of implanting a 3D AM part, replacing the damaged bone with metallic parts which are relatively of similar size and shape of the human natural organs.

2.5 Types of Bioadditive Manufacturing

3D structures for various applications, including tissue engineering and organ printing, are viable in various types of AM or 3D printing methods

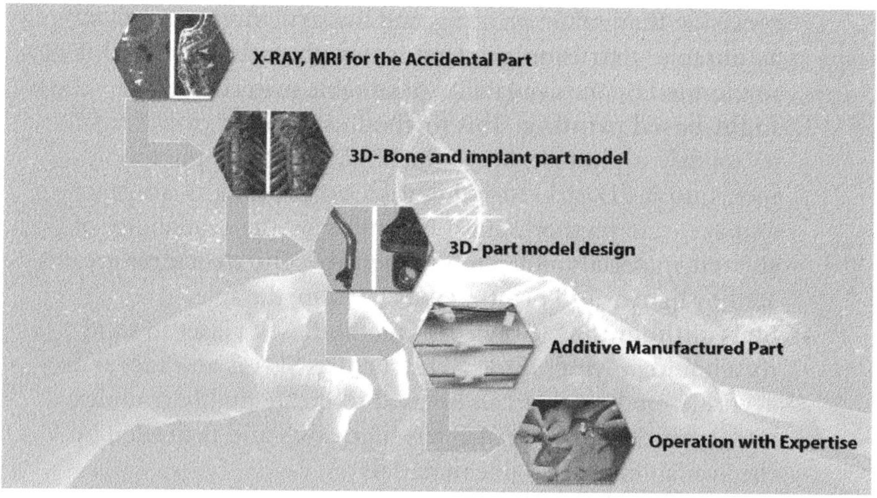

Figure 2.5 Accidental bone replacement by a 3D additive manufactured supportive part [31].

such as inkjet-based printing [32], extrusion-based printing [33], light-based printing [34], stereolithography [35], laser-assisted printing [36], and drop-on-demand (DOD) printing [37], all being bioadditive manufacturing techniques.

1. **Inkjet-based printing:** This method uses a nozzle to spray droplets of a biomaterial, such as a hydrogel, onto a substrate. Inkjet printing is a versatile and cost-effective method that can create 3D structures with high resolution and precision. However, the droplets may not bond well together, limiting the resolution. It is relatively inexpensive compared with other methods, is relatively slow, and requires a substrate to hold the structure. Inkjet printing is not photosensitive and, thus, not limited by the availability of suitable materials. The droplets may not bond well together, which can limit the resolution.
2. **Extrusion-based printing:** This method uses a nozzle to extrude a biomaterial, such as a hydrogel, in a continuous stream onto a substrate. Extrusion printing can create large structures quickly and efficiently, but it has a lower resolution and accuracy than inkjet printing. It is less expensive than other methods. However, it has a lower resolution and

precision than inkjet printing, and the structures may not be as durable. Extrusion printing is not photosensitive and is not limited by the availability of suitable materials.

3. **Light-based printing:** This method uses a laser or UV light to solidify a photosensitive material, such as a photopolymer, into a 3D structure. Light-based printing is an exact and accurate method, but it is limited by the availability of suitable materials and can be expensive. The structures created by light-based printing may be more durable.
4. **Stereolithography (SLA):** This method uses a laser to solidify a liquid polymer into a 3D structure. SLA is an exact and accurate method that can be used to create highly complex structures. However, it also has high costs and is limited by the availability of suitable materials.
5. **Laser-assisted printing:** This method uses a laser to assist in the deposition of materials onto a substrate. This method can create complex structures with high resolution and precision, but it is still in the research phase and has not yet been used to create functional, viable organs.
6. **Drop-on-demand (DOD) printing:** This method uses a nozzle to drop tiny droplets of a biomaterial, such as a hydrogel, onto a substrate. This can be used to create 3D structures that mimic the extracellular matrix of various tissues. It is a cost-effective method and can be used to create 3D structures with high resolution and precision, but it also has limitations in terms of the materials that can be used.

When it comes to the human body's reaction, all the mentioned techniques are still in the research phase and have not yet been used to create functional, viable organs. More research is needed to determine how the human body will respond to these structures. It is important to note that the choice of technique will depend on the specific application and the desired properties of the final 3D structure. Factors such as cost, speed, photosensitivity, durability, and the human body's reaction should be considered.

Most techniques are under the research and trial phase, and further investigation and research are under the early phase. Table 2.2 shows a comparative analysis of the AM bioapplication process, properties, and applications.

Table 2.2 Comparison of the additive manufacturing process for biomedical applications.

AM bioprocess	Speed	Cell density	Strength	Cost	Applications	Ref
Inkjet-based printing (IBP)	↑	↓	↓	↑	Skin, cartilage, bone	[38–40]
Extrusion-based printing (EBP)	↑	↑	↓	↑	Blood vessels, cartilage, bone, muscle, liver-on-a-chip	[41–45]
Light-based printing (LBP)	↑	↑	↓	↓	Vascular networks, skin, organ-on-a-chip	[46–48]
Stereolithography (SLA)	↓	↑	↓	↓	Prosthetics, surgical planning	
Laser-assisted printing (LAP)	↑	↑	↑	↓	Vascular networks, skin, organ-on-a-chip, sensors	[49, 50]
Drop-on-demand printing (DODP)	↓	↑	↑	↑	Biomedical sensors, scaffold for cartilage	[51]
Electrospinning	↓	↓	↑	↑	Wound healing, natural tissue, biomedical sensors, scaffold for cartilage	[52]

2.5.1 Classification of Organoids Using AM

Organoids are three-dimensional cell cultures that resemble the structure and function of specific tissues or organs in the body. They are created by growing cells in a controlled environment, typically in a nutrient-rich liquid medium, and allowing them to self-organize into a three-dimensional structure [52]. Organoids can be derived from various sources, including embryonic stem cells, induced pluripotent stem cells, or adult tissue. To model specific organs or tissues, they can also be made from different cell types, such as intestinal, liver, brain, or lung cells. The primary rewards of organoids are their ability to recapitulate the organization and functions of the tissue or organ they are modeled after and their potential to serve as a tool for disease modeling, drug discovery and toxicology, and regenerative medicine.

For example, organoids can be used to study the effects of genetic mutations or environmental factors on the development of diseases and to test new drugs and therapies more accurately than traditional cell cultures or animal models. Additionally, organoids can be used to engineer functional tissue constructs for regenerative medicine [53, 54]. Artificial recreation of human organs, also known as organoids or organ-on-a-chip technology, involves creating functional replicas of human organs using various techniques such as stem cell technology, 3D printing, and microfabrication.

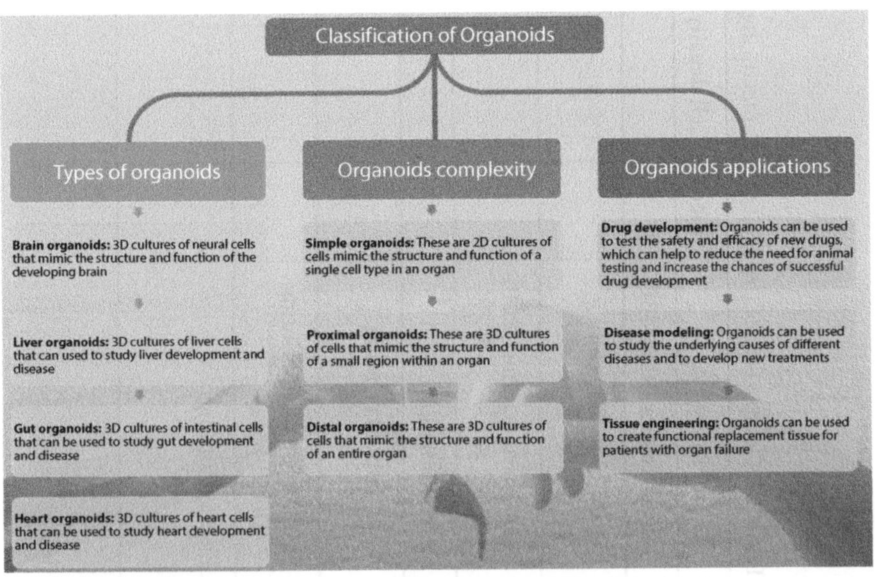

Figure 2.6 Classification, details, and applications of organoids.

The classification of organ-on-a-chip-based organoids is explained in detail below in Figure 2.6.

Overall, organ-on-a-chip and organoids are considered an exciting new frontier in the field of tissue engineering, with the potential to revolutionize drug development and precision medicine. The combination of organoids and additive manufacturing technology holds immense potential for the future of medicine and human body applications, offering new ways to understand biological processes, test new therapies, and advance medical treatments.

2.6 Conclusion

The field of artificial recreation of human organs is interdisciplinary, combining advances in additive manufacturing, bioimplantation, artificial intelligence programming, and cellular tissue mechanics. The goal is to create functional replicas of human organs that can be implanted in the body to replace damaged or diseased ones.

Additive manufacturing is a great technology that allows the creation of complex structures and shapes. Bioimplantation involves the insertion of artificial organs into the body. Artificial intelligence programming is used to optimize and control the 3D printing process and simulate and analyze the mechanical behavior of artificial organs. Cellular tissue mechanics studies the mechanical properties of biological tissues and how they respond to external loads and internal forces.

The future of artificial organ recreation holds much promise, with the potential to address the shortage of donor organs and improve the quality of life for millions of people. However, many challenges must be overcome, including developing biocompatible materials, integrating artificial organs with the human body, and regulating technology.

Overall, the integration of additive manufacturing, bioimplantation, artificial intelligence programming, and cellular tissue mechanics is a promising field with the potential to revolutionize regenerative medicine and provide new hope for people with damaged or diseased organs.

References

1. Jin, Z., Li, Y., Yu, K., Liu, L., Fu, J., Yao, X., Zhang, A., He, Y., 3D printing of physical organ models: Recent developments and challenges. *Adv. Sci.*, 8, 17, 2101394, 2021. https://doi.org/10.1002/advs.202101394.

2. Primo, G.A. and Mata, A., 3D patterning within hydrogels for the recreation of functional biological environments. *Adv. Funct. Mater.*, 31, 16, 2009574, 2021. https://doi.org/10.1002/adfm.202009574.
3. Melchels, F.P.W., Domingos, M.A.N., Klein, T.J., Malda, J., Bartolo, P.J., Hutmacher, D.W., Additive manufacturing of tissues and organs. *Prog. Polym. Sci.*, 37, 8, 1079–1104, 2012. https://doi.org/10.1016/j.progpolymsci.2011.11.007.
4. Murphy, S.V. and Atala, A., 3D bioprinting of tissues and organs. *Nat. Biotechnol.*, 32, 8, 773–785, 2014.
5. Wang, X., Bioartificial organ manufacturing technologies. *Cell Transplant.*, 28, 1, 5–17, 2018. https://doi.org/10.1177/0963689718809918.
6. Zadpoor, A.A. and Malda, J., Additive manufacturing of biomaterials, tissues, and organs. *Ann. Biomed. Eng.*, 45, 1, 1–11, 2016. https://doi.org/10.1007/s10439-016-1719-y.
7. Vaithilingam, J., Kilsby, S., Goodridge, R.D., Christie, S.D.R., Edmondson, S., Hague, R.J.M., Immobilisation of an antibacterial drug to Ti6Al4V components fabricated using selective laser melting. *Appl. Surf. Sci.*, 314, 642–654, 2014. https://doi.org/10.1016/j.apsusc.2014.06.014.
8. Giannopoulos, A.A., Mitsouras, D., Yoo, S.-J., Liu, P.P., Chatzizisis, Y.S., Rybicki, F.J., Applications of 3D printing in cardiovascular diseases. *Nat. Rev. Cardiol.*, 13, 12, 701–718, 2016. https://doi.org/10.1038/nrcardio.2016.170.
9. Haleem, A., Javaid, M., Suman, R., Singh, R.P., 3D printing applications for radiology: An overview. *Indian J. Radiol. Imaging*, 31, 1, 10–17, 2021. https://doi:10.1055/s-0041-1729129.
10. Bertolini, M., Rossoni, M., Colombo, G., Operative workflow from CT to 3D printing of the heart: Opportunities and challenges. *Bioengineering*, 8, 10, 130, 2021. https://doi.org/10.3390/bioengineering8100130.
11. Panja, N., Maji, S., Choudhuri, S., Ali, K.A., Hossain, C.M., 3D bioprinting of human hollow organs. *AAPS PharmSciTech*, 23, 5, 1–18, 139, 2022. https://doi.org/10.1208/s12249-022-02279-9.
12. Mohammadrezaei, D., Moghimi, N., Vandvajdi, S., Powathil, G., Hamis, S., Kohandel, M., Predicting and elucidating the post-printing behavior of 3D printed cancer cells in hydrogel structures by integrating *in-vitro* and *in-silico* experiments. *Sci. Rep.*, 13, 1, 2023. https://doi.org/10.1038/s41598-023-28286-9.
13. Barreiro Carpio, M., Dabaghi, M., Ungureanu, J., Kolb, M.R., Hirota, J.A., Moran-Mirabal, J.M., 3D bioprinting strategies, challenges, and opportunities to model the lung tissue microenvironment and its function. *Front. Bioeng. Biotechnol.*, 9, 1–24, 2021. https://doi.org/10.3389/fbioe.2021.773511.
14. Miller, J.S., The billion cell construct: Will three-dimensional printing get us there? *PloS Biol.*, 12, 6, e1001882, 2014. https://doi.org/10.1371/journal.pbio.1001882.

15. Gu, Z., Fu, J., Lin, H., He, Y., Development of 3D bioprinting: From printing methods to biomedical applications. *Asian J. Pharm. Sci.*, 15, 5, 529–557, 2019. https://doi.org/10.1016/j.ajps.2019.11.003.
16. Haleem, A., Javaid, M., Saxena, A., Additive manufacturing applications in cardiology: A review. *Egypt. Heart J.*, 70, 4, 433–441, 2018. https://doi.org/10.1016/j.ehj.2018.09.008.
17. Leberfinger, A.N., Dinda, S., Wu, Y., Koduru, S.V., Ozbolat, V., Ravnic, D.J., Ozbolat, I.T., Bioprinting functional tissues. *Acta Biomater.*, 95, 32–49, 2019. https://doi.org/10.1016/j.actbio.2019.01.009.
18. Zhang, B., Gao, L., Ma, L., Luo, Y., Yang, H., Cui, Z., 3D bioprinting: A novel avenue for manufacturing tissues and organs. *Engineering*, 5, 4, 777–794, 2019. https://doi.org/10.1016/j.eng.2019.03.009.
19. Ramiah, P., du Toit, L.C., Choonara, Y.E., Kondiah, P.P.D., Pillay, V., Hydrogel-based bioinks for 3D bioprinting in tissue regeneration. *Front. Mater.*, 7, 1–13, 2020. https://doi.org/10.3389/fmats.2020.00076.
20. Leung, C.M., de Haan, P., Ronaldson-Bouchard, K. *et al.*, A guide to the organ-on-a-chip. *Nat. Rev. Methods Primers*, 2, 33, 2022. https://doi.org/10.1038/s43586-022-00118-6.
21. Jung, Y., Ji, H., Chen, Z., Fai Chan, H., Atchison, L., Klitzman, B., Truskey, G., Leong, K.W., Scaffold-free, human mesenchymal stem cell-based tissue engineered blood vessels. *Sci. Rep.*, 5, 1, 1–9, 2015. https://doi.org/10.1038/srep15116.
22. Kim, S., Shah, S.B., Graney, P.L. *et al.*, Multiscale engineering of immune cells and lymphoid organs. *Nat. Rev. Mater.*, 4, 355–378, 2019. https://doi.org/10.1038/s41578-019-0100-9.
23. Song, H.-H.G., Rumma, R.T., Ozaki, C.K., Edelman, E.R., Chen, C.S., Vascular tissue engineering: Progress, challenges, and clinical promise. *Cell Stem Cell*, 22, 3, 340–354, 2018. https://doi.org/10.1016/j.stem.2018.02.009.
24. L'Heureux, N., Dusserre, N., Konig, G., Victor, B., Keire, P., Wight, T.N., Chronos, N.A.F., Kyles, A.E., Gregory, C.R., Hoyt, G., Robbins, R.C., McAllister, T.N., Human tissue-engineered blood vessels for adult arterial revascularization. *Nat. Med.*, 12, 3, 361–365, 2006.
25. Su, Z., Xing, Y., Wang, F. *et al.*, Biological small-calibre tissue engineered blood vessels developed by electrospinning and in-body tissue architecture. *J. Mater. Sci. Mater. Med.*, 33, 67, 2022. https://doi.org/10.1007/s10856-022-06689-w.
26. Lai, H.-Y., Wang, H.-Q., Lai, J.-C., Li, C.-H., A self-healing and shape memory polymer that functions at body temperature. *Molecules*, 24, 18, 3224, 2019. https://doi.org/10.3390/molecules24183224.
27. Dzobo, K., Motaung, K.S.C.M., Adesida, A., Recent trends in decellularized extracellular matrix bioinks for 3D printing: An updated review. *Int. J. Mol. Sci.*, 20, 18, 4628, 2019. https://doi.org/10.3390/ijms20184628.

28. Kabirian, F. and Mozafari, M.M., Decellularized ECM-derived bioinks: Prospects for the future. *Methods*, 171, 108–118, 2020. https://doi.org/10.1016/j.ymeth.2019.04.019.
29. Primo, G.A. and Mata, A., 3D patterning within hydrogels for the recreation of functional biological environments. *Adv. Funct. Mater.*, 31, 16, 2009574, 2021. https://doi.org/10.1002/adfm.202009574.
30. Salmi, M., Additive manufacturing processes in medical applications. *Materials*, 14, 1, 191, 2021. https://doi.org/10.3390/ma14010191.
31. Willemsen, K., Nizak, R., Noordmans, H.J., Castelein, R.M., Weinans, H., Kruyt, M.C., Challenges in the design and regulatory approval of 3D-printed surgical implants: A two-case series. *Lancet Digit. Health*, 1, 4, 163–171, 2019. https://doi: 10.1016/S2589-7500(19)30067-6.
32. Ngo, T.D., Kashani, A., Imbalzano, G., Nguyen, K.T.Q., Hui, D., Additive manufacturing (3D printing): A review of materials, methods, applications and challenges. *Compos. B Eng.*, 143, 172–196, 2018. https://doi.org/10.1016/j.compositesb.2018.02.012.
33. Placone, J.K. and Engler, A.J., Recent advances in extrusion-based 3D printing for biomedical applications. *Adv. Healthc. Mater.*, 7, 8, 1701161, 2017. https://doi.org/10.1002/adhm.201701161.
34. Kowsari, K., Lee, W., Yoo, S.S., Fang, N.X., Scalable visible light 3D printing and bioprinting using an organic light-emitting diode microdisplay. *iScience*, 24, 11, 103372, 2021.
35. Melchels, F.P.W., Feijen, J., Grijpma, D.W., A review on stereolithography and its applications in biomedical engineering. *Biomaterials*, 31, 24, 6121–6130, 2010. https://doi.org/10.1016/j.biomaterials.2010.04.050.
36. Dou, C., Perez, V., Qu, J., Tsin, A., Xu, B., Li, J., A state-of-the-art review of laser-assisted bioprinting and its future research trends. *ChemBioEng Rev.*, 8, 5, 517–534, 2021. https://doi.org/10.1002/cben.202000037.
37. Zhao, D., Zhou, H., Wang, Y., Yin, J., Huang, Y., Drop-on-demand (DOD) inkjet dynamics of printing viscoelastic conductive ink. *Addit. Manuf.*, 48, 102451, 2021. https://doi.org/10.1016/j.addma.2021.102451.
38. Skardal, A., Mack, D., Kapetanovic, E., Atala, A., Jackson, J.D., Yoo, J., Soker, S., Bioprinted amniotic fluid-derived stem cells accelerate healing of large skin wounds. *Stem Cells Transl. Med.*, 1, 11, 792–802, 2012. https://doi.10.5966/sctm.2012-0088.
39. Cui, X., Breitenkamp, K., Finn, M.G., Lotz, M., D'Lima, D.D., Direct human cartilage repair using three-dimensional bioprinting technology. *Tissue Eng. Part A*, 18, 11–12, 1304–12, 2012. https://doi.10.1089/ten.TEA.2011.0543.
40. Gao, G., Yonezawa, T., Hubbell, K., Dai, G., Cui, X., Inkjet-bioprinted acrylated peptides and PEG hydrogel with human mesenchymal stem cells promote robust bone and cartilage formation with minimal printhead clogging. *Biotechnol. J.*, 10, 10, 1568–77, https://doi.10.1002/biot.201400635.
41. Lee, V.K., Kim, D.Y., Ngo, H., Lee, Y., Seo, L., Yoo, S.S., Vincent, P.A., Dai, G., Creating perfused functional vascular channels using 3D bio-printing

technology. *Biomaterials*, 35, 28, 8092–102, 2014. https://doi.10.1016/j.biomaterials.2014.05.083.

42. Jia, W., Gungor-Ozkerim, P.S., Zhang, Y.S., Yue, K., Zhu, K., Liu, W., Pi, Q., Byambaa, B., Dokmeci, M.R., Shin, S.R., Khademhosseini, A., Direct 3D bioprinting of perfusable vascular constructs using a blend bioink. *Biomaterials*, 106, p58–68, 2016 Nov;106:58-68. 2016. https://doi.10.1016/j.biomaterials.2016.07.038.
43. Bhise, N.S., Manoharan, V., Massa, S., Tamayol, A., Ghaderi, M., Miscuglio, M., Lang, Q., Shrike Zhang, Y., Shin, S.R., Calzone, G., Annabi, N., Shupe, T.D., Bishop, C.E., Atala, A., Dokmeci, M.R., Khademhosseini, A., A liver-on-a-chip platform with bioprinted hepatic spheroids. *Biofabrication*, 8, 1, 014101, 2016. https://doi.10.1088/1758-5090/8/1/014101.
44. Kang, H.W., Lee, S., Ko, I. *et al.*, A 3D bioprinting system to produce human-scale tissue constructs with structural integrity. *Nat. Biotechnol.*, 34, 312–319, 2016. https://doi.org/10.1038/nbt.3413.
45. Choi, Y.J., Kim, T.G., Jeong, J., Yi, H.G., Park, J.W., Hwang, W., Cho, D.W., 3D cell printing of functional skeletal muscle constructs using skeletal muscle-derived bioink. *Adv. Healthc. Mater.*, 5, 20, 2636–2645, 2016. https://doi.10.1002/adhm.201600483.
46. Zhu, W., Qu, X., Zhu, J., Ma, X., Patel, S., Liu, J., Wang, P., Lai, C.S., Gou, M., Xu, Y., Zhang, K., Chen, S., Direct 3D bioprinting of prevascularized tissue constructs with complex microarchitecture. *Biomaterials*, 124, 106–115, 2017, https://doi.10.1016/j.biomaterials.2017.01.042.
47. Ma, X., Qu, X., Zhu, W., Li, Y.S., Yuan, S., Zhang, H., Liu, J., Wang, P., Lai, C.S., Zanella, F., Feng, G.S., Sheikh, F., Chien, S., Chen, S., Deterministically patterned biomimetic human iPSC-derived hepatic model via rapid 3D bioprinting. *Proc. Natl. Acad. Sci. U. S. A.*, 113, 8, 2206–2211, 2016. https://doi.10.1073/pnas.1524510113.
48. Gao, Q., He, Y., Fu, J.Z. *et al.*, Fabrication of shape controllable alginate microparticles based on drop-on-demand jetting. *J. Sol-Gel Sci. Technol.*, 77, 610–619, 2016. https://doi.org/10.1007/s10971-015-3890-2.
49. Koch, L., Brandt, O., Deiwick, A., Chichkov, B., Laser-assisted bioprinting at different wavelengths and pulse durations with a metal dynamic release layer: A parametric study. *Int. J. Bioprint.*, 3, 1, 001, 2017. https://doi.10.18063/IJB.2017.01.001.
50. Gao, Q., He, Y., Fu, J., Qiu, J., Jin, Y., Fabrication of shape controllable alginate microparticles based on drop-on-demand jetting. *J. Sol-Gel Sci. Technol.*, 77, 3, 610–619, 2016. https://doi.org/10.1007/s10971-015-3890-2.
51. Shim, J.-H., Lee, J.-S., Kim, J.Y., Cho, D.-W., Bioprinting of a mechanically enhanced three-dimensional dual cell-laden construct for osteochondral tissue engineering using a multi-head tissue/organ building system. *J. Micromech. Microeng.*, 22, 8, 085014, 2012. https://doi.org/10.1088/0960-1317/22/8/085014.

52. Vunjak-Novakovic, G., Ronaldson-Bouchard, K., Radisic, M., Organs-on-a-chip models for biological research. *Cell*, 184, 18, 4597–4611, 2021. https://doi.10.1016/j.cell.2021.08.005.
53. Kang, H.-W., Lee, S.J., Ko, I.K., Kengla, C., Yoo, J.J., Atala, A., A 3D bioprinting system to produce human-scale tissue constructs with structural integrity. *Nat. Biotechnol.*, 34, 3, 312–319, 2016. https://doi.org/10.1038/nbt.3413.
54. Ren, Y., Yang, X., Ma, Z., Sun, X., Zhang, Y., Li, W., Yang, H., Qiang, L., Yang, Z., Liu, Y., Deng, C., Zhou, L., Wang, T., Lin, J., Li, T., Wu, T., Wang, J., Developments and opportunities for 3D bioprinted organoids. *Int. J. Bioprint.*, 7, 3, 364, 1–36, 2021. https://doi.org/10.18063/ijb.v7i3.364.

3

Advances, Risks, and Challenges of 3D Bioprinting

Chinmaya Padhy[1]*, Manish Amin[2], Suhridh Sundaram[2] and Priyanka Paul[2]

[1]Department of Mechanical Eng., GITAM University, Hyderabad, India
[2]Avay Biosciences - IKP-EDEN, Bangalore, India

Abstract

3D bioprinting is an upcoming technique to produce patient-specific tissues for the development of precise, targeted, and customized treatments for creating artificial skin, tissues, and internal organs with the assembly of complex living and non-living biological products from raw materials such as living cells, extracellular matrices, and biomaterials. Fundamentally, these bioprinters have three axes for their motions—*x* (horizontal), *y* (vertical), and *z* (depth) —to print the predefined object layer-by-layer with accurately controlled pneumatic extrusion pressure. With each layer being deposited, it modifies to a solid form either by regulating temperature control on the print head, chemical reaction (UV cross-linking), or extrusion of different materials delivered by the extruder, and then new layers are extruded and deposited onto the existing layers forming a stable object. It differs from traditional 3D printing as it uses bioinks consisting of cells and other biomaterials which allow for the formation of functional tissues. Moreover, bioprinting consists of computational modeling, preparation of bioink and its deposition for its maturation, etc. Undoubtedly, this is an intricate complex process where the composition of the bioink, the process steps of printing, and the requirement of a particular bioprinter are essential. This technology has gained momentum in recent days for both *in vitro* and *in vivo* applications. Though a lot of advancement has been done in the field of 3D bioprinting, its application in human research has not been reached so far as expected due to several technical limitations and ethical challenges in place. Therefore, in this chapter, we tried to tabulate the present scenario along with the recent advancements, challenges, and future strategies of 3D bioprinting along with case studies at Avay Biosciences, Bangalore

**Corresponding author*: cpadhy@gitam.edu

Jay Prakash Srivastava, Dražan Kozak, Vinayak Ranjan, Pankaj Kumar, Ranjan Kumar and Shubham Tayal (eds.) Mechanical Engineering in Biomedical Applications: Bio-3D Printing, Biofluid Mechanics, Implant Design, Biomaterials, Computational Biomechanics, Tissue Mechanics, (43–76) © 2024 Scrivener Publishing LLC

(an R&D organization), a definitive aid for the bio-AM (additive manufacturing) community.

Keywords: Additive manufacturing, 3D-bioprinting, scaffolds, biomaterials, bioink, living tissues, living cells, MITO

3.1 Introduction

Presently, additive manufacturing (AM) is popularly known as additive layer manufacturing (ALM), where 3D objects can be manufactured by layer-upon-layer deposition of raw materials along with its predefined path as instructed by a computer-controlled program; interestingly, this is sometimes also called 3-dimensional (3D) printing. It has been proven that ALM techniques benefited from the limitations of conventional manufacturing to a large extent mostly for the fabrication of products with complex geometries, resulting in the reduction of overall production cost [1]. 3D printing has moved a step further from the production of prototyping (earlier) the manufacturing parts to the end use of the production capabilities. To date, this technology has been seen to have many advantages in various sectors, e.g., aerospace, automobile, architecture, energy, food, and medical. Recently, this technology has attracted many medical practitioners and researchers from biomedical engineering due to its ability to print human tissues and scaffolds for the desired replacements and repairs. This then opens up the channels for regenerative therapy, which helped many patients with their knee replacement, neck surgery, hip replacement, prolotherapy, etc. Earlier, from 3D printing of non-living constructs of groundbreaking discoveries, e.g., implants, surgical associates, and orthopedic devices to living tissues, cells, and hydrogels, bioinks have been seen to have a remarkable journey so far in biomedical engineering. However, 3D printing of living cells is not as straightforward as in non-living constructs. There are a lot of precautionary steps one needs to follow while dealing with 3D printing living cells. In the past, living cells were seeded during *in vitro* or *in vivo* applications, but this process has found some limitations, e.g., the cell distributions are not uniform and result in low seeding efficiency; however, few researchers came out with a solution by directly encapsulating living cells into a biomaterial (bioink) before the start of the actual printing process. However, this version of bioprinting demands strict biocompatible materials for the ink as well as the typical processing conditions to handle. With this, the paradigm of bioprinting has shifted to the development of a suitable bioink that could hold the tissue functionalities intact during the complete printing process [2]. In dditionn to the specific biomaterial in demand, 3D-bioprinting technologies largely depended on the printing speed, accuracies, and applications. Recently, acoustic and magnetic-based bioprinting techniques were under discussion and have evolved [3]. Similarly, with another attempt, volumetric bioprinting allows creating complex geometries in tissue engineering (TE), regenerative medicine (RM), and soft robotics and can be dealt with at an unprecedented speed [4]. There are a few emerging technologies enabling the fabrication of heterogeneous multicellular constructs [5]. Though 3D bioprinting has made significant progress so far, getting a cell distribution in the functional tissues at the micro as well as the macro level is still a matter of concern and always a challenge. Therefore, here, an attempt has been made to discuss and understand from the basics the complete workflow of three-dimensional (3D) bioprinting with an overview of the historical backgrounds of the biomaterial, bioinks, and various bioprinting techniques along with the recent advancements, risks, and challenges witnessed at Avay Biosciences (an R&D lab), Bangalore, India.

3.2 3D Bioprinting

3D bioprinting can be described as a broadened application of AM creating a tissue or organ layer-by-layer, mimicking the natural cellular architecture and functionality of complex tissues. Bioprinters follow the same mechanism as other 3D printers with one contrasting property where instead of printing materials such as plastic and polymers, ceramics, and certain metal alloys, the deposition of biomaterials happens in layers involving live cells to bioprint complex structures such as blood vessels or skin tissues. Cells in biomaterials are printed right away onto a substrate in a particular pattern to form the desired 3D construct [6]. The bioprinting approach entails knowing the biological sensitivity to various printing technical advances, supply of growth factors, and precise perfusion, among other things [7]. Gas and nutrient interchange, along with inter- and intracellular communications, is accomplished efficiently through the interconnected perforations in the construct during layer-wise deposition of the bioprinted tissues [8]. These 3D-bioprinted constructs have enhanced intercellular communication, thus providing a promising result to *in vivo* physiology. These results supply efficient data from preclinical trials since most of the time animal trials are not reliable for predicting human pathophysiological responses [9]. In general, 3D bioprinting can be segregated into three phases: the preparatory phase, the printing phase, and the post-processing phase [9].

Bioprinting technology is an upcoming technique to produce patient-specific tissues for the development of precise, targeted, and customized treatments [10]. The procedure of creating artificial skin, tissues, and internal organs is called *biofabrication*. Biofabrication can be defined as the assembly of complex living and non-living biological products from raw materials such as living cells, extracellular matrices, and biomaterials [11]. Among the different techniques available for 3D bioprinting, *extrusion-based* 3D bioprinting is one of the first modalities developed and is currently the most widely used technique for printing 3D tissue constructs. This technique of bioprinting has proved to be beneficial for the biofabrication of 3D scaffold tissues facilitating a high rate of translation in this field as it is cost-effective and has a higher rate of accessibility and capability to mimic the complex environment of tissues with a resolution of approximately 100 μm [12, 13]. In this technique, the material is loaded onto the extruder and extruded from the nozzle by pneumatic or mechanical movement in a layer-wise manner [14]. This indirectly helps in printing multiple materials onto the same structure resulting in composite models [15]. The major benefits of extrusion-based bioprinting compared with other traditional 3D printing methods are as follows: application of a wide number of biomaterials and cell types, along with hydrogel polymers in suspended cells for producing tissue scaffolds [16]; successfully designed layer-by-layer deposition of biomaterials with sufficient cell density [17]; comparatively less process-induced cell damage [18]; and possessing a high capability of monitoring stem cell growth and differentiation for various research applications [19]. In extrusion bioprinting, the printing process initiates with the dispensing head depositing the biomaterial onto the print bed through three different mechanisms: pneumatic, piston, or screw-based system [20].

Among the three different mechanisms of extrusion bioprinting, pneumatic extrusion bioprinting utilizes air pressure as the driving force, whereas in piston- and screw-based extrusion bioprinting, mechanical displacement and rotation are the driving factors for the extrusion of biomaterials through the nozzle [21]. Three main factors are to be considered for the printability of extrusion bioprinting: adaptability to the viscosity of the biomaterial, phase of the cell and biomaterial mixture before extrusion, and material-specific biofabrication arrangement [22]. The viscosity of the biomaterial can be dependent on temperature and thixotropic properties; thus, it needs to be adjusted for different printing techniques. Considering the limitations of conventional 3D-bioprinting technology, new advanced extrusion-based bioprinting techniques are being developed such as multidispenser, coaxial, and hybrid bioprinting [23].

3.2.1 Types of 3D Bioprinting

Apart from extrusion-based bioprinting, the major existing bioprinting techniques are inkjet, stereolithography-based, and laser-assisted bioprinting methods. These techniques have their respective pros and cons in the field of bioprinting.

1. *Inkjet bioprinting*:
   This type of bioprinting technology is based on the conventional inkjet printing procedure with the accommodation of desktop inkjet printers. It is a contactless procedure that deposits precise picoliter beads of bioink onto a specific hydrogel medium or petri plate based on computer inputs. Depending on the droplet control mechanism, this printing method might be further subdivided between thermal and piezoelectric activator procedures [75]. Printing can be done via a continuous or drop-on-demand (DOD) technique. In thermal inkjet printing technology, ink droplets are formed on heating, creating a bubble forcing the ink to eject out of the narrow nozzle onto a substrate. The temperature can rise very high in a few microseconds to create pulse pressure [76], but the main disadvantage is that the prepared droplets are mixed, non-aligned, and unequal in size. Although the technique is economical, because of repetitive nozzle clogging, printing is not smooth. Cell viability is frequently harmed by shear and heat stress [77].

   In piezoelectric inkjet technique, droplets are formed by transient pressure generated from a piezoelectric actuator. This method of inkjet bioprinting produces no additional heat or orifice blockage, allowing droplets to stay directed and uniform in size [78]. The overuse of this approach might result in membrane rupture and cell lysis [79]. More than 90% viability in mammalian cells such as human osteoblasts, fibroblasts, and chondrocytes has been reported using the piezoelectric inkjet technique [80].

   Inkjet printing has made significant advances in patterning biomolecules, cells, and organs, providing research data on printed DNA and studies of the oncogenesis of different organs [81]. This conventional bioprinting technique is advantageous because it is economical and a no-contact process, thus reducing the chances of contamination. Hydrogels are used as the primary biomaterials in inkjet bioprinting; however, they are too fragile to tolerate normal physiological circumstances [82]. Therefore, to overcome this challenge, new biomaterials with good biological function *in vivo* are being researched along with more advanced techniques in bioprinting. This study is being conducted for the human body in order to address the mechano-biological features of tissues in the body artificially.

2. *Laser-assisted bioprinting (LAB)*:
   This bioprinting technology utilizes a laser as the energy source for the deposition of biomaterials onto a substrate [77]. A pulsating laser projector, a string or band covered with hydrogel-based biological components surfaced on a metal film, and a receiving platform comprise the technology. The mechanism of this bioprinting technique initiates with the irradiation of the string or ribbon by the laser causing evaporation of the liquid biological materials and depositing them onto the substrate in droplet form. Based on the optical features of the laser and the bioink, a biopolymer or cell culture medium is already present on the substrate to stimulate cellular attachment and proliferation after cell transfusion from the band. The lasers used in this technology are nanosecond lasers with UV or near-UV wavelengths to print hydrogels, cells, and ceramic-based composites [83]. The resolution of this bioprinting technique varies from the complexity of the machine and usually provides micron-level accuracy of the order of cell sizes and is dependent upon many factors such as the thickness of the biological materials on the substrate, their rheological properties, the wavelength of the laser pulse, the porosity of the substrate, printing speed, and the constitutional makeup of the structure [84].

   Compared with other bioprinting techniques, LAB has quite a few advantages such as a nozzle-free contactless process, high activity and high resolution in the printing of cells, precise control of ink droplets, and deposition on the substrate [85]. Laser-based bioprinting technology showed prominent successful results in printing human dermal fibroblasts, mouse C2C12 myoblasts, bovine pulmonary artery endothelial cells (BPAECs), breast cancer (MCF-7) cells, and rat neural stem cells [86]. Researchers have developed new advanced methods in LAB, such as "absorbing film-assisted laser-induced forward transfer (AFA-LIFT)" or "biological laser processing (BioLP)" [87] and "matrix-assisted pulsed laser evaporation direct writing (MAPLEDW)" [88]. The fabrication of a soft freeform tissue with a high cell density *in vivo* was successfully printed by LAB having different combinations of cells onto the ECM [89].

3. *Stereolithographic-based 3D bioprinting (STL)*:
   This is a nozzle-free, light-based system that was created in the late 1980s [90]. Under illumination, a fluid photoreactive polymer solidifies. This technology uses computerized micromirror arrangements to adjust the intensity of light in order to produce crosslinked light-sensitive polymers. Its primary use is in the fabrication of constructs using reactive acrylics and epoxies. With more advancements in the field of bioprinting, the development of photocrosslinkable polymers along with multiple resins is required for a specific structure [91]. Digital laser projection oversees the printing procedure in this top-down system. As compared with other rigid freeform printing protocols, the STL file format properly depicts complex surface geometries with prominent accuracy, and a wide variety as well as quantity of materials may be employed in this technology. A major advantage of using the STL format has been historically used for light-sensitive polymers, and therefore, it is well-suited for an even hydrogel deposition in a stratified manner. The total processing time is determined by the structure's volume [92].

   There are several drawbacks, including a shortage of biocompatible polymers, adverse effects from leftover toxic photocuring agents, lack of technical skills to remove the supporting structure, and the complexity to create transverse gradients in 3D biofabricated structures [90]. UV-precise photoinitiators employed in this technology cause DNA damage, which can lead to skin cancer. Some of the most often utilized synthetic biomaterials for light-assisted polymerization of 3D tissue-engineered composites include polyethylene glycol (PEG) and its numerous counterparts such as PEG dimethacrylate (PEGDMA) and PEG diacrylate (PEGDA) [93]. This form of bioprinting is used in correlation with clinical imaging operations such as MRI or CT scan for the improvisation of diagnostic procedures and design, along with quality for the fabrication of prosthetics and medical inserts, and as a beneficial tool for complex surgeries.

STL is classified broadly into two categories: single-photon protocol and multiphoton protocol. The former protocol is subdivided into infrared stereolithography systems, visible radiation systems, conventional stereolithography, and stereothermal lithography systems. Light-projecting systems can be installed directly by laser scripting or by the use of mask projection mechanisms in either physical or digital form [94]. By employing photons from UV radiations, SLA, or "standard single-photon stereolithography apparatus," plays a vital role in crosslinking UV-responsive fluid oligomers into sol-gel polymeric frameworks. Polymerization using a photosensitive resin occurs in three steps: activation, elongation, and termination. A number of resins are being tested for SLA, including "biodegradable resins for non-toxicity, elastomeric resins for flexibility and high strength resins for analyzing mechanical strength" as cited by Barbeck *et al.* [95].

This method has been widely utilized in the field of tissue engineering for the development of biocompatible scaffold where resins help in preventing inflammatory responses that might occur during implantation. Moreover, sufficient degradability with harmless by-products generating renal clearance alongside tissue restoration is required.

4. *Coaxial extrusion-based bioprinting*:
This type of bioprinting is a widely known form of extrusion-based bioprinting, primarily applicable for printing microvasculature. The coaxial nozzles used in bioprinting are composed of concentric chambers for synchronous extrusion of a biomaterial as part of the outer shell and a crosslinker solution encapsulated within this shell to crosslink when the two materials combine at the extruding head. The crosslinking occurs at the nozzle aperture providing high printing accuracy since the crosslinked material retains its shape with increased fidelity [96]. This feature allows for

the fabrication of multilayer scaffolds even with bioinks having low viscosity during extrusion [97]. Structures such as hollow tubular or aggregated fibers are fabricated by cross-linking the extruded solution at the point of extrusion. A variety of cells and biomaterials can be combined in core-and-shell sections to form heterogeneous scaffolds in this configuration. It is a user-defined control system where the nozzle size, geometry, and size of the printed channels can be regulated along with other extrusion settings such as the pressure flow of the concentric chambers based on its application [98]. Multiple coaxial bioprinting stems have been formulated for printing continuous tubules in any required length, thus proving to be a beneficial technology.

5. *Embedded 3D bioprinting*:
During extrusion bioprinting, soft hydrogel bioinks (<100 kPa) are directly extruded onto the substrates without any support in the open air, thus causing gravity-induced sagging and reducing the complexity of the printed construct. To overcome this challenge, embedded 3D bioprinting was developed, wherein the extrusion occurs onto a physical support matrix. This reduces gravity-induced sagging and enhances an omnidirectional extrusion within the support matrix. The matrix being hydrated in nature promotes an aqueous media with optimum pH, temperature, and sterility required for providing cell-friendly environmental conditions during the printing process. The nozzle produces a space in its initial printing layer to promote the matrix. The bioprinted assisted matrix should ideally include shear-thinning and self-curing viscoelastic qualities for assisting the nozzle switch and covering the space to generate extrudate support [99]. Low viscosity and granular hydrogels are ideal support matrices for this type of bioprinting [100].

Granular-based hydrogels or biomaterials combine to form solidified scaffolds at low shear strain; however, they also can be achieved at higher shear pressures. Once the pressure resolves, the hydrogel can shrink back to its solid-like state. The perimeter of the extruded filaments may be controlled by regulating the nozzle tip momentum and rate of flow. In these support materials, complex structures like hollow tubes with dimensions of 100 μm have been successfully printed [101]. Therefore, these features allow the use of a wide number of various biomaterials for bioinks and support matrix for bioprinting tissue-specific scaffolds with microvasculature. Synthetic biomaterials such as polyethylene glycol (PEG), hyaluronic acid (HA), and natural biomaterials such as alginate are applicable for bioprinting in granular matrices.

3D embedded bioprinting has shown evident successful results in its utilization for microvascularization in tissue engineering. Physiological 3D structures with intrinsic complexity, such as the heart [102], cardiac patches [103], the brain [104], and vascular models, have been produced *in vitro* without any support using this technology [105]. This approach is especially important for *in vitro* drug trials and vascular modeling.

6. *Scaffold-free bioprinting*:
This form of bioprinting eliminates the usage of biomaterials as structural components in the creation of bioink. In this technique, the high density of cells is taken for deposition and production of their own ECM for forming their structure. Scaffold-free bioprinting permits the use of much higher cell densities than standard cell density, thus promoting a higher cell–cell interaction for the formation of spheroids with tissue/organ building blocks. Hanging drop [106] and micromolding [107] are two of the most common methods utilized for forming spheroids by scaffold-free bioprinting. Spheroids have imperfect printability and inability to sustain a bioprinted scaffold without collapse; therefore, they are frequently built inside hydrogels to facilitate orientation into 3D scaffolds. One challenge in this type of bioprinting is an accurate control over the placement of individual spheroids. The fusion of spheroids is dependent upon cell density [108], and irregular placement may cause inhomogeneous fusion [109].

Cell aggregate-based bioinks prove to be efficient tools for bioprinting vascularized tissues with relevantly high cell densities. The major disadvantage of this bioprinting is that it is a costly process, and a skilled, laborious cell culture is required for generating the required number of spheroids. Sacrificial writing into living tissue (SWIFT) is a newly created procedure that utilizes the technology of integrated 3D bioprinting. In this advanced technique, high concentrations of organ building blocks (OBBs) are crosslinked inside a temperature-sensitive matrigel/collagen scaffold fabricated for embedding and infusion of a sacrificial bioink. Based on their uses, OBBs can be described as cellular aggregates such as organoids and embryoid bodies. OBBs can be derived from induced pluripotent stem cells (iPSCs) for bioprinting specific constructs for patients. The OBB-based matrices are found to be self-healing, viscoplastic, and easy to use and have a high rate of recovery. Based on the print speed, structures can be bioprinted in the range of 400–1,000 μm in diameter by this technique.

Another disadvantage of scaffold-free bioprinting is the high density of cell aggregates, and their autogenous ECM hinders nutrient transfusion to the cells located positionally at the spheroid center, resulting in necrosis. Future research focuses on the fabrication of microvascular networks with functional and scaffold-free vascular building units for proper grid and structure [110].

3.3 Biomaterials and Bioinks

Normally, natural tissues comprise a systemic alignment of cells and cellular components performing some specific functions. Till now, researchers have performed experiments with cells and their corresponding growth factors and biomaterials along with some assisting components to fabricate functional constructs promoting the regeneration of natural diseases after an injury or disease [24]. One of the newest advanced approaches in the field of tissue engineering is the development of "tissue scaffolds." Tissue scaffolds can be defined as three-dimensional (3D) constructs composed of cells and biomaterials to assist and promote the regeneration of targeted tissues by cell differentiation, migration, and proliferation [25]. The scaffolds have high porosity with interconnected channels and have the potential to mimic the cellular environment resulting in higher metabolism of cells, along with providing suitable mechanical assistance to sustain the stability of the newly fabricated structures both *in vivo* and *in vitro* [26]. These scaffolds also possess a high rate of biodegradation which is exactly similar to that of cell and tissue regeneration. Biomaterials used in creating tissue scaffolds are described as materials that are designed to interact with biological systems in the human body, as well as several applications in the area of medical treatment or replacement of biological functions [27].

These biomaterials provide physical rigidity and functional morphology for cell and tissue interaction. Prior to the printing process, a quick investigation of the performance of these materials based on their pre-existing qualities is required. The most important property is the biocompatibility of the material passively or actively allowing the production of desirable effects on cells and tissues [28].

Along with biocompatibility, there are other factors based on which a biomaterial is selected for bioprinting such as adequate mechanical properties, superior physical and chemical qualities, sufficient wear and rust resistance, and a high rate of osseo-resistance (primarily applicable for bone implants) [29]. Biomaterials can be naturally derived or synthetic (man-made) in nature, and hydrogels are one type of biomaterial having good biocompatibility along with wide applications in the field of bioprinting currently. Biomaterials developed naturally come from a physiological source (mammalian or non-mammalian). These biomaterials are extracted using solvents or enzymatic absorption [30]. They are favored for their biocompatibility and inherent proangiogenic properties, thus providing a wide range of applications in the field of tissue engineering [31]. Naturally generated biomaterials may be further classified as polysaccharide-based or protein-based biomaterials. Some of the most important protein-based biomaterials are as follows:

Collagen: It is the main structural component of the mammalian extracellular matrix (ECM), playing a major role in homeostasis and tissue generation. Type 1 collagen protein is the most prevalent of the 29 identified collagen proteins, and type IV constitutes the scaffold component of the basement membrane enclosing blood vessels [32]. Collagen is composed of a triple helical structure and derived from different animal sources, primarily bovine. Type 1 collagen is an excellent biomaterial for therapeutic vascularization because it promotes an angiogenic microenvironment. Softer collagen gels generate narrower and more dense connections, whereas higher viscous collagen gels form broad and sparsely scattered microvessel channels [33]. The printability of collagen bioinks is determined by the loss moduli as well as storage quality prior to and following the printing process [34]. Collagen bioinks exhibit shear-thinning characteristics at molar concentrations ranging from 0.5% to 1.5% [35]. Factors such as gelation kinetics and moduli of collagen bioinks depend on the pH as well as temperature conditions. Additionally, the storage modulus should have a pH of roughly 8 at 37°C. Gelation of collagen occurs at an optimum temperature of 37°C with a pH of 8. Crosslinking compounds such as genipin [36] and tannic acid (TA) [37] can be utilized for better printability of collagen in extrusion bioprinting.

Gelatin: Derived from denatured collagen, this temperature-sensitive biomaterial is composed of polypeptides and forms a hydrogel below 37°C due to gelatin monomer buildup. Temperature above 37°C causes the dissociation of monomers resulting in a liquefied form of gelatin. Gelatin bioinks are composed of an adhesive peptide sequence of Arg-Gly-Asp amino acids, known as RGD peptides or protease-sensitive sites [38], leading to a useful biomaterial in restorative medicine and tissue engineering [39]. Crosslinking of gelatin can be enzymatically induced using microbial transglutaminase (mTG), but the gelation process is quite lengthy [40]. Sacrificial poly(N-isopropylacrylamide) (PNIPAM) microfibers were introduced to mTG-crosslinked gelatin hydrogels for faster gelation in order to fabricate 3D microvascular networks. Gelatin is commonly utilized as a sacrificial bioink because of its quick melting at 37°C, which is the typical temperature for incubation. It has great versatility for the biofabrication of the microvasculature. The biocompatibility and thermoreversibility of gelatin bioinks make them suitable for bioprinting of the vasculature within 3D constructions.

Gelatin methacryloyl (GelMA): It is a naturally derived semisynthetic hydrogel formed with an amalgamation of methacrylamide and methacrylate groups for better stability of bioprinted gelatin. The methacrylation process leads to the covalent crosslinking of gelatin macromers when exposed to a photoinitiator through photopolymerization [41]. GelMA-based bioinks are widely used in tissue engineering applications such as bone, cartilage, and cardiac tissues [42]. Similar to gelatin, GelMA also retains the RGD peptide sequence, thus assisting endothelial cell adhesion and microvascular network formation [43]. The bioink properties of GelMA, which include bioactivity and tunability, make it highly effective for the bioprinting of microvascular structures. Despite RGD peptide sites that adhere to cells, the vascular mechanism of GelMA can be increased by bioconjugation.

Some of the major polysaccharide-based naturally derived biomaterials are as follows:

Alginate: An anionic polysaccharide extracted from brown algae (Phaeophyceae) is a linear copolymer of (1,4)-linked β-D-mannuronate (M) and α-L-guluronate (G) residues. Morris *et al.* studied the crosslinking activity of alginate by ionic crosslinking of carboxylate groups in the G-block of alginate with divalent cations such as Ca^{+2}, Ba^{+2}, Mg^{+2}, and Sr^{+2} [44]. The molecular weight and M/G block ratio vary in the polymer chain of alginate bioinks depending on their source [45]. Alginate is an excellent biomaterial as it has high gelation kinetics and crosslinks immediately when divalent cations such as calcium chloride are present. In particular, sodium alginate has a shear-thinning property helping to reduce shear stress in cultured cells [46]. The accurate viscosity of alginate bioinks is dependent upon the pH, temperature, concentration, cell density, and molecular weight. Sun and Khalil were the primary researchers to create a "fabrication window" using alginate bioinks for bioprinting functional endothelial cells [47]. The rapid rate of gelation in alginate helps in blending other biomaterials of choice in it for upgraded gelation and shape stability during the printing procedure. These bioinks are highly printable and help in stimulating vascularization [48].

Hyaluronic acid (HA): Also known as hyaluronan, HA is a non-sulfated glycosaminoglycan predominantly present in human ECM. It plays a pivotal role in maintaining the structure and fluid balance in loose connective tissues [49], aiding in cell morphogenesis through direct binding to CD44, a cell surface receptor for hyaluronan, as well as other receptors [50]. The hydrophilicity of HA helps in establishing hydration-integrated networks that allow for signaling between cells [51]. Depending on the source, the molecular weight of HA varies from 10^3 to 10^4 kDa. HA oligosaccharides that have low molecular weight enhance angiogenesis, migration, and endothelial cell proliferation by cohering CD44 receptors onto endothelial cells [52]. A normal gelation concentration of HA results in frail hydrogel networks which have a high rate of degradation. Burdick and others developed a crosslinking protocol with methacrylated HA (MeHA), which is often used to promote light-reactive HA hydrogels with adjustable degradation and mechanical properties [53]. MeHA-based bioinks are applicable for supporting the differentiation and self-renewal of hESCs [54]. Crosslinking with acrylated synthetic polymers can lead to improvement in its rheological properties and applicability in the field of bioprinting [55].

Apart from the naturally derived biomaterials, there are some synthetic-based biomaterials that are prevalently used in the field of bioprinting for tissue engineering applications. Synthetic biomaterials are hydrated frameworks of polymers extracted using chemical mechanisms. The physical and chemical traits of synthetic polymers differ depending on the choice of monomers used and the nature of their crosslinking. These synthetic-based bioinks do not have the capability to promote cell adhesion but are capable of cell-mediated adhesion and breakdown, facilitated by bioactive peptides [56]. These hydrogels have a high printability rate as their physical and chemical properties can be altered as per the print requirements of any particular bioprinting protocol.

Some of these biomaterials are as follows:

Poly(ethylene glycol) [PEG]: PEG, or poly(ethylene glycol), is a commonly used artificial biomaterial with several biomedical purposes. PEG, a compound that may be characterized as a linear polyether, is recommended primarily because it is hydrophilic and resists protein and cell adsorption. The major advantage of using it in biomedical fields is because of its non-toxic nature and biocompatibility property, inducing minimal immune response [57]. PEG-based hydrogels are non-biodegradable and do not have cell attachment sites; thus, several approaches have been implemented to design proangiogenic PEG hydrogels. Proangiogenic signaling ligands like ephrinA1 promote endothelial cell migration and tubulogenesis [58]. In extrusion-based bioprinting, PEGs are mostly applicable for adjusting the mechanical integrity of extrudable bioinks. Primarily, PEG-based bioinks have UV-based crosslinking and are also a useful biomaterial for direct writing.

Poloxamers: This synthetic biomaterial, sometimes referred to as pluronics, is made up of tri-block copolymers made of ethylene oxide and propylene oxide in a linear chain of alternate PEO-PPO-PEO blocks. With regard to gelation, poloxamers have the unique trait of thermoreversibility—below their sol-gel point (about 20°C), and they generate micellar liquids and gelate at room temperature through the accumulation of micelles [59]. Higher polymer concentrations result in lower sol-gel temperatures, which affects the precise sol-gel point of poloxamers [60]. Poloxamers are often utilized for wound dressing [62] and as a strategic release vehicle for hydrophobic medicines [61]. Poloxamer 407 or Pluronic F127 is a commonly used FDA-approved biomaterial applicable in humans. Pluronics are cytocompatible; thus, they are not recommended for cell encapsulation. A 10% (w/w) concentration has shown some prominent negative effects on cell viability [63]; thus, the standard concentration for gelation in mammalian cell culture of Pluronic F127 was standardized at 14% [64]. This group of biomaterials is primarily used as sacrificial bioinks because of their thermoreversible gelation property. These bioinks have a high printing quality and are applicable for constructing networks of sacrificial vasculature with precise form integrity [65].

3.4 Applications of 3D Bioprinting

3D bioprinting has helped in enhancing the standard and prognosis of ongoing disease modeling, disease diagnosis, and treatment and also plays a pivotal role in different methods of tissue reconstruction. Some of its applications are as follows:

1. Role of bioprinting in disease modeling and drug testing *in vitro*:
   Traditionally, laboratory tissue models are restricted to microfluidic devices and 2D structures. These methods were only useful for the collection of inexpensive data but could not provide detailed information on complex and physiologically relevant tissues. Yet, in complicated interactions between cells, growth factors, and the extracellular matrix, 3D constructs showed superiority in *in vivo* settings [111]. Conventionally, microengineering protocols had limited applications in multiple cell types and ECM environments mimicking the *in vivo* environment [112]. Biomaterials and cells may be precisely positioned via 3D bioprinting, which also maintains perfect control over the production of spatiotemporal components in the 3D build [7]. Bioprinting is a cutting-edge method that may be used in drug testing and *in vitro* disease modeling research.
2. Role of bioprinting in the field of regeneration and tissue engineering:
   The final application is the replacement of whole tissue and organs in the area of tissue engineering; microvasculature demonstrates to be an important step in nutrient transfusion in complex intricate tissues [113]. Microvasculature also plays an essential role in accurate therapeutic designed structures' *in vivo* vascular integration [114]. Proangiogenic communications between 3D matrices and cells, fabricating complex vascular structures with proper diffusion and capillary-size vasculature, and functional anastomosis *in vivo* [116] are the main difficulties encountered in this application [115]. To overcome these challenges, bioprinting has

a significant contribution to advanced tissue-engineered structures by bioprinting an accurate microvasculature. Hormones, growth factors, and neurotransmitters must be sent to the bone through the microvasculature in order to monitor hematopoiesis [117]. Bioprinted 3D bone structures have shown positive results in inducing prevascularization of bone constructs. Due to the intricacy of its vascularization and significance in sustaining the vasculature of teeth, microvasculature also plays a key part in the bioconstruction of dental pulp [118]. Bioprinting of dental pulp helps in controlling the structural integrity of dental tissues and initiates a synergistic mechanism of different parts in dental tissues. The fabrication of kidneys, skin, skeletal muscle, and heart tissue is all made possible through bioprinting.

Handling, assembling, and operation of a 3D bioprinter: A 3D bioprinter has three axes for its motion—*x* (horizontal), *y* (vertical), and *z* (depth)—to print an object layer-by-layer with accurate pressure by a process known as pneumatic extrusion. A blueprint or a digital file is created using inbuilt modeled software for printing the 3D object. Once the blueprint is loaded onto the printer along with the chosen material, the printer head goes up and down, side to side and forward and back, forming successive layers of the desired material to form the final product. With each layer being printed, it modifies to a solid form either by regulating temperature control on the print head, chemical reaction (rosslynking), or extrusion of different materials delivered by one extruder. New layers are extruded onto the existing layers forming a stable object. A complete detailed software design and the models obtained directly from the scan reports of a patient lead the precision printer heads to settle the cells at the correct point, and over a certain span of time, an organic object is fabricated comprising a large number of many thin layers [10].

The 3D bioprinter "MITO" (refer Figure 3.1) manufactured at Avay Biosciences, located in Bangalore, India, is capable of 3D bioprinting of cells and tissues with the pneumatic extrusion mechanism. The bioprinter comprises two extruders according to customer requisition onto which the syringe is loaded for extrusion. Pressure is moderated with the pressure adjusting knob and pressure display with the help of an air compressor connected to the pressure port on the printer. According to the different biomaterials chosen for testing, appropriate air pressure is set in tandem with the viscosity of the material among other parameters. A touchscreen is provided at the bottom on one side of the printer for monitoring printing

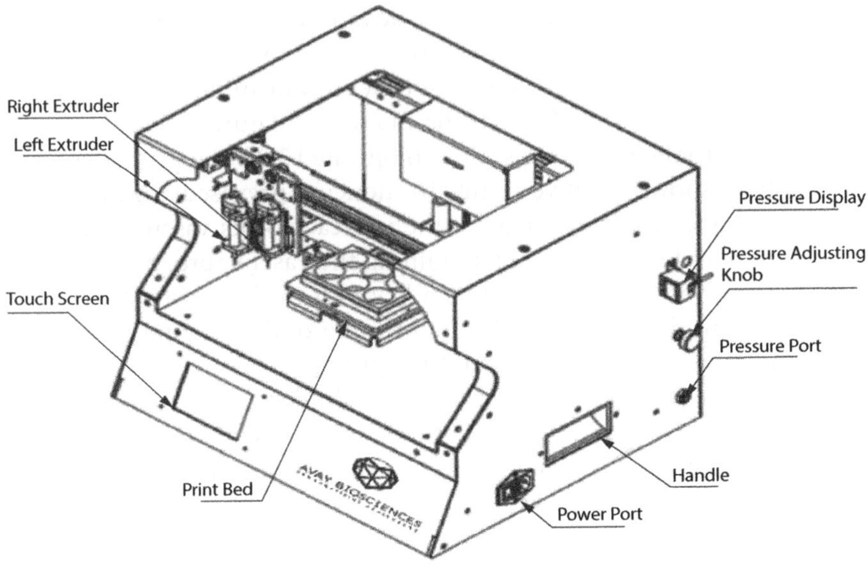

Figure 3.1 A schematic diagram of a 3D bioprinter named "MITO," Avay Biosciences, Bangalore, India.

conditions such as temperature control, crosslinking activity, and multimaterial extrusion and also regulating the motion of the bioprinter. A power port is also provided on the side for loading the blueprint onto the printer for printing the 3D construct. Avay's proprietary software is used for slicing a 3D model by converting it into an STL file before printing with refined adjustments on the extrusion speed, infill density and angle, layer number and height, extrusion width and extrusion delay, curing settings, and also some optional settings. A calibration page is provided in the software for needle calibration, done both manually and with the help of a sensor installed in the bioprinter. Accurate calibration helps in the precise layer printing of the 3D construct. The print bed is designed in such a way to hold a petri dish, six-well plate, 12-well plate, and even a 24-well plate along with a special feature of temperature control on the bed. This feature adds to maintaining the shape and density of the 3D object. Disposable syringes are used for loading biomaterials into the extruders for extrusion, thus providing sterility to the object. Handles are provided on both sides of the bioprinter for convenient transport of the bioprinter, and different nozzles ranging from narrow to wider diameters are also provided for extrusion from the print head. A safety hood is used to cover the bioprinter for preventing major contamination during the research experiments. Once the bioprinter is installed inside a biosafety cabinet, wireless control can be operated for monitoring the printing process.

Assembling a bioprinter comes with a lot of challenges that need to be addressed for the proper functioning of the bioprinter. The main challenge is providing sterility to the bioprinter when working inside the biosafety cabinets in order to avoid contamination. The other challenges remain such as with different biomaterials maintaining extrusion at a specified speed along with crosslinking reaction in between layers during the process of printing. Maintaining the correct UV wavelength during the entire extrusion is very important. The precise control of the motion system in the bioprinter is necessary for controlling the hardware as well as software applications. There are a few risks that are monitored with adequate quality control during the manufacturing of a bioprinter. Primarily, this would involve calibration of several hardware and software factors before the printing process is initiated and selection of the correct extrusion nozzle along with the print head option for the proper extrusion of the 3D construct, and lastly, detailed proper knowledge is required for all the individual components being fabricated into the bioprinter. Overcoming the challenges and risks involved in the fabrication of a bioprinter, some of the significant advancements in MITO are as follows: having a range of temperature adjustments during the ongoing printing protocol according to the choice of biomaterials either in single extrusion or in combination, implicating non-contact needle referencing, and in-process print monitoring, as well as if the user requires the print bed, it can be modified, making the machine experiment-specific by implementing advanced software applications and hardware components. One advancement that MITO is undergoing is the drop-on-demand printing technology for less viscous biomaterials. Another frequently requested feature is integrating hard tissue printing in the bioprinter along with imparting an incubator sterile microenvironment while printing.

3.5 A Case Study

(a) Considerations Before Scaffold Bioprinting

A reliable biocompatible substance should be capable of tolerating long-term transplantation, which means that its rate of breakdown must be sufficient to allow the scaffold to be gradually replaced by ECM proteins released by the related cells over time. As a result, the rate of disintegration of the material should be compatible with the pace of ECM secretion and result in only the non-toxic breakdown of by-products that can be quickly or easily eliminated from the body. A bioprintable substance must have a precise fluidic consistency for controlled deposition and the right crosslinking technique to properly solidify the desired 3D construct with enough structural integrity [7]. Cells and biomaterial solutions are combined in proportion for tissue scaffold bioprinting, thus being permeable and non-toxic and having a neutral pH, mimicking an aqueous environment for the high metabolism of cells with adequate cell viability. Based on the above factors, the selection of an accurate biomaterial before the initiation of the bioprinting process is essential as it is indirectly dependent upon the performance of the 3D-printed scaffolds.

The amalgamation of cells with the biomaterial solution and the subsequent loading of the solution into the bioprinter are the last steps in the bioprinting process. Extrusion-based bioprinting allows a broad variety of cell densities to be mixed into the biomaterial solutions (cell density of 108–109 cells/ml), which is one of the main benefits of using it to fabricate 3D tissue constructs. This characteristic aids in shortening the time needed for cell development and proliferation prior to scaffold implantation.

(b) The Printing Process of MITO

For the purpose of creating 3D-printed tissue scaffolds in a regulated stratified pattern, MITO uses an extrusion-based bioprinting technology. An autonomous robotic system and a fluid-dispensing head make up the extrusion bioprinting technology [66]. When the printing process initiates after loading the STL file, the dispensing head of the extruder moves across the *x*- and *y*-axes that the robotic system controls to accumulate bioink in a predesigned shape onto the bed of the bioprinter, and moves up and down along the *z*-axis to deposit different layers and fabricate the scaffold as directed by the digital code. The pneumatic extrusion with the help of an air compressor helps in a controlled volume extrusion of bioinks from a syringe at an appropriate pressure onto the bed of the bioprinter, thus maintaining the structural integrity and viscosity of the bioink. For higher viscosity suspensions of bioinks, a significant pressure drop from the inlet of the syringe to the output of the nozzle is beneficial. Various crosslinking methods such as the change in temperature and UV curing can be implemented during the printing process. Thermal crosslinked hydrogel polymers (e.g., collagen, gelatin) are fluid-based above a certain temperature but crosslink into a hydrogel at a particular temperature (refer Figure 3.2).

Naturally derived biomaterials (e.g., gelatin methacryloyl (GelMA)) require UV crosslinking at a particular wavelength on layer-by-layer deposition during extrusion to form a stable 3D structure (refer Figure 3.3).

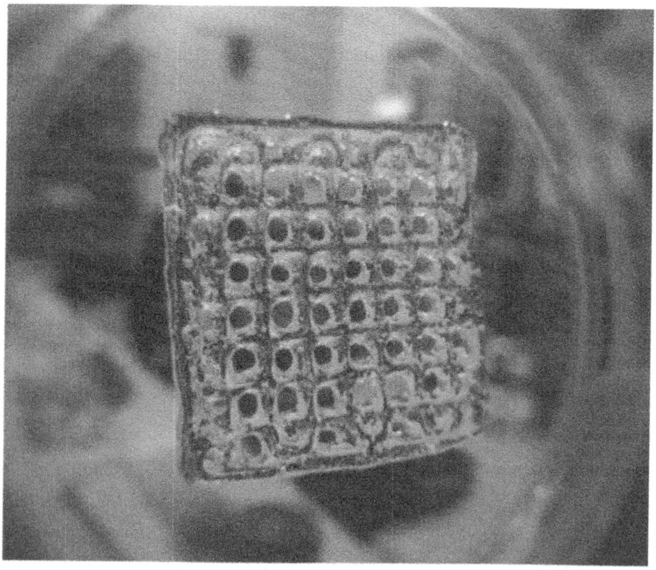

Figure 3.2 A 3D-bioprinted gelatin scaffold extruded at an optimum temperature of 22°C during extrusion.

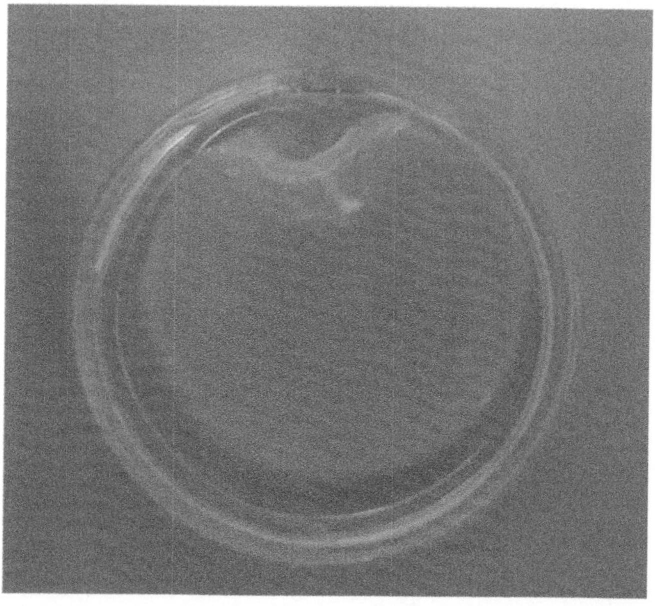

Figure 3.3 A solid layer of GelMA formed on UV crosslinking at a wavelength of 365 nm for a time exposure of around 2–3 min. On pouring warm water at 37°C overnight, the layer did not dissolve ensuring an accurate crosslinked GelMA scaffold.

Naturally derived biomaterial such as alginate shows efficient crosslinking when mixed with calcium chloride in appropriate proportions.

During printing (refer Figure 3.4), several modifications can be implemented in MITO such as the extrusion of two different bioinks simultaneously enabling the dual extruder printing system, refilling of bioink for forming a complete 3D object, and temperature modifications in different layers during printing and pressure regulations. While lower viscosity materials are more suited for biopathways, higher viscosity biomaterials are better suited to provide tissue scaffolds with structural stability.

Testing the optimum concentration and stringing property of the biomaterial before printing posed a big challenge in the field of bioprinting. Therefore, to overcome this challenge, a bioink tester is constructed from a designed STL file and used for testing different biomaterials. Higher or lower concentrations of biomaterials will create a crest or trough with increasing distance gaps in the tester, thus proving to be a beneficial tool for standardizing the concentration of materials.

From Figure 3.5, it was determined that the optimum concentration of gelatin required for bioprinting is 12% and that of alginate is 10% for a firm 3D scaffold.

For the printing of hard tissues, pore size distribution and structure of the 3D-printed tissue must maintain a modulus of elasticity in order to sustain a healthy rate of cell growth while implantation is conducted [67].

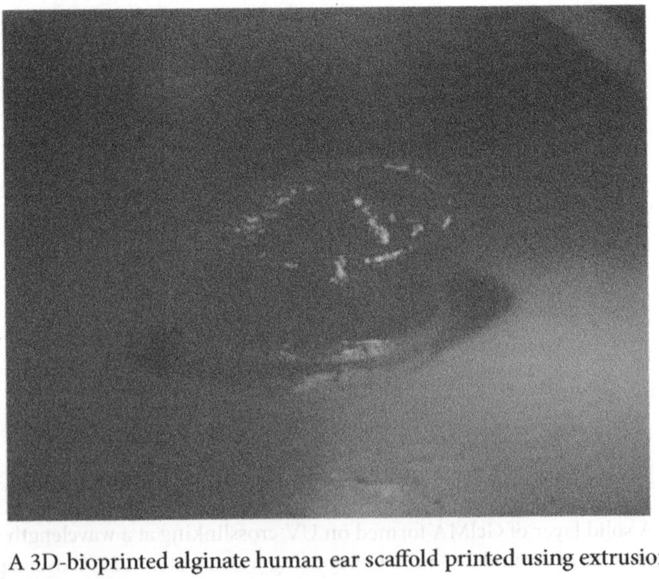

Figure 3.4 A 3D-bioprinted alginate human ear scaffold printed using extrusion bioprinting.

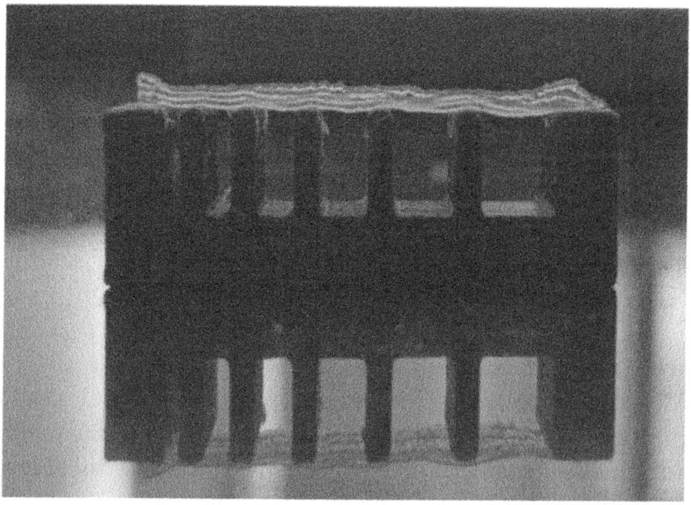

Figure 3.5 Three to four consecutive layers of biomaterials are extruded from a syringe onto a bioink tester for optimum concentration.

During the printing process, the layer height of the 3D construct is also important to be noted for proper structural integrity. Another risk of post-printing is scaffold deformation of newly constructed tissues, which can occur if appropriate mechanical support and structural stability requirements are not fulfilled by the scaffold. For the newly 3D-bioprinted construct, proper vascularization *in vivo* is important for the proper transfer of cell growth factors, oxygen, and nutrients and the removal of waste. Challenges on the part of biomaterials include the selection of a suitable binder and monitoring the life span of the biomaterials before printing; thus, 3D fabrication is a process that involves several steps that have to work cohesively with each other.

Using extrusion bioprinting helped in fabricating tissue scaffolds with various cell types such as primary cells and stem cells. Primary cell types can be classified as osteocytes, chondrocytes, and keratinocytes derived from animals and humans and have been used in tissue scaffolds for the 3D fabrication of bone, cartilage, and skin [68]. Extrusion bioprinting has shown successful results in biofabricating human lung components [69], functional skeletal muscle [70], and restoration of the human liver [71]. In some cases, the isolation of primary cells from living tissues is challenging; thus, in the process of bioprinting, stem cells are employed in place of primary cells to get around this problem [72]. Stem cells have the capacity to self-renew and differentiate into any particular cell type, thus showing a range of applications in adipose tissue, bone marrow tissue, gingival tissue, etc. [73]. Owing to the advancements in MITO, new biomaterials and modified bioprinting techniques are being implemented to print tissue constructs with comparable biological and mechanical qualities made from a variety of materials and cells with increased resolution of deposition. Modifications of bioinks are also under research for improving their biocompatibility. Lab-on-a-chip is also an advanced method that is being implemented in the fabrication of scaffolds with complex cross-sections.

(c) Postprinting Process

After the 3D scaffold construct has been bioprinted in MITO, a cell suspension of optimum cell density (>70%) is microinjected into the scaffold construct, and some culture medium is placed inside the CO_2 incubator under optimum conditions for the binding of cells to the scaffold and the growth of cells on it over a period of time. Once the entire 3D-bioprinted construct has replicated the morphology and structure of the desired native organ, a set of clinical trials is performed for its viability, morphology, cytotoxicity, and rheology. Once all the experiments are validated, the next step involves the implantation of the 3D construct into the target site of the patient's body. The major challenge in the postprinting process involves preventing contamination of the bioconstruct in a sterile environment.

(d) Considerations for Preventing Cell Damage in the Bioprinting Process

During the process of bioprinting, cells experience extensional and shear pressure when forced to extrude through the nozzle attached to the syringe in the extruder causing deformation of cells and rupture of the cell membrane [74]. Cell damage and malfunction brought on by cell membrane rupture significantly reduce cell viability. As a result, a proper analysis of cell damage caused during bioprinting is important to be noted for inducing cell survival during the fabrication of scaffolds. One of the key requirements in the bioprinting process is to keep cell viability high. Applying a low emitting pressure can help in reducing stresses to the cells, but it will lead to inefficient deposition of viscous bioinks. Modeling an object before loading it into the printer is considered a realistic approach to overcome the risk for the accurate physical and biological performance of bioprinted constructs. Optimal designs and control based on the models may be easily accomplished when validation is carried out through a series of tests.

3.6 Conclusions

Bioadditive manufacturing experts are able to create simple structures, such as a single layer of cells or a small piece of tissue. One of the key challenges in 3D bioprinting is to ensure that the printed structures are biocompatible, meaning they are able to integrate with the body's natural tissues without causing an immune response or other adverse reactions. In some cases, this may involve using natural materials that are already found in the body, such as collagen or hyaluronic acid, to create a scaffold or support structure for the printed cells. Researchers might also have used specialized bioprinting techniques, such as extrusion or inkjet printing, that can help in minimizing the damage to the cells and ensure that they remain viable after printing. Once the structure has been printed, it may undergo additional processing, such as incubation or exposure to growth factors, to encourage the cells to grow and differentiate into the desired tissue type. Researchers also conducted extensive testing to ensure that the printed structure is safe and biocompatible, both *in vitro* and in animal models. Despite the many challenges involved in cellular resources, biomaterials, and bioprinting methods, 3D bioprinting has progressed in recent times where researchers have been able to create more complex structures, such as 3D-printed organs and tissues that imitate the natural structure and function of human organs. While 3D bioprinting has made significant progress in recent years, there is still a lot of room or challenges for improvement in terms of both the technology and its applications, e.g., printing larger structures, materials with multicell, multifunctional properties of natural tissues and organs, and some form of advanced imaging and modeling techniques to help and guide the printing process to ensure that the printed structures have the desired properties and functionality. Undoubtedly, bioinks are specialized materials that are used to print cells and other biological materials, and they play a critical role in ensuring that the printed structures are biocompatible and functional. One of the key challenges in developing bioinks is ensuring that they are biocompatible and able to support the growth and differentiation of cells. This requires careful selection of materials, as well as optimization of the printing process to ensure that the cells remain viable and functional. In addition to biocompatibility, the properties of the bioink can also affect the mechanical and biological properties of the printed structure. Therefore, with the development of more advanced and specialized bioinks, researchers will be able to print increasingly complex structures, which will open up new possibilities in biomedical engineering.

While progress has been made in bioprinting complex tissue structures, most studies to date have focused on simpler tubular structures and millimeter-scale vascular constructs with basic cylindrical or branched geometry. More delicate functional capillary networks and intra-organ vasculature for the fabrication of tissues and organs, which allows for proper blood flow and nutrient exchange, are still under research. With the advancement in more modified bioinks and inventive bioprinting techniques, there is rapid progress in the field of bioprinting for vascularized tissues and organs with precise structural integrity. 3D bioprinting provides a unique opportunity to artificially recreate any tissues, thus avoiding the risk of immunological graft rejection or issues related to donor scarcity. The application of 3D bioprinting could provide personalized treatment for the patient yielding a better clinical outcome with less risk to life. Bioinks formulated with biomaterials should closely mimic the native microenvironment within ECM, thus providing an interlink between matrix biology and tissue construct. Overall, bioprinting technology is rapidly evolving and will undoubtedly mark a major contribution to healthcare.

In this chapter, we presented the basics of bioprinting, its techniques, different processes, bioinks, notable research studies, and more importantly a typical modern bioprinter (the proprietor being Avay Biosciences, Bangalore), along with its advantages and limitations in a detailed manner. Despite the many challenges including high cost, difficulty in reproducing the complex microenvironment of living tissues, and limited availability of suitable bioinks, 3D bioprinting has the potential to revolutionize the field of regenerative medicine and clinical rendition and to lead to the development of new rehabilitation for various diseases in the days to come.

References

1. Frazier, W.E., Metal additive manufacturing: A review. *J. Mater. Eng. Perform.*, 23, 1917–1928, 2014.
2. Fang, Y. and Guo, Y., 3D printing of cell-laden microgel-based biphasic bioink with heterogeneous microenvironment for biomedical applications. *Adv. Funct. Mater.*, 32, 13, 2109810, 2021.
3. Huang, P., Tian, Z., Yang, S., Wave number-spiral acoustic tweezers for dynamic and reconfigurable manipulation of particles and cells. *Sci. Adv.*, 5, eaau6062, 2019.
4. Kim, D.Y., Hong, H., Seo, Y.B., Digital light processing 3D printed silk fibroin hydrogel for cartilage tissue engineering. *Biomaterials*, 232, 119679, 2020.

5. Wu, Y., Su, H., Li, M., Xi, H., Digital light processing-based multi-material bioprinting: Processes, applications, and perspectives. *J. Biomed. Mater. Res. A*, 111, 4, 527–542, 2022. https://doi.org/10.1002/jbm.a.37473.
6. Xiongfa, J., Hao, Z., Liming, Z., Jun, X., Recent advances in 3D bioprinting for the regeneration of functional cartilage. *Regen. Med.*, 13, 1, 73–87, 2018.
7. Murphy, S.V. and Atala, A., 3D bioprinting of tissues and organs. *Nat. Biotechnol.*, 32, 8, 773–785, 2014.
8. Knowlton, S., Anand, S., Shah, T., Tasoglu, S., Bioprinting for neural tissue engineering. *Trends Neurosci.*, 41, 1, 31–46, 2018.
9. Resnik, D.B., The clinical investigator-subject relationship: A contextual approach. *Philos. Ethics Humanit. Med.*, 4, 16, 2009.
10. Melissa, L. and Gordan, W., Printing the future:3D bioprinters and their uses. *Aust. Acad. Sci.*, 2016. https://www.science.org.au/curious/people-medicine/bioprinting
11. Zhang, Y.S., Haghiashtiani, G., Hübscher, T. *et al.*, 3D extrusion bioprinting. *Nat. Rev. Methods Primers*, 1, 75, 2021.
12. Fisch, P., Holub, M., Zenobi-Wong, M., Improved accuracy and precision of bioprinting through progressive cavity pump-controlled extrusion. *Biofabrication*, 13, 1, 015012, 2020.
13. Matai, I., Kaur, G., Seyedsalehi, A., McClinton, A., Laurencin, C.T., Progress in 3D bioprinting technology for tissue/organ regenerative engineering. *Biomaterials*, 226, 119536, 2020.
14. Mironov, V., Printing technology to produce living tissue. *Expert Opin. Biol. Ther.*, 3, 5, 701–704, 2003.
15. Jiang, T., Munguia-Lopez, J.G., Flores-Torres, S., Kort-Mascort, J., Kinsella, J.M., Extrusion bioprinting of soft materials: An emerging technique for biological model fabrication. *Appl. Phys. Rev.*, 6, 1, 011310, 2019.
16. Pati, F., Jang, J., Ha, D.H., Won Kim, S., Rhie, J.W., Shim, J.H., Cho, D.W., Printing three-dimensional tissue analogues with decellularized extracellular matrix bioink. *Nat. Commun.*, 5, 1, 3935, 2014.
17. Mironov, V., Trusk, T., Kasyanov, V., Little, S., Swaja, R., Markwald, R., Biofabrication: A 21st century manufacturing paradigm. *Biofabrication*, 1, 2, 022001, 2009.
18. Ozbolat, I.T. and Hospodiuk, M., Current advances and future perspectives in extrusion-based bioprinting. *Biomaterials*, 76, 321–343, 2016.
19. Chen, D.X., Extrusion bioprinting of scaffolds: An introduction, in: *Extrusion Bioprinting of Scaffolds for Tissue Engineering Applications*, pp. 1–13, 2019.
20. Ning, L. and Chen, X., A brief review of extrusion-based tissue scaffold bio-printing. *Biotechnol. J.*, 12, 8, 1600671, 2017.
21. Derakhshanfar, S., Mbeleck, R., Xu, K., Zhang, X., Zhong, W., Xing, M., 3D bioprinting for biomedical devices and tissue engineering: A review of recent trends and advances. *Bioact. Mater.*, 3, 2, 144–156, 2018.

22. He, Y., Yang, F., Zhao, H., Gao, Q., Xia, B., Fu, J., Research on the printability of hydrogels in 3D bioprinting. *Sci. Rep.*, 6, 1, 29977, 2016.
23. Pati, F., Jang, J., Lee, J.W., Cho, D.W., Extrusion bioprinting, in: *Essentials of 3D Biofabrication and Translation*, pp. 123–152, Academic Press, USA, 2015.
24. Skardal, A. and Atala, A., Biomaterials for integration with 3-D bioprinting. *Ann. Biomed. Eng.*, 43, 3, 730–746, 2015.
25. Zhang, L., Yang, X., Yue, Y., Ye, J., Yao, Y., Fu, Y., Gong, P., Cyclic mechanical stress modulates neurotrophic and myelinating gene expression of Schwann cells. *Cell Prolif.*, 48, 1, 59–66, 2015.
26. Larson, C. and Shepherd, R., 3D bioprinting technologies for cellular engineering, in: *Microscale Technologies for Cell Engineering*, pp. 69–89, 2016.
27. Ng, I.C., Pawijit, P., Tan, J., Yu, H., Anatomy and physiology for biomaterials research and development, in: *Encyclopedia of Biomedical Engineering*, pp. 225–236, Elsevier, 2019. https://doi.org/10.1016/B978-0-12-801238-3.99876-3.
28. Ferris, C.J., Gilmore, K.G., Wallace, G.G., in het Panhuis, M., Biofabrication: An overview of the approaches used for printing of living cells. *Appl. Microbiol. Biotechnol.*, 97, 4243–4258, 2013.
29. Bharadwaj, A., An overview on biomaterials and its applications in medical science, in: *IOP Conference Series: Materials Science and Engineering*, vol. 1116, IOP Publishing, p. 012178, 2021.
30. Ha, T.L.B., Quan, T.M., Vu, D.N., Naturally derived biomaterials: Preparation and application, in: *Regenerative Medicine and Tissue Engineering*, Intech Open, United Kingdom, 2013.
31. Liu, F., Chen, Q., Liu, C., Ao, Q., Tian, X., Fan, J., Wang, X., Natural polymers for organ 3D bioprinting. *Polymers*, 10, 11, 1278, 2018.
32. Pöschl, E., Schlötzer-Schrehardt, U., Brachvogel, B., Saito, K., Ninomiya, Y., Mayer, U., Collagen IV is essential for basement membrane stability but dispensable for initiation of its assembly during early development. *Development*, 131, 7, 1619–28, Apr 2004.
33. Yamamura, N., Sudo, R., Ikeda, M., Tanishita, K., Effects of the mechanical properties of collagen gel on the *in vitro* formation of microvessel networks by endothelial cells. *Tissue Eng.*, 13, 7, 1443–1453, 2007.
34. Diamantides, N., Wang, L., Pruiksma, T., Siemiatkoski, J., Dugopolski, C., Shortkroff, S., Bonassar, L.J., Correlating rheological properties and printability of collagen bioinks: The effects of riboflavin photocrosslinking and pH. *Biofabrication*, 9, 3, 034102, 2017.
35. Lai, G., Li, Y., Li, G., Effect of concentration and temperature on the rheological behavior of collagen solution. *Int. J. Biol. Macromol.*, 42, 3, 285–291, 2008.
36. Kim, Y.B., Lee, H., Kim, G.H., Strategy to achieve highly porous/biocompatible macroscale cell blocks, using a collagen/genipin-bioink and an optimal 3D printing process. *ACS Appl. Mater. Interfaces*, 8, 47, 32230–32240, 2016.
37. Lee, J., Yeo, M., Kim, W., Koo, Y., Kim, G.H., Development of a tannic acid cross-linking process for obtaining 3D porous cell-laden collagen structure. *Int. J. Biol. Macromol.*, 110, 497–503, 2018.

38. Van den Steen, P.E., Dubois, B., Nelissen, I., Rudd, P.M., Dwek, R.A., Opdenakker, G., Biochemistry and molecular biology of gelatinase B or matrix metalloproteinase-9 (MMP-9). *Crit. Rev. Biochem. Mol. Biol.*, 37, 6, 375–536, 2002.
39. Devillard, C.D., Mandon, C.A., Lambert, S.A., Blum, L.J., Marquette, C.A., Bioinspired multi-activities 4D printing objects: A new approach toward complex tissue engineering. *Biotechnol. J.*, 13, 12, 1800098, 2018.
40. Yang, G., Xiao, Z., Ren, X., Long, H., Qian, H., Ma, K., Guo, Y., Enzymatically crosslinked gelatin hydrogel promotes the proliferation of adipose tissue-derived stromal cells. *Peer J.*, 4, e2497, 2016.
41. Ramón-Azcón, J., Ahadian, S., Obregón, R., Camci-Unal, G., Ostrovidov, S., Hosseini, V., Matsue, T., Gelatin methacrylate as a promising hydrogel for 3D microscale organization and proliferation of dielectrophoretically patterned cells. *Lab. Chip*, 12, 16, 2959–2969, 2012.
42. Yue, K., Trujillo-de Santiago, G., Alvarez, M.M., Tamayol, A., Annabi, N., Khademhosseini, A., Synthesis, properties, and biomedical applications of gelatin methacryloyl (GelMA) hydrogels. *Biomaterials*, 73, 254–271, 2015.
43. Aubin, H., Nichol, J.W., Hutson, C.B., Bae, H., Sieminski, A.L., Cropek, D.M., Khademhosseini, A., Directed 3D cell alignment and elongation in microengineered hydrogels. *Biomaterials*, 31, 27, 6941–6951, 2010.
44. Grant, G.T., Morris, E.R., Rees, D.A., Smith, P.J., Thom, D., Biological interactions between polysaccharides and divalent cations: The egg-box model. *FEBS Lett.*, 32, 1, 195–198, 1973.
45. Tønnesen, H.H. and Karlsen, J., Alginate in drug delivery systems. *Drug Dev. Ind. Pharm.*, 28, 6, 621–630, 2002.
46. Ma, J., Lin, Y., Chen, X., Zhao, B., Zhang, J., Flow behavior, thixotropy and dynamical viscoelasticity of sodium alginate aqueous solutions. *Food Hydrocoll.*, 38, 119–128, 2014.
47. Khalil, S. and Sun, W., Bioprinting endothelial cells with alginate for 3D tissue constructs. *J. Biomech. Eng.*, 131, 11, 111002, 2009 Nov.
48. Peters, M.C., Isenberg, B.C., Rowley, J.A., Mooney, D.J., Release from alginate enhances the biological activity of vascular endothelial growth factor. *J. Biomater. Sci. Polym. Ed.*, 9, 12, 1267–1278, 1998.
49. Fraser, J.R.E., Laurent, T.C., Laurent, U.B.G., Hyaluronan: Its nature, distribution, functions and turnover. *J. Int. Med.*, 242, 1, 27–33, 1997.
50. Day, A.J. and Prestwich, G.D., Hyaluronan-binding proteins: Tying up the giant. *J. Biol. Chem.*, 277, 7, 4585–4588, 2002.
51. Toole, B.P., Hyaluronan: From extracellular glue to pericellular cue. *Nat. Rev. Cancer*, 4, 7, 528–539, 2004.
52. Genasetti, A., Vigetti, D., Viola, M., Karousou, E., Moretto, P., Rizzi, M., Passi, A., Hyaluronan and human endothelial cell behavior. *Connect. Tissue Res.*, 49, 3-4, 120–123, 2008.

53. Burdick, J.A., Chung, C., Jia, X., Randolph, M.A., Langer, R., Controlled degradation and mechanical behavior of photopolymerized hyaluronic acid networks. *Biomacromolecules*, 6, 1, 386–391, 2005.
54. Gerecht, S., Burdick, J.A., Ferreira, L.S., Townsend, S.A., Langer, R., Vunjak-Novakovic, G., Hyaluronic acid hydrogel for controlled self-renewal and differentiation of human embryonic stem cells. *Proc. Natl. Acad. Sci.*, 104, 27, 11298–11303, 2007.
55. Kiyotake, E.A., Douglas, A.W., Thomas, E.E., Nimmo, S.L., Detamore, M.S., Development and quantitative characterization of the precursor rheology of hyaluronic acid hydrogels for bioprinting. *Acta Biomater.*, 95, 176–187, 2019.
56. Lutolf, M.P. and Hubbell, J.A., Synthetic biomaterials as instructive extracellular microenvironments for morphogenesis in tissue engineering. *Nat. Biotechnol.*, 23, 1, 47–55, 2005.
57. Yang, Q. and Lai, S.K., Anti-PEG immunity: Emergence, characteristics, and unaddressed questions. *Wiley Interdiscip. Rev. Nanomed. Nanobiotechnol.*, 7, 5, 655–677, 2015.
58. Saik, J.E., Gould, D.J., Keswani, A.H., Dickinson, M.E., West, J.L., Biomimetic hydrogels with immobilized ephrinA1 for therapeutic angiogenesis. *Biomacromolecules*, 12, 7, 2715–2722, 2011.
59. Malmsten, M. and Lindman, B., Self-assembly in aqueous block copolymer solutions. *Macromolecules*, 25, 20, 5440–5445, 1992.
60. Fakhari, A., Corcoran, M., Schwarz, A., Thermogelling properties of purified poloxamer 407. *Heliyon*, 3, 8, e00390, 2017.
61. Batrakova, E.V. and Kabanov, A.V., Pluronic block copolymers: Evolution of drug delivery concept from inert nanocarriers to biological response modifiers. *J. Control. Release*, 130, 2, 98–106, 2008.
62. Nalbandian, R.M., Henry, R.L., Balko, K.W., Adams, D.V., Neuman, N.R., Pluronic F-127 gel preparation as an artificial skin in the treatment of third-degree burns in pigs. *J. Biomed. Mater. Res.*, 21, 9, 1135–1148, 1987.
63. Khattak, S.F., Bhatia, S.R., Roberts, S.C., Pluronic F127 as a cell encapsulation material: Utilization of membrane-stabilizing agents. *Tissue Eng.*, 11, 5-6, 974–983, 2005.
64. Matthew, J.E., Nazario, Y.L., Roberts, S.C., Bhatia, S.R., Effect of mammalian cell culture medium on the gelation properties of Pluronic® F127. *Biomaterials*, 23, 23, 4615–4619, 2002.
65. Barrs, R.W., Jia, J., Silver, S.E., Yost, M., Mei, Y., Biomaterials for bioprinting microvasculature. *Chem. Rev.*, 120, 19, 10887–10949, 2020.
66. Mironov, V., Reis, N., Derby, B., Bioprinting: A beginning. *Tissue Eng.*, 12, 4, 631–634, 2006.
67. Hollinger, J.O., Brekke, J., Gruskin, E., Lee, D., Role of bone substitutes. *Clin. Orthop. Relat. Res®.*, 324, 55–65, 1996.
68. Castilho, M., Rodrigues, J., Pires, I., Gouveia, B., Pereira, M., Moseke, C., Vorndran, E., Fabrication of individual alginate-TCP scaffolds for bone

tissue engineering by means of powder printing. *Biofabrication*, 7, 1, 015004, 2015.

69. Horváth, L., Umehara, Y., Jud, C., Blank, F., Petri-Fink, A., Rothen-Rutishauser, B., Engineering an *in vitro* air-blood barrier by 3D bioprinting. *Sci. Rep.*, 5, 1, 1–8, 2015.
70. Choi, Y.J., Kim, T.G., Jeong, J., Yi, H.G., Park, J.W., Hwang, W., Cho, D.W., 3D cell printing of functional skeletal muscle constructs using skeletal muscle-derived bioink. *Adv. Healthc. Mater.*, 5, 20, 2636–2645, 2016.
71. Wang, X., Yan, Y., Zhang, R., Rapid prototyping as a tool for manufacturing bioartificial livers. *Trends Biotechnol.*, 25, 11, 505–513, 2007.
72. Dimri, G.P., Lee, X., Basile, G., Acosta, M., Scott, G., Roskelley, C., Pereira-Smith, O., A biomarker that identifies senescent human cells in culture and in aging skin *in vivo*. *Proc. Natl. Acad. Sci.*, 92, 20, 9363–9367, 1995.
73. Mehrban, N., Teoh, G.Z., Birchall, M.A., 3D bioprinting for tissue engineering: Stem cells in hydrogels. *Int. J. Bioprinting*, 2, 1, 1–14, 2016.
74. Gupta, R., Truong, L., Bear, D., Chafik, D., Modafferi, E., Hung, C.T., Shear stress alters the expression of myelin-associated glycoprotein (MAG) and myelin basic protein (MBP) in Schwann cells. *J. Orthop. Res.*, 23, 5, 1232–1239, 2005.
75. Boland, T., Xu, T., Damon, B., Cui, X., Application of inkjet printing to tissue engineering. *Biotechnol. J.*, 1, 9, 910–917, 2006.
76. Cui, X., Dean, D., Ruggeri, Z.M., Boland, T., Cell damage evaluation of thermal inkjet printed chinese hamster ovary cells. *Biotechnol. Bioeng.*, 106, 6, 963–969, 2010.
77. Jana, S. and Lerman, A., Bioprinting a cardiac valve. *Biotechnol. Adv.*, 33, 8, 1503–1521, 2015.
78. Wayne, J.S., McDowell, C.L., Shields, K.J., Tuan, R.S., *In vivo* response of polylactic acid–alginate scaffolds and bone marrow-derived cells for cartilage tissue engineering. *Tissue Eng.*, 11, 5-6, 953–963, 2005.
79. Seetharam, R. and Sharma, S.K. (Eds.), *Purification and Analysis of Recombinant Proteins, Vol. 12*, CRC Press, USA, 1991.
80. Saunders, R.E., Gough, J.E., Derby, B., Delivery of human fibroblast cells by piezoelectric drop-on-demand inkjet printing. *Biomaterials*, 29, 2, 193–203, 2008.
81. Okamoto, T., Suzuki, T., Yamamoto, N., Microarray fabrication with covalent attachment of DNA using bubble jet technology. *Nat. Biotechnol.*, 18, 4, 438–441, 2000.
82. Hockaday, L.A., Kang, K.H., Colangelo, N.W., Cheung, P.Y.C., Duan, B., Malone, E., Butcher, J.T., Rapid 3D printing of anatomically accurate and mechanically heterogeneous aortic valve hydrogel scaffolds. *Biofabrication*, 4, 3, 035005, 2012.
83. Trombetta, R., Inzana, J.A., Schwarz, E.M., Kates, S.L., Awad, H.A., 3D printing of calcium phosphate ceramics for bone tissue engineering and drug delivery. *Ann. Biomed. Eng.*, 45, 23–44, 2017.

84. Williams, C.G., Malik, A.N., Kim, T.K., Manson, P.N., Elisseeff, J.H., Variable cytocompatibility of six cell lines with photoinitiators used for polymerizing hydrogels and cell encapsulation. *Biomaterials*, 26, 11, 1211–1218, 2005.
85. Ali, M., Pages, E., Ducom, A., Fontaine, A., Guillemot, F., Controlling laser-induced jet formation for bioprinting mesenchymal stem cells with high viability and high resolution. *Biofabrication*, 6, 4, 045001, 2014.
86. Ringeisen, B.R., Kim, H., Barron, J.A., Krizman, D.B., Chrisey, D.B., Jackman, S., Spargo, B.J., Laser printing of pluripotent embryonal carcinoma cells. *Tissue Eng.*, 10, 3-4, 483–491, 2004.
87. Barron, J.A., Ringeisen, B.R., Kim, H., Spargo, B.J., Chrisey, D.B., Application of laser printing to mammalian cells. *Thin Solid Films*, 453, 383–387, 2004.
88. Patz, T.M., Doraiswamy, A., Narayan, R.J., He, W., Zhong, Y., Bellamkonda, R., Chrisey, D.B., Three-dimensional direct writing of B35 neuronal cells. *J. Biomed. Mater. Res. B Appl. Biomater.*, 78, 1, 124–130, 2006.
89. Guillotin, B., Souquet, A., Catros, S., Duocastella, M., Pippenger, B., Bellance, S., Laser assisted bioprinting of engineered tissue with high cell density and microscale organization. *Biomaterials*, 31, 28, 7250–7256, [Internet] 2010.
90. Melchels, F.P., Feijen, J., Grijpma, D.W., A review on stereolithography and its applications in biomedical engineering. *Biomaterials*, 31, 24, 6121–6130, 2010.
91. Chia, H.N. and Wu, B.M., Recent advances in 3D printing of biomaterials. *J. Biol. Eng.*, 9, 1, 1–14, 2015.
92. Wang, Z., Abdulla, R., Parker, B., Samanipour, R., Ghosh, S., Kim, K., A simple and high-resolution stereolithography-based 3D bioprinting system using visible light crosslinkable bioinks. *Biofabrication*, 7, 4, 045009, 2015.
93. Zhu, J., Beamish, J. A., Tang, C., Kottke-Marchant, K., Marchant, R. E., Extracellular matrix-like cell-adhesive hydrogels from RGD-containing poly(ethylene glycol) diacrylate. *Macromolecules*, 39, 4, 1305–1307, 2006. https://doi.org/10.1021/ma052333s
94. Bártolo, P.J., Stereolithographic processes, in: *Stereolithography: Materials, Processes and Applications*, pp. 1–36, 2011.
95. Kačarević, Ž. P., Rider, P.M., Alkildani, S., Retnasingh, S., Smeets, R., Jung, O., Barbeck, M., An introduction to 3D bioprinting: Possibilities, challenges and future aspects. *Materials*, 11, 11, 2199, 2018.
96. Costantini, M., Colosi, C., Święszkowski, W., Barbetta, A., Co-axial wet-spinning in 3D bioprinting: State of the art and future perspective of microfluidic integration. *Biofabrication*, 11, 1, 012001, 2018.
97. Colosi, C., Shin, S.R., Manoharan, V., Massa, S., Costantini, M., Barbetta, A., Khademhosseini, A., Microfluidic bioprinting of heterogeneous 3D tissue constructs using low-viscosity bioink. *Adv. Mater.*, 28, 4, 677–684, 2016.
98. Gao, Q., He, Y., Fu, J.Z., Liu, A., Ma, L., Coaxial nozzle-assisted 3D bioprinting with built-in microchannels for nutrients delivery. *Biomaterials*, 61, 203–215, 2015.

99. Grosskopf, A.K., Truby, R.L., Kim, H., Perazzo, A., Lewis, J.A., Stone, H.A., Viscoplastic matrix materials for embedded 3D printing. *ACS Appl. Mater. Interfaces*, 10, 27, 23353–23361, 2018.
100. Jin, Y., Compaan, A., Bhattacharjee, T., Huang, Y., Granular gel support-enabled extrusion of three-dimensional alginate and cellular structures. *Biofabrication*, 8, 2, 025016, 2016.
101. Bhattacharjee, T., Zehnder, S.M., Rowe, K.G., Jain, S., Nixon, R.M., Sawyer, W.G., Angelini, T.E., Writing in the granular gel medium. *Sci. Adv.*, 1, 8, e1500655, 2015.
102. Skylar-Scott, M.A., Uzel, S.G., Nam, L.L., Ahrens, J.H., Truby, R.L., Damaraju, S., Lewis, J.A., Biomanufacturing of organ-specific tissues with high cellular density and embedded vascular channels. *Sci. Adv.*, 5, 9, eaaw2459, 2019.
103. Noor, N., Shapira, A., Edri, R., Gal, I., Wertheim, L., Dvir, T., 3D printing of personalized thick and perfusable cardiac patches and hearts. *Adv. Sci.*, 6, 11, 1900344, 2019.
104. Hinton, T.J., Jallerat, Q., Palchesko, R.N., Park, J.H., Grodzicki, M.S., Shue, H.J., Feinberg, A.W., Three-dimensional printing of complex biological structures by freeform reversible embedding of suspended hydrogels. *Sci. Adv.*, 1, 9, e1500758, 2015.
105. Hinton, T.J., Hudson, A., Pusch, K., Lee, A., Feinberg, A.W., 3D printing PDMS elastomer in a hydrophilic support bath via freeform reversible embedding. *ACS Biomater. Sci. Eng.*, 2, 10, 1781–1786, 2016.
106. Foty, R., A simple hanging drop cell culture protocol for generation of 3D spheroids. *J. Vis. Exp.*, 51, e2720, 2011.
107. Bernard, A.B., Lin, C.C., Anseth, K.S., A microwell cell culture platform for the aggregation of pancreatic β-cells. *Tissue Eng. Part C: Methods*, 18, 8, 583–592, 2012.
108. Yang, X., Mironov, V., Wang, Q., Modeling fusion of cellular aggregates in biofabrication using phase field theories. *J. Theor. Biol.*, 303, 110–118, 2012.
109. Ozbolat, I.T. and Yu, Y., Bioprinting toward organ fabrication: Challenges and future trends. *IEEE Trans. Biomed. Eng.*, 60, 3, 691–699, 2013.
110. Barrs, R.W., Jia, J., Silver, S.E., Yost, M., Mei, Y., Biomaterials for bioprinting microvasculature. *Chem. Rev.*, 120, 19, 10887–10949, 2020.
111. Esch, E.W., Bahinski, A., Huh, D., Organs-on-chips at the frontiers of drug discovery. *Nat. Rev. Drug Discovery*, 14, 4, 248–260, 2015.
112. Lee, H. and Cho, D.W., One-step fabrication of an organ-on-a-chip with spatial heterogeneity using a 3D bioprinting technology. *Lab. Chip*, 16, 14, 2618–2625, 2016.
113. Jain, R.K., Au, P., Tam, J., Duda, D.G., Fukumura, D., Engineering vascularized tissue. *Nat. Biotechnol.*, 23, 7, 821–823, 2005.
114. Chang, W.G. and Niklason, L.E., A short discourse on vascular tissue engineering. *NPJ Regen. Med.*, 2, 1, 7, 2017.

115. Bersini, S., Yazdi, I.K., Talò, G., Shin, S.R., Moretti, M., Khademhosseini, A., Cell-microenvironment interactions and architectures in microvascular systems. *Biotechnol. Adv.*, 34, 6, 1113–1130, 2016.
116. Paulsen, S.J. and Miller, J.S., Tissue vascularization through 3D printing: Will technology bring us flow ? *Dev. Dyn.*, 244, 5, 629–640, 2015.
117. Filipowska, J., Tomaszewski, K.A., Niedźwiedzki, Ł., Walocha, J.A., Niedźwiedzki, T., The role of vasculature in bone development, regeneration and proper systemic functioning. *Angiogenesis*, 20, 291–302, 2017.
118. Huang, C.C., Narayanan, R., Warshawsky, N., Ravindran, S., Dual ECM biomimetic scaffolds for dental pulp regenerative applications. *Front. Physiol.*, 9, 495, 2018.

4

Laser-Induced Forward Transfer for Biosensor Application

Ankit Das[1], Samarpan Deb Majumder[2], Drazan Kozak[3] and Chien-Fang Ding[1]*

[1]*Department of Biomechatronics Engineering, National Taiwan University, Taipei, Taiwan*
[2]*Department of Civil and Environmental Engineering, University of Wisconsin Madison, Wisconsin, United States*
[3]*Faculty of Mechanical Engineering, University of Slavonski Brod, Slavonski Brod, Croatia*

Abstract

Additive manufacturing is the future of advanced manufacturing. One of the additive manufacturing techniques that can be successfully, efficiently, and widely utilized is the laser-induced forward transfer (LIFT) process. In recent years, LIFT has been exploited extensively in printing thin films, and it has also been incorporated into manufacturing and printing sensor materials. Additionally, with the rise in miniaturization, the use of thin films in sensor applications has significantly increased. Recently, biosensors are a comprehensively studied area of research, for their easy, rapid, low-cost, highly sensitive, and highly selective properties. Biosensors have shown significant potential in areas of analytical biotechnology such as healthcare. Furtherance of biosensors for next-generation medicines and healthcare has been predominantly observed. Furthermore, biosensors have the ability to provide real-time information in healthcare applications. One of the major concerns over the use of biosensors is its expensive and time-consuming manufacturing processes. LIFT as a one-step process poses to be an efficient solution by collectively saving cost and time. This chapter imparts an elaborate description and overview on the LIFT process used for biosensor fabrication. First, biosensors are introduced in brief. Subsequently, a number of manufacturing processes used in biosensor fabrication along with their shortcomings

**Corresponding author*: cfding@ntu.edu.tw

Jay Prakash Srivastava, Dražan Kozak, Vinayak Ranjan, Pankaj Kumar, Ranjan Kumar and Shubham Tayal (eds.) Mechanical Engineering in Biomedical Applications: Bio-3D Printing, Biofluid Mechanics, Implant Design, Biomaterials, Computational Biomechanics, Tissue Mechanics, (77–118) © 2024 Scrivener Publishing LLC

are discussed. Following the biosensor discussion, LIFT is introduced. The process parameters and descriptions are reviewed in detail to provide a better insight into the LIFT process. Finally, considering the limitations offered by conventional biosensor manufacturing processes, the potential and advantages of the LIFT process are presented. It has been seen that the accuracy, cost, and time effectiveness of the LIFT process enable the efficient fabrication of biosensors, thereby proving itself as a potential manufacturing option for the future.

Keywords: Additive manufacturing, laser-induced forward transfer (LIFT), biosensors

4.1 Introduction

Additive manufacturing is one of the rapidly developing manufacturing processes that is being incorporated into our daily lives. Additive manufacturing has a variety of synonyms in technological aspects, such as 3D printing, rapid prototyping, and layered manufacturing [1]. Additive manufacturing has broken the chains of limitations for traditional manufacturing, thereby making manufacturing easier, cost-effective, accurate, and rapid. Some of the widely used AM processes are stereolithography (SLA), selective laser melting (SLM), selective laser sintering (SLS), 3D printing, coating technologies, etc. In light of the advantages offered by additive manufacturing, one such process is laser-induced forward transfer (LIFT) which contributes widely to the sensor industry, discussed subsequently.

LIFT is a high-resolution printing/transfer process used for depositing a smidgen of materials on a substrate. Either solid or liquid phase media can be used in this process. In short, the desired material (to be deposited) is transferred from a donor material onto a receiver substrate material using pulsed laser irradiation. Various research studies have been made to utilize LIFT for material-specific applications like DNA [2], polymers [3, 4], nanotubes [5], biomaterials [6, 7], metals [8, 9], particles [10–13], and inks [14, 15]. Nearly all materials can be deposited using the LIFT process, thereby providing evidence for sensitive material printing in a one-step solvent-free method. The other applications of LIFT are utilized as an alternative manufacturing process for the fabrication of transistors [16–18], MEMS [19, 20], organic LEDs [21–23], sensors [24–27], etc.

Human senses receive and process information from the surrounding environment; however, human senses have limitations [28]. Therefore, there is a need for practical sensors, which overcome the limitations posed by nature [28]. For example, a device with an output having a digital number or some other light intensity, fabricated using a current generator, voltage amplifier, and signal converter (analog to digital), is also termed a sensor [29]. LIFT has been widely utilized for sensor fabrication and has gained popularity recently. The advantage of the LIFT process outperforms the disadvantage of debris generation. An advantage offered by the LIFT process is the ability to deposit sensitive materials, thereby providing the potential to improve sensor selectivity [30]. Subsequently, the fact that LIFT can utilize a wide range of materials, thereby enabling the development of highly sensitive sensors, is the other advantage, making the process versatile for a wide range of applications [31]. The research on LIFT has been vastly made; therefore, this review only focuses on biosensors manufactured/fabricated using LIFT. First, biosensors are described in detail; subsequently, the LIFT process is introduced. Focus is made on the electrochemical biosensors to make to review critical and succinct. Afterward, a summary is made on biosensors fabricated using the LIFT process. Finally, encouraging discussions are made to provide effective solutions to the limitation of biosensors through LIFT.

4.2 Biosensor

4.2.1 History/Background

In the early 1960s to 1970s, biosensors were more generalized as some sort of probe, due to their close association with pH and ion-selective electrodes [32]. In those early days, biosensors were often referred to as bioelectrodes and biocatalytic enzyme electrodes [33]. According to the literature, the first biosensor which detected oxygen was introduced by Clark in 1956 [34], and he began to be known as the "father of biosensors." In 1962, Clark and Lyons developed a biosensor, more often referred to as enzyme electrodes, and following that, Guilbault and Montalvo, Jr [35] developed the first potentiometric biosensor for urea detection in 1969. In 1984, Christopher R. Lowe [36] published an article stating the rapid growth and commercial potential of biosensors. Since then, various biological systems and processes have initiated the development of sensor technologies. Therefore, biosensors revolutionized the areas of analytical biotechnology (e.g., healthcare monitoring and control) [32].

An instrument that utilized a biological sensing element integrated with a transducer is known as a biosensor [37]. According to Eggins [38], biosensors consist of sensing elements that respond, when exposed to biological substances. To account for this response, a transducer is attached, which provides a visual response like the voltage and change in color. Some of the components usually present in a biosensor are as follows: (i) analyte: a material of interest which is to be detected, e.g., in glucose biosensors, and glucose is the "analyte"; (ii) bioreceptor: molecules which are responsible for recognizing the analyte, and the subsequent signal generated due to their interaction is termed biorecognition; (iii) transducer: this device converts the biorecognition signal to a quantitative signal; (iv) electronics: transducer signals are processed for display, visual, and qualitative analysis, and this is done by the electronics; and (v) display: the measured signals and results are displayed through this component of the biosensor. The display can illustrate numbers, graphs, images, etc. as per requirements. A basic illustration of a biosensor having all the aforementioned elements/components is shown in Figure 4.1. However, it must be noted that the above components are usually present in a biosensor, but not a compulsion. The latest designs of biosensors are capable of eliminating some of the components to further enhance the performance and cut costs for biosensors.

Before diving into the details and types of biosensors, one must know the basic performance parameters of a biosensor as described below:

Selectivity: This property enables the biosensor to distinguish between various substrates. Identifying the specific analyte of interest among a mix of other substances is important. Therefore, selectivity is highly important.

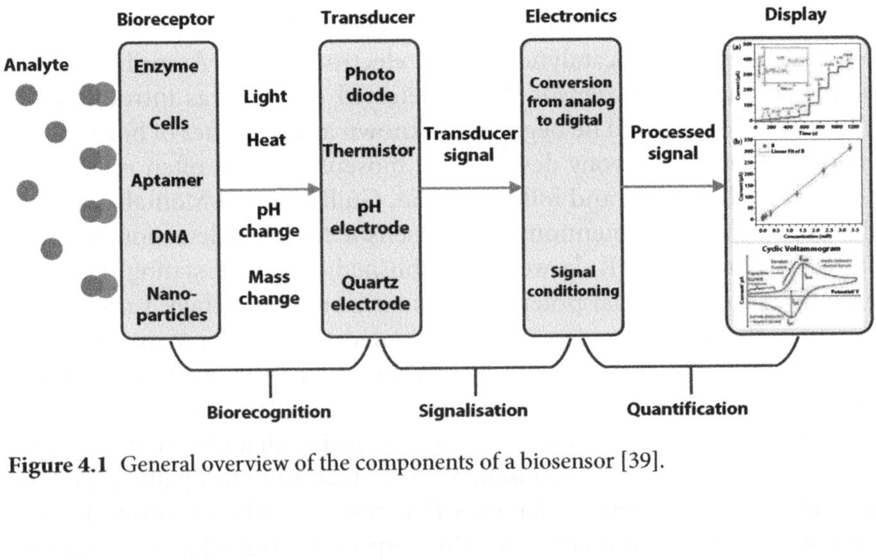

Figure 4.1 General overview of the components of a biosensor [39].

Reproducibility: This is the ability to produce identical response signals for a duplicated experimental setup [39]. The reproducibility of a biosensor is highly dependent on the accuracy and precision of the embedded transducer and electronics. High reliability and robustness are achieved through reproducible signals in a biosensor.

Stability: This refers to the degree of susceptibility or vulnerability to various surrounding disturbances experienced by the biosensing system. Such disturbances have the ability to alter the output signals during sensing resulting in errors hampering the biosensor precision and accuracy.

Sensitivity range: It marks the ability of the biosensor's sensing in terms of the substrate concentration. Sensitivity ranges are recommended to near submillimolar and can go up to femtomolar order.

Linearity: This illustrates the accuracy of the measured response to a straight line. According to the straight-line equation, $y = mx + c$, where c is the analyte concentration, y is the measured signal, and m is the biosensor sensitivity. Resolution and range can have a significant influence on the linearity of a biosensor.

In addition to the above, there are numerous other factors that have a significant influence on the biosensor performance. Some of these are precision, accuracy, resolution, response time, recovery time, working lifetime, linearity range, nature of the solution, etc. Precision is defined as the sensor's ability to produce similar responses each time a specimen is measured. Moreover, accuracy is a measure that shows the closeness of the mean value to the true/real value during multiple measurements (recommended accuracy ~5%). Resolution is responsible for capturing and providing a change in biosensor response when the required small changes in the concentration of an analyte take place. The linear range is associated with linearity; this is the analyte concentration's range that provides linear biosensor response. Time is an important asset; therefore, response time, recovery time, and working lifetime must be determined. Response time is the time taken to generate a new signal when the analyte concentration is changed. Simultaneously, the duration of the response current to attain a new and steady-state value post-disturbance (by the presence of a toxic component) is termed recovery time [40]. Finally, the solution nature (pH, ionic strength, and temperature) must be carefully considered as they can affect biosensor response.

4.2.2 Types of Biosensors

Since the 1960s, a number of biosensors have been fabricated according to their intended application. Majority of biosensors include electrochemical biosensors, optical biosensors, optoelectronic biosensors, piezoelectric biosensors, field-effect transistor (FET-based) biosensors, surface acoustic wave-based biosensors, thermal method-based biosensors, biosensors based on ion-selective electrodes (ISEs), plasmonic biosensors, immune biosensors, microbial biosensors, metabolism sensors, and affinity biosensors. Karunakaran *et al.* [41] widely classified biosensors on the basis of bioreceptors or transducers used as shown in Figure 4.2. However, it must be noted that a variety of biosensors have been and are being developed constantly.

A number of biosensors have been introduced above; however, to make the chapter succinct, electrochemical biosensors will be discussed henceforth. Ever since glucose biosensors have been developed, electrochemical sensing has gained attention as a potential option in commercial point-of-care diagnostic systems [42]. One of the subsets of chemical sensors is electrochemical biosensors [43]. Enhanced sensitivity and real-time analysis capabilities of electrochemical biosensors have led to the development of various advanced designs [42]. Most electrochemical sensors use the three-electrode setup in an electrolyte medium, composed of a working electrode (WE), a counter electrode (CE), and a reference electrode (RE) [44]. WE is one of the prime electrodes of interest as the reaction occurs

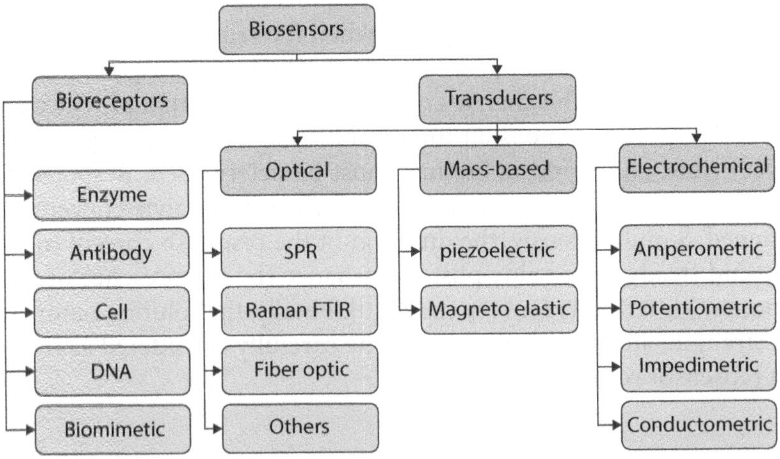

Figure 4.2 Classification of biosensors based on various types of bioreceptors and transducers [41].

on its interface and functions as a transducer for electron transfer [45]. The RE is used to measure the half-cell potentials and is commonly made up of Ag/AgCl. The CE (also named as an auxiliary electrode) is used to complete the circuit and is made significantly larger than the WE to ensure that current limitations are avoided in the electrochemical circuit [44, 46].

Further classification of electrochemical biosensors is based on transduction mechanisms. These are potentiometric, conductometric, and voltammetric/amperometric. In potentiometric sensors, the potential difference (between WE and RE) is altered by the analyte. In voltammetric or amperometric sensors, the current response at a fixed potential is captured, which is due to the redox reactions at the interface. Conductometric sensors as the name suggests capture the changes in conductivity due to the variation in the mobility of the charge carriers or the variation in density due to analyte interactions [47, 48]. Impedometric sensors on the other hand follow either impedance (Z) or its component's resistance (R) and capacitance (C). Conductometric and impedometric sensors are closely related as their properties are just inversely related to each other [49].

4.2.2.1 Potentiometric Biosensors

Potentiometric biosensors come from the concept of potentiometry, which provides information on the ions in an electrochemical reaction [50, 51]. Potentiometric biosensors measure the charge potential accumulated across the WE and RE, provided that no significant (near zero) current flow is present [52–54]. The primary constituents of potentiometric biosensors are ion-selective electrodes (ISE) and ion-sensitive field effect transistors (ISFET) [49]. ISEs are often utilized for determining the lowest detection limits. Ions which are accumulated on the interface of the ion-selective membrane are responsible for the initial response. The Nernst equation is one of the most important equations in this type of biosensor, where a relationship is established between potential and concentration.

$$EMF \ or \ E_{cell} = E_{0\,cell} - \frac{RT}{nF} lnQ \tag{4.1}$$

where E_{cell} is the electromotive force (EMF) or the cell potential observed when current = 0 A, E_{0cell} is the constant potential contribution to the cell, R is the universal gas constant, T is the absolute temperature (K), n is the charge number of the electrode reaction, F is the Faraday constant, and Q is the ion concentration ratio $\left(Q = \frac{[anode]}{[cathode]}\right)$ [55].

Some of the notable uses are in areas where there are low concentrations and very few specimen volumes as they do not chemically influence the specimen. In addition, this can be used for potentiometric titration for endpoint measurement using variations in potentials resulting from concentration variation of the target ion. However, the limit of detection (LOD) is limited for a variety of ions. Therefore, further studies on LOD for different ions must be made.

4.2.2.2 Amperometric Biosensors

This type of electrochemical sensor continuously measures the current generated via oxidation and reduction of certain electroactive species [52, 56]. Amperometric biosensors are thus based on amperometry (current measured at a constant potential). The bulk analyte (electronic species) concentration is directly proportional to the current peak, which is measured over a linear potential range [52, 53, 57]. Amperometric biosensors are quite sensitive and favored over potentiometric sensors for mass production [58]. These biosensors can be further classified into single use, intermittent use, and continuous use. Biorecognition components over a screen-printed layer, noble metals, or carbon paste with enzyme embedded (economic option) are usually chosen for WE of an amperometric biosensor [59, 60]. These biosensors accommodate both two-electrode (RE and WE) or three-electrode (RE, WE, and CE) configurations. The limited potential control for higher WE surface currents makes the linear range short, in the two-electrode configuration. Therefore, to overcome this limitation, the three-electrode system with a CE is used. In such configuration, the voltage is applied between the RE and WE, and current flows through the WE and CE.

There are some limitations to amperometric biosensors. Due to the incapability of some protein analytes to serve as redox partners, mediators are to be used for suitable transduction (direct/indirect) [53, 56]. However, despite the limitations, amperometric systems are preferred over potentiometric ones due to their ability to maintain sensitivity [56, 61]. A glucose biosensor is a common amperometric biosensor that shows amperometric detection of hydrogen peroxide (H_2O_2). Some of the analytes which commonly use amperometric biosensors on a large scale are glucose, lactate [62], and sialic acid [63]. Other applications include biological agents [64], serological diagnosis [65], pharmacological studies [66], and pesticide-nerve agent detention [67].

4.2.2.3 *Impedimetric Biosensors*

Impedimetric biosensors are established on the principle of impedance (Z) or resistance (R) and capacitance (C) [49]. The general equation used to represent impedance is given as

$$Z^2 = R^2 + \frac{1}{(2fC)^2} \tag{4.2}$$

It is to be noted that inductance is not considered since it has a negligible influence in a typical electrochemical setup. The setup consists of two electrodes and has an alternating voltage with amplitudes up to 100 mV. Generally, the biosensor imitates a part of the Wheatstone bridge. Impedimetric biosensors are not popular as the potentiometric and amperometric biosensors; however, some promising research evidence shows their potential. Reciprocal of resistance is known as conductance; hence, there are research studies made on conductometric biosensors, discussed subsequently.

4.2.2.4 *Conductometric Biosensors*

This type of biosensor works on the analyte's or medium's ability (e.g., electrolyte solutions, nanowires) to conduct electricity between the electrodes. As mentioned above, these biosensors are a subdivision of impedimetric sensors. The conductometric biosensors have a strong association with enzymes, as conductivity resulting from ionic strength is dependent on the enzymatic reactions. Therefore, enzymatic reactions causing changes in ionic concentration can be monitored and studied using such sensors [68]. Subsequently, variation in conductance due to immobilization of analytes onto the surface of the electrodes can be utilized for suitable monitoring.

Field-effect transistor (FET) devices in association with multianalyte and polymer-based conductive biosensors have been developed [69, 70]. Recently, conductometric biosensors along with nanostructures and nanowires are becoming popular [71, 72]. Although there has been a successful application of conductometric biosensors in practical applications like the detection of drugs, urine, pollution, toxicity (using cells as biorecognition elements), etc. [73, 74], they have not been extensively used as per their potential [52].

4.2.3 Biosensor Manufacturing Processes

Electrochemical biosensors dominate over others due to their low cost, militarization capabilities, and automotive and easy-to-use properties [75]. A number of manufacturing processes have been used to fabricate electrochemical biosensors, of which printing technologies have been the most popular techniques. These include screen printing, inkjet printing, and roll-to-roll printing (gravure printing, roll-to-roll nanoimprinting, and flexographic printing) [76, 77].

In the 1980s, the MediSense team adapted screen printing, and at present, it is utilized in most disposable glucose sensors [78] due to its reproducibility, reliability, low-cost, and mass production capabilities [75]. The other properties of screen-printed electrodes for biosensors include the ability to develop ductile devices, which can be made of different materials (enzymes, antibodies, DNA, etc.) in various shapes. Screen printing has paved the way for numerous applications such as enzyme-based biosensors (nanomaterial, wearable and paper-based microfluidic devices) [75], DNA biosensors [79], and biosensors for monitoring metal pollutants [80] and detection of biomarkers (cardiac, cancer, inflammatory) [80].

Inkjet printing is the widely used and most promising method for mass production due to its wide material range and mask-less and non-contact mode of operation. Inkjet printing consists of two main parts: (i) an ink chamber and (ii) a channel connecting the ink chamber to the nozzles [81]. Since the first concept in 1878 [82], thermal and piezoelectric printers have been the most popular types of inkjet printing. However, both these have their respective limitations. Thermal printing introduces limitations such as high-temperature generation [83], thermal stability, and thermal degradation of organic/biological materials [84], and these can be overcome by piezoelectric printing. However, piezoelectric methods offer complicated mechanisms, less environment-friendly ink solvents, and higher ink viscosities when compared with the thermal methods [84, 85]. Additional limitations of inkjet printing lie in its resolution at nanoscales (limited to microscales). In such cases of nanoscale-level resolution requirements, scanning probe lithographic techniques are beneficial, but again, they are limited to the industry-level application due to its time-consuming nature. One process which saves time is microcontact printing, but it is limited due to the mask requirement. Lithography techniques are slow and therefore not preferred for mass production, especially e-beam lithography (even slower than photolithography). Albeit, MAPLE permits printing of multilayers; however, it is only suitable for low volume, prototyping, and mask production or repair [77]. To summarize, lithographic techniques mentioned above are not employed for the mass production of biosensors, but they only act as preliminary steps to manufacturing. The plasma polymerization method offers advantages like improved enzyme immobilization (by substrate functionalization) and limitations like high cost of operation (due to its vacuum requirements), limiting its industrial use. If only this technique is used in complementary to other processes for improvements in enzymatic adhesion, it might be plausible.

Gravure printing utilizes a rotating printing cylinder used to engrave patterns. Conducting polymers [86, 87] and ITO nanoparticle dispersions [87, 88] are some materials that have been used. This process has not been popular due to high-pressure requirements and the resulting low resolution of the printed patterns hampering the development of high-precision devices. Furthermore, this kind of process is long and utilizes the cost of cylinders, which otherwise is expensive for short runs. In flexographic printing, the printing roll is coated through a flexography rubber having raised portions coated with a uniform film of ink. Although this printing technique is a little better than gravure printing in terms of pressure and resolution, it is only used for short and medium runs. Another flexographic technique that enables the patterning of submicron features is roll-to-roll nanoimprinting (with speeds up to 20 m/min). This process transfers the pattern to the substrate using pressure, usually at high temperatures.

Other techniques include microcontact printing, soft lithographic method (microfluidic network, (μFN), laminar flow, and metal transfer (MT) printing) [89, 90], probe-based techniques (dip-pen nanolithography [91, 92], nanografting, local oxidation nanolithography), lithographic techniques (photolithography, electron beam, and laser-assisted (matrix-assisted pulsed laser evaporation (MAPLE)) deposition), plasma-based patterning (plasma polymerization) [93–98], vapor deposition, and subsequent laser ablation.

There are a number of limitations that are encountered in the aforementioned biosensor fabrication techniques and are discussed subsequently. Inkjet and lithographic techniques enable direct [99, 100] and indirect [101, 102] deposition of enzymes. Simultaneously, the extreme conditions required in plasma polymerization do not support direct deposition [103]. Furthermore, direct deposition of enzymes/proteins is possible for dip-pen nanolithography and local oxidation, whereas indirect deposition is supported by nanografting and microcontact printing.

In conclusion, from the aforementioned limitations, it is evident that existing techniques cannot provide a short time, high precision, and nanometer-scale patterns at the same time. Therefore, in the past, manufacturing processes were decided according to the application, i.e., which is more important to the final device resolution: time effectiveness or precision. Upon deciding the process specific to the final device, optimizations are made to achieve the best results. Therefore, the need for a single-step process addressing the aforementioned shortcomings is inevitable. This leads us to a process called laser-induced forward transfer (LIFT), with enormous potential in the development of biosensors, due to its fast, precise, and high-resolution capabilities. LIFT offers unique advantages over inkjet methods, due to its wide range of printable viscosities and no important composition and material rheology restrictions [104]. Furthermore, LIFT is a non-contact process that prevents contamination. Lastly, the nozzle-free LIFT technique eliminates problems like clogging and drying which hampers the performance of printing heads for inkjet printing. LIFT is discussed in the subsequent sections, and successful applications in biosensor fabrication are presented.

4.3 Laser-Induced Forward Transfer (LIFT)

4.3.1 History and Process Description

In 1986, Bohandy *et al.* [106] first introduced laser-induced forward transfer or LIFT. The LIFT process was motivated by the research of Levene *et al.* [105] where the Nd:YAG laser was used to transfer black ink onto a substrate. The first application of LIFT deposited copper (Cu) metal on fused silica and silicon [106]. Following the invention of LIFT, Adrian *et al.* [107] from Bohandy's group studied the metal deposition mechanism for the first time in 1987. In the early days after the introduction of the LIFT process, droplets and condensed vapor were proposed to be the mechanisms for the process of transfer and deposition of the material [108]. A model based on partial evaporation of the film material was introduced. According to this model, a gas–molten particle mixture was responsible for the material transfer from the donor support toward the target or receiver substrate [108].

The direct-write technique is analogous to digital printing, which is a combination of processes used to deposit functional materials onto a substrate with digitally specified locations [109]. The printing press invention which utilized masks and molds, allowing mass production methods, lacks the possibilities for customization and adaptability to any alterations in production change. In comparison to mass production-based methods, the direct-write technique offers customization for every product. Similar to the direct-write technique, LIFT is a type of digital printing technique that transfers materials from the donor film onto the receiver substrates [25, 104]. A pulsed laser beam is employed to project material from the donor film onto the receiving substrate, where the material is deposited as a voxel. A graphical illustration of the LIFT process is shown in Figure 4.3. The LIFT can operate on both solid and liquid donor films, which makes it versatile for various applications such as droplet printing, graphics production, printing of functional materials, bioprinting, and printed electronics with a wide range of ink viscosities and loading particles [110]. LIFT can be simply described in three main steps [6]. In the first step, the film material is ablated from the donor (initial) base support. Subsequently, transfer of the ablated material takes place, i.e., from the donor to the receiver (target) substrate. The ejection velocity of ablated solid materials/pixels/voxels is in the range of 200 to 1,200 m/s (approx. more than three times the speed of sound in air) [111]. Finally, in the third step, the flyer (as voxel) material is deposited.

One LIFT can be broadly classified on the basis of the phase of its working materials [111]. Simultaneously, the process and the mechanisms involved are different for different phases: LIFT of (i) solids, (ii) liquids [112], (iii) paste [113], and (iv) solid in liquid. In addition to the above classification, LIFT has a wide range of variants, briefly summarized in Figure 4.4. It is evident to note that the phase involved in the LIFT process largely depends on the process and material parameters and is sometimes difficult to distinguish. A number of LIFT-related processes and their abbreviations are summarized as follows: **SS-LIFT**: solid-state LIFT; **LDT**: laser decal transfer; **BLAST LIFT**: ballistic laser-assisted solid transfer;

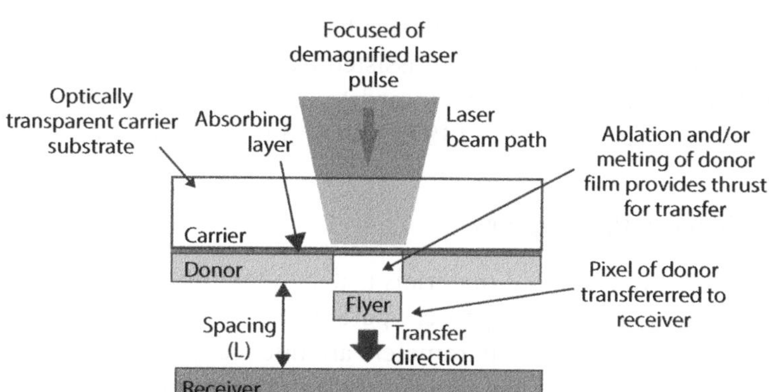

Figure 4.3 Graphical illustration of the LIFT process [110].

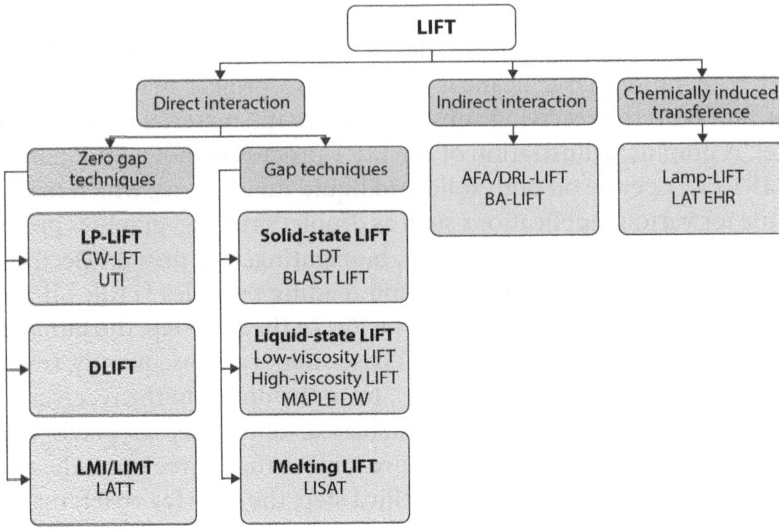

Figure 4.4 Classification of LIFT-related processes.

LS-LIFT: liquid-state LIFT; **MAPLE DW**: matrix-assisted pulsed laser evaporation direct write; **LIMJ**: laser-induced microjet transference; **TIN-LIFT**: thermal-induced nozzle LIFT; **LITV**: laser-induced thermal voxels; **LISAT**: laser-induced self-alloying transfer; **LIIFT**: laser-interference-induced forward transfer; **LIIBT**: laser-interference-induced backward transfer; **LP-LIFT**: long pulsed LIFT; **CW-LIFT**: continuous wave LIFT; **LITI**: laser-induced thermal imaging; **DLIFT**: laser-induced forward transfer doping; **LTD**: laser transfer doping; **LMI**: laser molecular implantation/ **LIMIT**: laser-induced molecular implantation; **DRL-LIFT**: dynamic release layer LIFT/**AFA-LIFT**: absorbing film-assisted LIFT; **BA-LIFT**: blister-actuated LIFT/**BB-LIFT**: blister-based LIFT; **LIFT**: lamp-induced forward transfer; **LAT HER**: laser-assisted transfer by explosive hydrogen release; **LIBT**: laser-induced backward transfer; **HILIRT**: high-intensity laser-induced reverse transfer; **LIST**: laser-induced side transfer.

4.3.2 Process Parameters

The quality of LIFT-printed materials is dependent on various major process parameters. The major process parameters can affect the LIFT process to a great extent. These parameters are fluence (J/cm^2), film–acceptor substrate distance (μm), scan rate (mm/s), and focus position of the laser beam. However, before diving further into the details of the LIFT process parameters, a few important fundamental relations must be understood. Therefore, knowledge of the fluence, peak power, and intensity in a laser process is a must. The following equations (Eqs. (4.3), (4.4), and (4.5)) are widely used:

$$Fluence\ (J/cm^2) = \frac{Laser\ pulse\ Energy\ (J)}{Effective\ focal\ spot\ area\ (cm^2)} \tag{4.3}$$

$$Peak\ Power\ (W) = \frac{Laser\ pulse\ Energy\ (J)}{Pulse\ duration\ (s)} \tag{4.4}$$

$$Intensity\ (W/cm^2) = \frac{Laser\ peak\ power\ (W)}{Effective\ focal\ spot\ area\ (cm^2)} \tag{4.5}$$

4.3.2.1 *Fluence of Lasers*

Fluence is the energy of the laser pulse over the effective focal spot area or simply the energy per unit area. Appropriate fluences are important for the ablation of material from donor support. Past research shows that ablation can take place not only through evaporation (for bulk materials) but also in the solid phase (for thin films) when exposed to certain laser fluence [108, 114]. At low fluences (in the order of mJ/cm^2), ablation starts, and above this, the film material takes off completely from the donor support without any collateral damage to the substrates. If the fluence is maintained at certain magnitudes, then the ablation of the material takes place without any melting (in the solid phase). The limitation of the solid-phase LIFT is its poor adhesion properties between the deposited voxels. At higher fluences, melting of the film material takes place, leading to liquid phase transfer. However, this leads to messy ablation and incomplete removal due to the resolidification of material droplets on the donor support. Upon further increase in fluence magnitudes, evaporation of the film material takes place, and the materials are deposited on the receiver (target) substrate through condensation, leading to higher material removal and transfer efficiency. A single pulse with high fluences (above 1 J/cm^2) evaporates the entire layer of the material, resulting in complete ablation. However, several disadvantages are encountered at such extreme fluences like enlargement of the ablated area, debris, and thermal damage [69, 115, 133]. Therefore, although adherence between the deposited voxels is low

Figure 4.5 LIFT deposition characteristics at different fluence magnitudes [116]: (a) low fluence, (b) optimized fluence, and (c) high fluence.

in the solid phase (low fluence), still there is the propitious possibility of continuous and high-quality LIFT-deposited patterns.

Sano *et al.* [45] extensively described the problems associated due to improper fluence. Figure 4.5 illustrates the effects of fluence on the LIFT deposited print quality [116]. Optimum fluence leads to high size accuracy and good adherence with the substrate with desired film thickness. Bad appearance due to incomplete material removal and spreading of the deposited structure due to big shocks was observed from the lower and higher fluence magnitudes, respectively. Furthermore, fluence also has significant effects on the flyer velocity [111].

4.3.2.2 Film–Acceptor Substrate Distance

The film–acceptor substrate distance (L) is one of the important parameters that affect the LIFT process. The ablated material velocity during transfer decreases with the increase in L [117]. Additionally, L affects the cooling (radiation and convection) of the flyer material. This can in turn affect the deposition adhesion. Therefore, the distance L is crucial in controlling the transfer yield and transfer efficiency [117]. One of the earliest investigations made by Kántor *et al.* [115] attempted to study the effects of film to substrate distance on LIFT in 1992. They provided evidence showing the highest transfer efficiency for low film substrate distance at fluence generating boiling point temperatures.

Sano *et al.* [116] described how deposition spread depends on the film–receiver distance (L). Figure 4.6(a) shows the deposited pixel (from the Ni thin film, thickness: 180 nm, fluence: 1.5 J/cm^2, L: 20 μm). Material spread is denoted by a length d, which is a function of L (as seen in Figure 4.6(b)). Experimental results show a linear relationship between d and L for a few hundred micrometers, and this is due to the absence of an external force on the transfer material. However, for $L \leq 100$ μm, d starts to increase in a non-linear fashion, which is due to the influence of external force derived

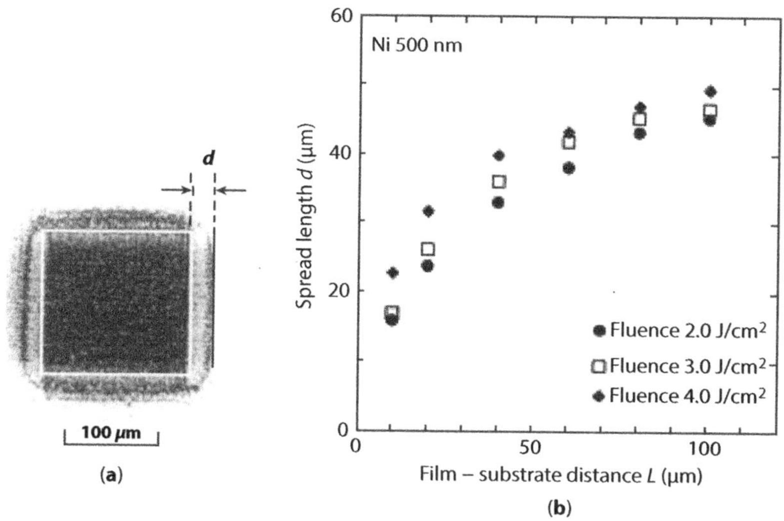

Figure 4.6 (a) Deposited structure and (b) the relationship between *d* and *L* [116].

from the flowing gas in the region. At optimum fluence, short *L* leads to enhanced size accuracy in the deposited structures.

Dinca *et al.* [118] studied PBS and glycerol solution transfer using a 10-nm-thick Au sacrificial layer and concluded that no significant variation in deposits occurred (distance varied from 20 μm to 1 mm). Subsequently, Palla-Papavlu *et al.* [119] studied the effects of the target–substrate distance using triazene polymers (TP) over Au or Ti layers as sacrificial layers. TP shows no debris formation (rare for other materials [120]), and the size–shape of the transferred patterns was not affected by the changes in the target–substrate distance (from 50 μm to 2 mm). Additionally, the droplet size was affected by the variation in the target–substrate distance (for 150- and 350-nm-thick TP films).

4.3.2.3 *Material Selection*

Material selection is a characteristic of the application. However, one must know which materials require low fluence and high fluence values for proper ablation, transfer, and adhesion. Some clean ablated materials at low fluences are V [121], Sn [115], Ti [115, 122], Cu [123–125], W [126], and Ge [127] (substrates: glass or quartz). Past evidence shows that for achieving a clear ablation, the onset of the film melting marks the upper limit of the fluence domain [6]. Subsequently, for optimizing the lower limit, the thermomechanical properties of the metal film and its supporting substrate along with their interface are considered. Additionally, a material (metal) with a high melting point can offer a wider clean ablation domain. For example, tungsten has a wide clean ablation domain due to its high melting point (100 nm film: solid-phase ablation with a fluence of 40 to 200 mJ/cm^2) [13]. Furthermore, strong adhesion is expected between the deposited material and the target (receiver) substrate when flyer materials are transferred at high temperatures (due to the transport of higher energy).

4.3.2.4 *Pulse Characteristics of Lasers*

A lot of efforts have been made in the past to study the effects of pulse characteristics on LIFT [117, 128, 129]. According to Sano *et al.* [116], Kantor *et al.* [117, 128, 129] studied long-pulsed (100–1,000 μs) and short-pulsed (nanosecond) lasers and compared them in detail for the LIFT process. They concluded that short-pulsed lasers show complete ablation but limited adherence with the acceptor/target substrate. However, when using long-pulsed lasers, enhanced adherence was achieved for tungsten micrometer-sized deposits, due to solid-phase transfer. Further investigations like time-resolved transmittance, temperature calculations, and reflectance measurements revealed that ablation is not dependent on pulse width and pulse shape and starts at a price temperature for long-pulsed lasers.

The LIFT process initially employed nanosecond lasers; however, several advancements and changes like femtosecond lasers result in positive effects on reducing heating effects and precise material processing [130, 131]. Klini *et al.* [132] demonstrated the effects of temporally shaped femtosecond laser pulses on LIFT. It was reported that temporal shape and the time scale of the excitation pulse affect the size and morphology of the deposits [132]. Past literature suggests that spatial characteristics of the transferred films can be controlled through the variation in pulse shapes by accessing the time scales of the aforementioned fundamental processes. Subsequently, these femtosecond laser pulses can be effectively utilized for controlling the size and morphology of micron-sized metallic structures.

Recently, numerous research has been made to transfer nanoscale features using picosecond and femtosecond lasers [58, 125]. Wills *et al.* [133] deposited microdroplets using nanosecond pulses. Nanosecond pulses are limited by the focal spot size of laser [125, 134, 135]. Subsequently, fluence and laser pulses have little interrelations. For nanosecond pulses, the transferred material area is greater or equal to the laser spot. The flyer velocity is not significantly affected by the pulse duration (when changed between ps and μs). However, it is interesting to note that when the order of the pulse duration is reduced to femtosecond, the flyer velocity is reduced (to about 34 m/s) with negligible shock wave generation (material: PZT; pulse duration: 100 fs) [136]. Furthermore, controlling the pulse duration has additional advantages; the possibility of a heat-affected zone (HAZ) and thermal menace can be minimized by using ultrashort pulses. Nevertheless, the ultrashort pulses may induce high stresses. Short pulse duration in the femtosecond order also restricts the formation of a uniform top hat energy beam profile which is an essential requirement for most printing applications. Therefore, in such cases where a uniform beam profile is a necessity, ns and ps beams are used.

In LIFT where transfer takes place by liquid droplets, it shows different characteristics with pulse duration. A short laser pulse generates liquid jets and cavitation bubbles; however, since cavitation is a thermally driven mechanism, short pulses do not affect the LIFT process. On the contrary, when ultrashort pulses are used, printing of tiny droplets is favored by limiting thermal diffusion within the film thickness [111]. After discussion on the short and ultrashort laser pulses, longer pulse durations are discussed subsequently. For μm-sized thin tungsten patterns with long (100–1,000 μs) laser pulses, several facts [115] were observed as follows: (i) regardless of the LIFT pulse shape and duration, the onset of characteristic processes is a well-defined constant temperature, and (ii) on-site annealing of deposited patterns takes place by proper shaping of the subsequent (second) part of the irradiating pulse, thereby achieving good adherence with the target substrate. Additionally, with longer pulses, the conventional description of LIFT is not applicable (like distinct differences between material removal, transfer, and deposition). Past research suggests that using long pulse durations resulted in high debris formation when transferring liquid metal drops from the solid metal film [137, 138]. However, this limitation was addressed by Zenou *et al.* [139, 140], where they adjusted the parameters (400 ps pulse; ~100 nm; fluence) to melt through the total film thickness. Additionally, the double pulse approach has also been seen to be effective in countering such problems. In conclusion, it can be inferred that μm-sized patterns cannot be significantly optimized by altering (a few tens of) ns. Additionally, optimizing the pulse energy and duration can lead to a successful deposition with the required properties, through proper ablation and conservation of high film temperature (just below the melting temperature) during and after the material transfer [115].

4.3.2.5 *Laser Spot Size*

Laser spot size can be altered to control the feature size [141]. For lasers with similar fluences, incident energy and deposit thickness are lower for the smaller spot size. However, the deposit thickness is not significantly affected (reduced) at higher fluences with a reduction in spot sizes. Printing quality is predominantly affected by the spreading effect, which again is not significantly affected for higher viscosity droplets. Therefore, printing quality can be improved by variation in spot sizes. However, this may hamper productivity due to additional layering. Albeit efficient transfer of materials takes place at higher fluences (since energy per pulse = laser fluence × spot size), cell damage is more likely to occur [142, 143]. Large spot sizes are sometimes essential as well, particularly for thick structures, as small spot sizes might lead to buckling and collapse of the structure. Furthermore, for printing biomaterials like oligonucleotide, spot size along with laser fluence affects the coverage (%) of the receiver substrate (membrane) [144]. An increase in the spot size increases the coverage percentage.

4.4 Laser-Induced Forward Transfer for Biosensor Manufacturing

LIFT, its process parameters, and its advantages have been emphasized in the sections above. Subsequently, efforts have been made to describe biosensors and its types and limitations in the sections above. In this section, an effort is made to provide evidence that LIFT can be employed to successfully develop/fabricate/manufacture biosensors. Table 4.1 summarizes studies on manufacturing biosensors using LIFT through the years.

4.5 Outlook and Conclusion

One of the most active research areas of analytical chemistry is biosensors as it diminishes the requirement of sample preparation [43]. Over the past 50 years, biosensors have evolved continuously. The most successful biosensor is for lateral flow pregnancy 1tests and electrochemical glucose sensing, and only a handful have achieved retail-level success on a global scale. This is mainly due to the challenges in developing academic research into industrial/commercial products. The challenges are broadened due to the limitations in clinical applications (complex regulations) and the lack of research professionals from various backgrounds for multidisciplinary

Table 4.1 Summary of studies on manufacturing biosensors using LIFT through the years.

Investigators	Biosensor type	Important highlights and prominent results/inferences	Notable limitations
Serra *et al.* [2] (2004)	Functional DNA microarray biosensor	• Buffer solution spotted at different laser pulse energies. • Uniform spots of ~40 μm (diameter) deposited. • Full functionality of the microarray was demonstrated (proving the potential of LIFT for DNA microarray production). • Resulted hybridization is specific in nature (allowing multiple different gene discrimination).	• Spot diameter has not been optimized. Smaller droplets can be deposited by tuning laser focus, liquid film thickness, etc.
Zergioti *et al.* [144] (2005)	DNA/protein microarray biosensor	• The high directionality (ultrafast laser) results in high quality and spatial resolution (10 μm). • Ultrafast laser pulses prevent melting and vaporization and deposits with minimum spread, narrow angular divergence, and high spatial resolution. • Superior printing quality achieved at ultrafast pulses <1 ps due to minimum thermal effects and negligible damage to the deposited material. • Adverse thermal effects prevented, thereby not damaging labile biomaterials (for < ns pulse width regime). • Simple method to deposit biomaterials on substrates (without requiring any transfer matrix material). • Efficient patterning of protein microarrays and DNA with a single-step process, with repetitive and reproducible properties. • Supports *in vitro* studies (protein function at genome-wide levels).	• Negligible notable limitations associated with LIFT.

(*Continued*)

Table 4.1 Summary of studies on manufacturing biosensors using LIFT through the years. (*Continued*)

Investigators	Biosensor type	Important highlights and prominent results/inferences	Notable limitations
Serra *et al.* [145] (2006)	DNA microarray biosensor	• Material transfers through microdroplet formation (liquid ejection). • Micrometric patterns of biomolecules produced. • Liquid solvent is the transport vector of biomolecules. • Negligible biomolecule decomposition and bioactivity loss (transferred material remains in the liquid state). • Micron-sized droplets with high spatial resolution. • DNA microarray (biosensor) has the ability to sense specific DNA strands and discriminate between two different human genes. • Microarray signal proves liquid film is comparable to the DNA concentration of the transferred material (no biomolecule damage).	• Results show some loss in the shape accuracy of the larger deposited structures.
Fernández-Pradas *et al.* [146] (2007)	Functional miniaturized biosensors	• Lasers are adequate tools for high spatial resolution. • LIFT transfers biological solutions onto solid substrates (μm biomolecule patterns). • Miniaturized biosensors are more viable through LIFT, providing enhanced integration capabilities and preventing issues related to contamination and clogging. • Droplet deposition with diameters <10 μm. • LIFT is an effective method for protein microarray deposition, preserving biomolecule bioactivity.	• Comparative studies show similar fluorescence intensity and unspecific hybridization levels for both LIFT and pin microspotting.

(*Continued*)

Table 4.1 Summary of studies on manufacturing biosensors using LIFT through the years. (*Continued*)

Investigators	Biosensor type	Important highlights and prominent results/inferences	Notable limitations
Duocastella *et al.* [147] (2010)	DNA/protein miniaturized biosensor	• Enhanced reproducibility along with resolution can be achieved. • Microarray development (proteins and DNA) shows the potential of LIFT for miniaturized biosensor manufacturing. • User-friendly, uniform, reproducible, well-defined microdroplet deposition. • Selective protein and DNA detection, suitable for miniaturized biosensor preparation. • Contamination risks are minimized through LIFT.	• Deep laser focusing is a limitation that results in no material deposition.
Palla-Papavlu *et al.* [119] (2011)	Liposome micropatterning-based biosensor	• Liposomes employed in micro- and nano-sized biosensors. • LIFT enables μm-sized well-defined patterns with no splashing (at 193 nm irradiation; sacrificial layer: TP). • Sacrificial layer thickness >150 nm (for clean, debris-free patterns). • Liposome integrity and chemical composition maintained (40 to 60 mJ/cm^2). • Target–substrate distance does not affect micropattern morphology. • Economical and time-effective technique for liposome array production. • LIFT shows the potential application in *in vitro* and *in vivo* studies.	• Au or Ti donor layers produce debris. • Results show incomplete TP (sacrificial layer) removal for fluence range with regular patterns. • The shape and size (transferred patterns) dependence for the TP DRL layer with laser fluence was not established.

(*Continued*)

Table 4.1 Summary of studies on manufacturing biosensors using LIFT through the years. (*Continued*)

Investigators	Biosensor type	Important highlights and prominent results/inferences	Notable limitations
Boutopoulos *et al.* [27] (2011)	Amperometric biosensor	• Photosynthetic material deposited on low-cost non-functionalized SPEs. • Thylakoids are directly immobilized on the electrodes through high kinetic energy of the transfer material. • Efficient electrochemical contact (protein-electrode: stable photosynthetic biomolecule) and efficient transport of electrons. • Rough electrode surface demonstrates complete wetting. • High photocurrent signals (335 nA), high signal-to-noise ratio, reduced response time (5 min), negligible diffusion limitations, and higher lifetime (48 h).	• Response time of 5 min can be improved, considering the current advancements of the process.

(*Continued*)

Table 4.1 Summary of studies on manufacturing biosensors using LIFT through the years. (*Continued*)

Investigators	Biosensor type	Important highlights and prominent results/inferences	Notable limitations
Touloupakis *et al.* [148] (2012)	Electrochemical biosensors	• Immobilization of photosynthetic biomaterial on SPE shows efficient electron transfer through LIFT. • Enhanced efficiency of energy production of a photosynthetic system. • Efficient and sensitive detection of herbicides (~nM concentration). • Biological molecules, supramolecules, and living cells combined with artificial supports show enhanced performance. • LIFT enhances energy conversion in electrochemical sensors. • High-spatial resolution (liquid biomaterials: ~μm). • Efficient electrochemical contact (between proteins and electrodes). • No damage to materials and functions during interfacing (with solid substrates). Extreme solid junctions were created (with electrodes). • High stable response signal (335 ± 13 nA) and excellent signal-to-noise ratio for the low photosynthetic mass of 28 ± 4 ng. • Au SPEs show the best performance.	• Half-life of 24 h. • Storage requirements for immobilized electrodes (−20°C: maintained properties for 3 months). • Expensive SPE material (gold, carbon paste, and carbon nanotubes).

(*Continued*)

Table 4.1 Summary of studies on manufacturing biosensors using LIFT through the years. (*Continued*)

Investigators	Biosensor type	Important highlights and prominent results/inferences	Notable limitations
Tsekenis *et al.* [149] (2012)	Micro-membrane array capacitive DNA biosensor	• Successful immobilization of thiol-modified oligonucleotides on low-temperature oxide (LTO) silicon (Si) surfaces using Au and 3-glycidoxypropyl-tri-methoxy silane (GOPTS). • The functional layer thickness and roughness along with the probe and target concentrations affect the sensor sensitivity by increasing the surface stress developed upon probe immobilization and target hybridization. • GOPTS suited as a functionalization layer (enhanced hybridization efficiencies and relative stability over time and ability to regenerate after analyte recognition). • LIFT uniformly deposited probe oligonucleotides at the μm scale on GOPTS-functionalized surfaces.	• GOPTS functionalized surfaces do not show high stability at room temperatures, therefore, limiting its use storage in normal conditions.
Di Pietrantonio *et al.* [150] (2013)	Surface acoustic wave biosensor	• LIFT employed for wtpOBPs transfer onto SAW resonators for odor detection applications. • Excellent surface coverage (overlap of ~50% for droplets) (20% glycerol solution; laser fluence: 650 mJ/cm^2). • Lower detecting agent consumption. • Good repeatability and linear response curves were obtained. • Sensitivity of 20.7 and 13.8 Hz/ppm for octenol and carvone, respectively. • Detection limit of 0.48 and 0.72 ppm for octenol and carvone, respectively.	• Low individual printed droplet resolution. • Sensitivities are not exceptionally advantageous when compared with other methods [151].

(*Continued*)

Table 4.1 Summary of studies on manufacturing biosensors using LIFT through the years. (*Continued*)

Investigators	Biosensor type	Important highlights and prominent results/inferences	Notable limitations
Chatzipetrou *et al.* [152] (2013)	Label-free capacitive biosensor	• Biosensor for the detection of the DNA hybridization process. • Deposition of thiol-modified oligonucleotides onto (LTO)/Si membranes. • Interaction forces deflect ultrathin LTO membranes, resulting in a capacitance shift (hybridization of target oligonucleotides with their immobilized complementary strands). • Printing parameters affected oligonucleotide conditions, the maximum deflection of the membranes leading to variation in capacitance. • Biomaterial deposition without bioactivity loss. • Percentage coverage of the sensor's membranes. • Printing multiple biomaterials without mixing. • LIFT ensured high-resolution deposition of oligonucleotides, enabling multi-analyte detection. • The fabricated biosensors can be reused.	• Percentage coverage of the membrane was limited to a maximum of ~80% (usually between 60% and 80%).

(*Continued*)

Table 4.1 Summary of studies on manufacturing biosensors using LIFT through the years. (*Continued*)

Investigators	Biosensor type	Important highlights and prominent results/inferences	Notable limitations
Touloupakis *et al.* [153] (2014)	Amperometric biosensor	• Improved sensitivity and detection limit (LOD). • Laccase enzyme directly immobilized without any toxic reagents interfering/attenuating the signal. • Stable and highly sensitive biosensor. • Laccase (active) successfully immobilized on graphite SPE. • Biologic materials are directly immobilized onto rough supports (without any functionalization). • Fair reproducibility (RSD: 11.7%) and detection limit (150 nM).	• Reproducibility may be improved. • Compromised response stability to catechol (at 4°C) after 35 days (90% of the initial response).
Palla-Papavlu *et al.* [25] (2014)	Surface acoustic wave (SAW) odor biosensor	• Successful development of surface acoustic wave (SAW) biosensors for odor detection in the food industry. • Wild-type bovine odorant binding protein (wtbOBP) solutions containing 20% and 50% glycerol are used to coat SAW biosensors. • High resolution, individual droplet printing of 50% glycerol solution. • High uniformity (over an active surface) and coffee ring effect suppression, despite low resolutions for 20% glycerol-printed droplets. • Pipette surface density overshadowed by LIFT deposits four times. • Fair sensitivity (5 Hz/ppm) and detection limit (low ppm range) were obtained, with low detecting agent consumption.	• Resolution has to be compromised to achieve high uniformity and coverage over the active area. • No significant improvement in sensitivity and detection limit for LIFT-printed sensors.

(*Continued*)

Table 4.1 Summary of studies on manufacturing biosensors using LIFT through the years. (*Continued*)

Investigators	Biosensor type	Important highlights and prominent results/inferences	Notable limitations
Tsekenis *et al.* [154] (2015)	Capacitive micromechanical sensor	• DNAzyme strand immobilized on the sensor surface and hybridized with the substrate strand through laser printing. • Change in surface stress leads to capacitance change. • 10 M of Pb^{2+} detected selectively. • LIFT enabled the achievement of high spatial resolution in biological material spotting and different DNAzymes printing for multiple heavy metal water pollutant detection simultaneously. • The fabricated sensor can be modified into a microfluidic device which can lead to a portable miniaturized sensor for onsite detection of Pb^{2+} or other heavy metals.	• Negligible notable limitations associated with LIFT.
Skotadis *et al.* [155] (2016)	Nanoparticle-based electrochemical biosensor	• Fast, easy, highly specific and label-free optimized biosensor for in situ detection of DNA hybridization. • Response range covers four orders of magnitude, LOD of 1 nM and the ability to detect a single base pair mismatch between the probe and complementary DNA. • LIFT utilized to deposit ssDNA probes. • Increased structural and electrical response at the nanoparticle surface coverage of 46%. • Inter-nanoparticle distance adjusted through deposition time. • Extremely flexible biosensor has multipurpose scenarios such as environmental monitoring and specific disease detection and is integrated with other sensing devices.	• Negligible notable limitations associated with LIFT.

(*Continued*)

Table 4.1 Summary of studies on manufacturing biosensors using LIFT through the years. (*Continued*)

Investigators	Biosensor type	Important highlights and prominent results/inferences	Notable limitations
Khairullina [123] (2021)	Electrochemical biosensor	• Electrodes suitable for non-enzymatic detection of biologically essential analytes, e.g., glucose. • High-rate metallization of various dielectrics. • Promising results for Cu–Au electrode structures manufactured on glass ceramics. • Sensitivity of 3,060 μA mM^{-1} cm^{-2} glucose sensing. • Linear range of glucose detection (0.3 and 1,000 μM). • High selectivity and long-term stability of electrodes. • Cu electrodes manufactured quickly and efficiently on various dielectric substrates. • Cu and Cu–Au electrodes on glass ceramics (i.e., Cu surface with Au nanoparticles) significantly increase the sensitivity (3,060 *vs.* 911 μA mM^{-1} cm^{-2}), enhance the detection limit, and increase the linear range of detection of the electrode toward dopamine sensing. • Additionally, Cu and Cu–Au sensors have good selectivity and excellent stability.	• Use of Au may be expensive when the biosensor is considered for mass production. • The long-run stability for non-enzymatic glucose sensing is limited to 15 days (~93%–96% of the initial electrocatalytic activity was retained).

research. Therefore, biosensor research should be application- and product-based.

Through a practical approach, the variety of limitations offered by the conventional techniques for sensor fabrication has led to the rise in LIFT. Although LIFT has not yet been widely used and limited literature could be extracted from the past, the advantages offered by LIFT are various, and LIFT-based biosensors show its potential to be utilized as a sensor manufacturing method on an industrial scale.

References

1. Bose, S., Sarkar, N., Vahabzadeh, S., Ke, D., Bandyopadhyay, A., Additive manufacturing of ceramics, in: *Additive Manufacturing*, Second Edition, pp. 183–231, CRC Press, Boca Raton, Florida, United States, 2019.
2. Serra, P., Colina, M., Fernández-Pradas, J.M., Sevilla, L., Morenza, J.L., Preparation of functional DNA microarrays through laser-induced forward transfer. *Appl. Phys. Lett.*, 85, 9, 1639–1641, 2004.
3. Thomas, B., Alloncle, A.P., Delaporte, P., Sentis, M., Sanaur, S., Barret, M., Collot, P., Experimental investigations of laser-induced forward transfer process of organic thin films. *Appl. Surf. Sci.*, 254, 4, 1206–1210, 2007.
4. Palla-Papavlu, A., Dinca, V., Luculescu, C., Shaw-Stewart, J., Nagel, M., Lippert, T., Dinescu, M., Laser induced forward transfer of soft materials. *J. Opt.*, 12, 12, 124014, 2010.
5. Arutyunyan, N.R., Komlenok, M.S., Kononenko, T.V., Dezhkina, M.A., Popovich, A.F., Konov, V., II, Printing of single-wall carbon nanotubes via blister-based laser-induced forward transfer. *Laser Phys.*, 29, 2, 026001, 2019.
6. Zergioti, I., Karaiskou, A., Papazoglou, D.G., Fotakis, C., Kapsetaki, M., Kafetzopoulos, D., Time resolved schlieren study of sub-pecosecond and nanosecond laser transfer of biomaterials. *Appl. Surf. Sci.*, 247, 1–4, 584–589, 2005.
7. Doraiswamy, A., Narayan, R.J., Lippert, T., Urech, L., Wokaun, A., Nagel, M., Hopp, B., Dinescu, M., Modi, R., Auyeung, R.C.Y., Chrisey, D.B., Excimer laser forward transfer of mammalian cells using a novel triazene absorbing layer. *Appl. Surf. Sci.*, 252, 13, 4743–4747, 2006.
8. Berg, Y., Zenou, M., Dolev, O., Kotler, Z., Temporal pulse shaping for smoothing of printed metal surfaces. *Opt. Eng.*, 54, 1, 011010, 2015.
9. Grant-Jacob, J.A., Mills, B., Feinaeugle, M., Sones, C.L., Oosterhuis, G., Hoppenbrouwers, M.B., Eason, R.W., Micron-scale copper wires printed using femtosecond laser-induced forward transfer with automated donor replenishment. *Opt. Mater. Express*, 3, 6, 747–754, 2013.
10. Florian, C., Caballero-Lucas, F., Fernández-Pradas, J.M., Artigas, R., Ogier, S., Karnakis, D., Serra, P., Conductive silver ink printing through the

laser-induced forward transfer technique. *Appl. Surf. Sci.*, 336, 304–308, 2015.

11. Kim, H.R.C.Y., Auyeung, R.C.Y., Lee, S.H., Huston, A.L., Piqué, A., Laser forward transfer of silver electrodes for organic thin-film transistors. *Appl. Phys. A*, 96, 2, 441–445, 2009.
12. Boutopoulos, C., Kalpyris, I., Serpetzoglou, E., Zergioti, I., Laser-induced forward transfer of silver nanoparticle ink: Time-resolved imaging of the jetting dynamics and correlation with the printing quality. *Microfluid. Nanofluidics*, 16, 3, 493–500, 2014.
13. Rapp, L., Ailuno, J., Alloncle, A.P., Delaporte, P., Pulsed-laser printing of silver nanoparticles ink: Control of morphological properties. *Opt. Express*, 19, 22, 21563–21574, 2011.
14. Palla-Papavlu, A., Dinca, V., Paraico, I., Moldovan, A., Shaw-Stewart, J., Schneider, C.W., Kovacs, E., Lippert, T., Dinescu, M., Microfabrication of polystyrene microbead arrays by laser induced forward transfer. *J. Appl. Phys.*, 108, 3, 033111, 2010.
15. Kononenko, T.V., Alloncle, P., Konov, V., II, Sentis, M., Laser transfer of diamond nanopowder induced by metal film blistering. *Appl. Phys. A*, 94, 3, 531–536, 2009.
16. Rapp, L., Diallo, A.K., Alloncle, A.P., Videlot-Ackermann, C., Fages, F., Delaporte, P., Pulsed-laser printing of organic thin-film transistors. *Appl. Phys. Lett.*, 95, 17, 171109, 2009.
17. Makrygianni, M., Verrelli, E., Boukos, N., Chatzandroulis, S., Tsoukalas, D., Zergioti, I., Laser printing and characterization of semiconducting polymers for organic electronics. *Appl. Phys. A*, 110, 3, 559–563, 2013.
18. Rapp, L., Serein-Spirau, F., Lère-Porte, J.P., Alloncle, A.P., Delaporte, P., Fages, F., Videlot-Ackermann, C., Laser printing of air-stable high performing organic thin film transistors. *Org. Electron.*, 13, 10, 2035–2041, 2012.
19. Birnbaum, A.J., Kim, H., Charipar, N.A., Piqué, A., Laser printing of multi-layered polymer/metal heterostructures for electronic and MEMS devices. *Appl. Phys. A*, 99, 4, 711–716, 2010.
20. Wang, J., Auyeung, R.C., Kim, H., Charipar, N.A., Piqué, A., Three-dimensional printing of interconnects by laser direct-write of silver nanopastes. *Adv. Mater.*, 22, 40, 4462–4466, 2010.
21. Fardel, R., Nagel, M., Nüesch, F., Lippert, T., Wokaun, A., Fabrication of organic light-emitting diode pixels by laser-assisted forward transfer. *Appl. Phys. Lett.*, 91, 6, 061103, 2007.
22. Shaw-Stewart, J.R.H., Mattle, T., Lippert, T.K., Nagel, M., Nüesch, F.A., Wokaun, A., The fabrication of small molecule organic light-emitting diode pixels by laser-induced forward transfer. *J. Appl. Phys.*, 113, 4, 043104, 2013.
23. Kattamis, N.T., McDaniel, N.D., Bernhard, S., Arnold, C.B., Ambient laser direct-write printing of a patterned organo-metallic electroluminescent device. *Org. Electron.*, 12, 7, 1152–1158, 2011.

24. Tsuboi, Y., Furuhata, Y., Kitamura, N., A sensor for adenosine triphosphate fabricated by laser-induced forward transfer of luciferase onto a poly (dimethylsiloxane) microchip. *Appl. Surf. Sci.*, 253, 20, 8422–8427, 2007.
25. Palla-Papavlu, A., Patrascioiu, A., Di Pietrantonio, F., Fernandez-Pradas, J.M., Cannata, D., Benetti, M., D'Auria, S., Verona, E., Serra, P., Preparation of surface acoustic wave odor sensors by laser-induced forward transfer. *Sens. Actuators B Chem.*, 192, 369–377, 2014.
26. Di Pietrantonio, F., Benetti, M., Cannatà, D., Verona, E., Palla-Papavlu, A., Dinca, V., Dinescu, M., Mattle, T., Lippert, T., Volatile toxic compound detection by surface acoustic wave sensor array coated with chemoselective polymers deposited by laser induced forward transfer: Application to sarin. *Sens. Actuators B Chem.*, 174, 158–167, 2012.
27. Boutopoulos, C., Touloupakis, E., Pezzotti, I., Giardi, M.T., Zergioti, I., Direct laser immobilization of photosynthetic material on screen printed electrodes for amperometric biosensor. *Appl. Phys. Lett.*, 98, 9, 093703, 2011.
28. Powner, E.T. and Yalcinkaya, F., From basic sensors to intelligent sensors: Definitions and examples. *Sens. Rev.*, 15, 4, 19–22, 1995.
29. D'Amico, A. and Di Natale, C., A contribution on some basic definitions of sensors properties. *IEEE Sens. J.*, 1, 3, 183–190, 2001.
30. Di Pietrantonio, F., Benetti, M., Cannatà, D., Verona, E., Palla-Papavlu, A., Dinca, V., Dinescu, M., Mattle, T., Lippert, T., Volatile toxic compound detection by surface acoustic wave sensor array coated with chemoselective polymers deposited by laser induced forward transfer: Application to sarin. *Sens. Actuators B Chem.*, 174, 158–167, 2012.
31. Mattle, T., Hintennach, A., Lippert, T., Wokaun, A., Laser induced forward transfer of SnO2 for sensing applications using different precursors systems. *Appl. Phys. A*, 110, 2, 309–316, 2013.
32. Kissinger, P.T., Biosensors—A perspective. *Biosens. Bioelectron.*, 20, 12, 2512 2516, 2005.
33. Arnold, M.A. and Meyerhoff, M.E., Ion-selective electrodes. *Anal. Chem.*, 56, 5, 20–48, 1984.
34. Heineman, W.R. and Jensen, W.B., Leland c. Clark jr. (1918–2005). *Biosens. Bioelectron.*, 8, 21, 1403–1404, 2006.
35. Guilbault, G.G. and Montalvo Jr., J.G., Urea-specific enzyme electrode. *J. Am. Chem. Soc.*, 91, 8, 2164–2165, 1969.
36. Lowe, C.R., Biosensors. *Trends Biotechnol.*, 2, 3, 59–65, 1984.
37. Turner, A., Karube, I., Wilson, G.S., *Biosensors: Fundamentals and applications*, Oxford University Press, Oxford, New York: Oxford University Press, p. 770, Boca Raton, Florida, United States, 1987.
38. Eggins, B.R., *Biosensors: An introduction*, Springer-Verlag, Copyright @ 1996 by John Wiley & Sons Ltd. (West Sussex, England) and B.G. Teubner (Stuttgart, Germany), 2013.
39. Bhalla, N., Jolly, P., Formisano, N. *et al.*, Introduction to biosensors. *Essays Biochem.*, 60, 1, 1–8, 2016.

40. Stein, N.E., Hamelers, H.V., Buisman, C.N., The effect of different control mechanisms on the sensitivity and recovery time of a microbial fuel cell based biosensor. *Sens. Actuators B Chem.*, 171, 816–821, 2012.
41. Karunakaran, C., Bhargava, K., Benjamin, R., *Biosensors and bioelectronics*, Elsevier, Amsterdam, Netherlands, 2015.
42. Veloso, A.J., Cheng, X.R., Kerman, K., Electrochemical biosensors for medical applications, in: *Biosensors for Medical Applications*, pp. 3–40, Woodhead Publishing, Sawston, Cambridge, England, 2012.
43. Ronkainen, N.J., Halsall, H.B., Heineman, W.R., Electrochemical biosensors. *Chem. Soc. Rev.*, 39, 5, 1747–1763, 2010.
44. Zoski, C.G. (Ed.), *Handbook of Electrochemistry*, Elsevier, Amsterdam, Netherlands, 2006.
45. Schmickler, W. and Santos, E., *Interfacial electrochemistry*, Springer Science & Business Media, First edition by Oxford University Press, New York, 1996.
46. Yosypchuk, B. and Novotný, L., Reference electrodes based on solid amalgams. *Electroanalysis*, 16, 3, 238–241, 2004.
47. Fraden, J. and King, J.G., *Handbook of modern sensors: Physics, designs, and applications*, vol. 3, Springer, New York, 2010.
48. Grieshaber, D., MacKenzie, R., Vörös, J., Reimhult, E., Electrochemical biosensors-sensor principles and architectures. *Sensors*, 8, 3, 1400–1458, 2008.
49. Pohanka, M. and Skládal, P., Electrochemical biosensors–principles and applications. *J. Appl. Biomed.*, 6, 2, 57–64, 2008.
50. Grieshaber, D., MacKenzie, R., Vörös, J., Reimhult, E., Electrochemical biosensors-sensor principles and architectures. *Sensors*, 8, 3, 1400–1458, 2008.
51. Bakker, E. and Pretsch, E., Potentiometric sensors for trace-level analysis. *Trac-Trends Anal. Chem.*, 24, 3, 199–207, 2005.
52. Eggins, B., *Chemical sensors and biosensors; Analytical Techniques in the Sciences*, John Wiley & Sons, West Sussex, 2002.
53. Chaubey, A. and Malhotra, B.D., Mediated biosensors. *Biosens. Bioelectron.*, 17, 6-7, 441–456, 2002.
54. D'Orazio, P., Biosensors in clinical chemistry. *Clin. Chim. Acta*, 334, 1-2, 41–69, 2003.
55. Buerk, D., *Biosensors: Theory and applications*, Technomic Publishing, Lancaster, 1993.
56. Luppa, P.B., Sokoll, L.J., Chan, D.W., Immunosensors - principles and applications to clinical chemistry. *Clin. Chim. Acta*, 314, 1-2, 1–26, 2001.
57. Thevenot, D.R., Toth, K., Durst, R.A., Wilson, G.S., Electrochemical biosensors: Recommended definitions and classification. *Biosens. Bioelectron.*, 16, 1-2, 121–131, 2001.
58. Ghindilis, A.L., Atanasov, P., Wilkins, M., Wilkins, E., Immunosensors: Electrochemical sensing and other engineering approaches. *Biosens. Bioelectron.*, 13, 1, 113–131, 1998.
59. Wang, J., Amperometric biosensors for clinical and therapeutic drug monitoring: A review. *J. Pharm. Biomed. Anal.*, 19, 1-2, 47–53, 1999.

60. Cui, X., Liu, G., Lin, Y., Amperometric biosensors based on carbon paste electrodes modified with nanostructured mixed-valence manganese oxides and glucose oxidase. *Nanomedicine*, 1, 130–135, 2005.
61. Wang, J., Electrochemical biosensors: Towards point-of-care cancer diagnostics. *Biosens. Bioelectron.*, 21, 10, 1887–1892, 2006.
62. Ohnuki, H., Saiki, T., Kusakari, A., Endo, H., Ichihara, M., Izumi, M., Incorporation of glucose oxidase into languir-blodgett films based on prussian blue applied to amperometric glucose biosensor. *Langmuir*, 23, 4675–4681, 2007.
63. Marzouk, S.A., Ashraf, S.S., Al Tayyari, K.A., Prototype amperometric biosensor for sialic acid determination. *Anal. Chem.*, 79, 4, 1668–1674, 2007.
64. Yemini, M., Levi, Y., Yagil, E., Rishpon, J., Specific electrochemical phage sensing for bacillus cereus and mycobacterium smegmatis. *Bioelectrochemistry*, 70, 180–184, 2007.
65. Pohanka, M., Jun, D., Kuča, K., Mycotoxin assays using biosensor technology–A review. *Drug Chem. Toxicol.*, 30, 3, 253–261, 2007.
66. Pohanka, M., Jun, D., Kuča, K., Amperometric biosensor for evaluation of competitive cholinesterase inhibition by the reactivator HI-6. *Anal. Lett.*, 40, 2351–2359, 2007.
67. Liu, G. and Lin, Y., Biosensor based on self-assembling acetylcholinesterase on carbon nanotubes for flow injection/amperometric detection of organophosphate pesticides and nerve agents. *Anal. Chem.*, 78, 3, 835–843, 2006.
68. D'Orazio, P., Biosensors in clinical chemistry. *Clin. Chim. Acta*, 334, 1-2, 41–69, 2003.
69. Cullen, D.C., Sethi, R., Lowe, C.R., Multianalyte miniature conductance biosensor. *Anal. Chim. Acta*, 231, 1, 33–40, 1990.
70. Contractor, A.Q., Sureshkumar, T.N., Narayanan, R., Sukeerthi, S., Lal, R., Srinivasa, R.S., Conducting polymer-based biosensors. *Electrochim. Acta*, 39, 8-9, 1321–1324, 1994.
71. Patolsky, F., Zheng, G.F., Lieber, C.M., Nanowire-based biosensors. *Anal. Chem.*, 78, 13, 4260–4269, 2006.
72. Stadler, B., Solak, H.H., Frerker, S., Bonroy, K., Frederix, F., Voros, J., Grandin, H.M., Nanopatterning of gold colloids for label-free biosensing. *Nanotechnology*, 18, 15, 6, 2007.
73. Yagiuda, K., Hemmi, A., Ito, S., Asano, Y., Fushinuki, Y., Chen, C.Y., Karube, I., Development of a conductivity-based immunosensor for sensitive detection of methamphetamine (stimulant drug) in human urine. *Biosens. Bioelectron.*, 11, 8, 703–707, 1996.
74. Chouteau, C., Dzyadevych, S., Chovelon, J.M., Durrieu, C., Development of novel conductometric biosensors based on immobilised whole cell chlorella vulgaris microalgae. *Biosens. Bioelectron.*, 19, 9, 1089–1096, 2004.
75. Arduini, F., Micheli, L., Moscone, D., Palleschi, G., Piermarini, S., Ricci, F., Volpe, G., Electrochemical biosensors based on nanomodified screen-printed

electrodes: Recent applications in clinical analysis. *TrAC Trends Anal. Chem.*, 79, 114–126, 2016.

76. Maddipatla, D., Narakathu, B.B., Atashbar, M., Recent progress in manufacturing techniques of printed and flexible sensors: A review. *Biosensors*, 10, 12, 199, 2020.
77. Gonzalez-Macia, L., Morrin, A., Smyth, M.R., Killard, A.J., Advanced printing and deposition methodologies for the fabrication of biosensors and biodevices. *Analyst*, 135, 5, 845–867, 2010.
78. Turner, A.P., Biosensors: Sense and sensibility. *Chem. Soc. Rev.*, 42, 8, 3184–3196, 2013.
79. Yamanaka, K., Vestergaard, M.D.C., Tamiya, E., Printable electrochemical biosensors: A focus on screen-printed electrodes and their application. *Sensors*, 16, 10, 1761, 2016.
80. Honeychurch, K.C., Screen-printed electrochemical biosensors and sensors for Monitoring metal pollutants. *Insciences J.*, 2, 1, 1–51, 2012.
81. Hussain, A., Abbas, N., Ali, A., Inkjet printing: A viable technology for biosensor fabrication. *Chemosensors*, 10, 3, 103, 2022.
82. Strutt, J.W. and Rayleigh, L., On the instability of jets. *Proc. London Math. Soc.*, 10, 4, 4–13, 1878.
83. Parashkov, R., Becker, E., Riedl, T., Johannes, H.H., Kowalsky, W., Large area electronics using printing methods. *Proc. IEEE*, 93, 7, 1321–1329, 2005.
84. Setti, L. *et al.*, An amperometric glucose biosensor prototype fabricated by thermal inkjet printing. *Biosens. Bioelectron.*, 20, 10, 2019–2026, 2005.
85. de Gans, B.J. and Schubert, U.S., Inkjet printing of polymer micro-arrays and libraries: Instrumentation, requirements, and perspectives. *Macromol. Rapid Commun.*, 24, 11, 659–666, 2003.
86. Zhu, J. *et al.*, Nano-structured TiO2 film fabricated at room temperature and its acoustic properties. *J. Phys. D Appl. Phys.*, 41, 16, 162001, 2008.
87. Tuomikoski, M. *et al.*, *Organic Optoelectronics and Photonics II*, pp. 19204–19204, 2006.
88. Santa-Nokki, H. *et al.*, Dynamic preparation of TiO2 films for fabrication of dye-sensitized solar cells. *J. Photochem. Photobiol. A Chem.*, 182, 2, 187–191, 2006.
89. Parashkov, R., Becker, E., Riedl, T., Johannes, H.H., Kowalsky, W., Large area electronics using printing methods. *Proc. IEEE*, 93, 7, 1321–1329, 2005.
90. Kane, R.S. *et al.*, Patterning proteins and cells using soft lithography. *Biomaterials*, 20, 23-24, 2363–2376, 1999.
91. Tang, Q. and Shi, S.Q., Preparation of gas sensors via dip-pen nanolithography. *Sens. Actuators B Chem.*, 131, 2, 379–383, Mar. 2008.
92. Lim, J.H. and Mirkin, C.A., Electrostatically driven dip-pen nanolithography of conducting polymers. *Adv. Mater.*, 14, 20, 1474–1477, Oct. 2002.
93. Kampfrath, G. and Hintsche, R., Plasma-polymerized thin films for enzyme immobilization in biosensors. *Anal. Lett.*, 22, 11-12, 2423–2431, 1989.

94. Yoshimura, K., Kitade, T., Kitamura, K., Hozumi, K., Preparation of immobilized glucose oxidase membrane by the plasma polymerization technique. *Microchem. J.*, 43, 2, 133–142, 1991.
95. Yoshimura, K. and Hozumi, K., Response characteristics of a glucose electrode with a sensing membrane prepared by plasma polymerization. *Microchem. J.*, 53, 4, 404–412, 1996.
96. Liu, M. and Liu, G.Y., Hybridization with nanostructures of single-stranded DNA. *Langmuir*, 21, 5, 1972–1978, Feb. 2005.
97. Muguruma, H., Plasma-polymerized films for biosensors II. *TrAC Trends Anal. Chem.*, 26, 5, 433–443, 2007.
98. Li, J., Wang, H., Deng, T., Wu, Z., Shen, G., Yu, R., A plasma-polymerized film for capacitance immunosensing. *Biosens. Bioelectron.*, 20, 4, 841–847, Nov. 2004.
99. Roda, A., Guardigli, M., Russo, C., Pasini, P., Baraldini, M., Protein microdeposition using a conventional ink-jet printer. *Biotechniques*, 28, 3, 492–496, Mar. 2000.
100. Ringeisen, B.R., Callahan, J., Wu, P.K., Pique, A., Spargo, B., McGill, R.A., Bucaro, M., Kim, H., Bubb, D.M., Chrisey, D.B., Novel laser-based deposition of active protein thin films. *Langmuir*, 17, 11, 3472–3479, May 2001.
101. Britland, S., Perez-Arnaud, E., Clark, P., McGinn, B., Connolly, P., Moores, G., Micropatterning proteins and synthetic peptides on solid supports: A novel application for microelectronics fabrication technology. *Biotechnol. Prog.*, 8, 2, 155–160, Mar. 1992.
102. Hasenbank, M.S., Edwards, T., Fu, E., Garzon, R., Kosar, T.F., Look, M., Mashadi-Hossein, A., Yager, P., Demonstration of multi-analyte patterning using piezoelectric inkjet printing of multiple layers. *Anal. Chim. Acta*, 611, 1, 80–88, 2008.
103. Muguruma, H., Kase, Y., Uehara, H., Nanothin ferrocene film plasma polymerized over physisorbed glucose oxidase: High-throughput fabrication of bioelectronic devices without chemical modifications. *Anal. Chem.*, 77, 20, 6557–6562, 2005.
104. Serra, P., Duocastella, M., Fernández-Pradas, J.M., Morenza, J.L., Laser-induced forward transfer: A laser-based technique for biomolecules printing, in: *Cell and Organ Printing*, pp. 53–80, Springer, Dordrecht, 2010.
105. Levene, M.L., Scott, R.D., Siryj, B.W., Material transfer recording. *Appl. Opt.*, 9, 10, 2260–2265, 1970.
106. Bohandy, J., Kim, B.F., Adrian, F.J., Metal deposition from a supported metal film using an excimer laser. *J. Appl. Phys.*, 60, 4, 1538–1539, 1986.
107. Adrian, F.J., Bohandy, J., Kim, B.F., Jette, A.N., Thompson, P., A study of the mechanism of metal deposition by the laser-induced forward transfer process. *J. Vac. Sci. Technol. B*, 5, 5, 1490–1494, 1987.
108. Ehrlich, D.J., Higashi, G.S., Oprysko, M.M., Laser and particle-beam chemical processing for microelectronics, in: *Conference Record of the 1987*

International Symposium on Microelectronics, Materials Research Society, Pittsburgh, PA, USA, pp. 187–196, 1988.

109. Serra, P. and Piqué, A., Laser-induced forward transfer: Fundamentals and applications. *Adv. Mater. Technol.*, 4, 1, 1800099, 2019.
110. Feinaeugle, M., Gregorčič, P., Heath, D.J., Mills, B., Eason, R.W., Time-resolved imaging of flyer dynamics for femtosecond laser-induced backward transfer of solid polymer thin films. *Appl. Surf. Sci.*, 396, 1231–1238, 2017.
111. Delaporte, P. and Alloncle, A.P., Laser-induced forward transfer: A high resolution additive manufacturing technology. *Opt. Laser Technol.*, 78, 33–41, 2016.
112. Wu, P.K. *et al.*, The deposition, structure, pattern deposition, and activity of biomaterial thin-films by matrix-assisted pulsed-laser evaporation (MAPLE) and MAPLE direct write. *Thin Solid Films*, 398, 607–614, 2001.
113. Piqué, A. *et al.*, A novel laser transfer process for direct writing of electronic and sensor materials. *Appl. Phys. A*, 69, 1, S279–S284, 1999.
114. Jadin, A. *et al.*, Plasma formation during excimer laser irradiation of thin selenium films in air, in: *Laser-Assisted Processing II*, vol. 1279, pp. 112–117, 1990.
115. Kántor, Z. *et al.*, Laser induced forward transfer: The effect of support-film interface and film-to-substrate distance on transfer. *Appl. Phys. A*, 54, 2, 170–175, 1992.
116. Sano, T. *et al.*, Experimental investigation of laser induced forward transfer process of metal thin films. *Appl. Surf. Sci.*, 186, 1-4, 221–226, 2002.
117. Kántor, Z. *et al.*, Metal pattern deposition by laser-induced forward transfer. *Appl. Surf. Sci.*, 86, 1-4, 196–201, 1995.
118. Dinca, V. *et al.*, Patterning parameters for biomolecules microarrays constructed with nanosecond and femtosecond UV lasers. *Thin Solid Films*, 516, 18, 6504–6511, 2008.
119. Palla-Papavlu, A. *et al.*, Liposome micropatterning based on laser-induced forward transfer. *Appl. Phys. A*, 102, 3, 651–659, 2011.
120. Rapp, L. *et al.*, Characterization of organic material micro-structures transferred by laser in nanosecond and picosecond regimes. *Appl. Surf. Sci.*, 255, 10, 5439–5443, 2009.
121. Chakraborty, S. *et al.*, Laser-induced forward transfer technique for maskless patterning of amorphous V2O5 thin film. *Appl. Surf. Sci.*, 254, 2, 638–643, 2007.
122. Mogyorósi, P. *et al.*, Pulsed laser ablative deposition of thin metal films. *Appl. Surf. Sci.*, 36, 1-4, 157–163, 1989.
123. Khairullina, E.M., Panov, M.S., Andriianov, V.S., Ratautas, K., Tumkin, I., II, Račiukaitis, G., High rate fabrication of copper and copper–gold electrodes by laser-induced selective electroless plating for enzyme-free glucose sensing. *RSC Adv.*, 11, 32, 19521–19530, Apr. 2021.
124. Zergioti, I., Mailis, S., Vainos, N.A., Papakonstantinou, P., Kalpouzos, C., Grigoropoulos, C.P., Fotakis, C., Microdeposition of metal and oxide

structures using ultrashort laser pulses. *Appl. Phys. A Mater. Sci. Process.*, 66, 4, 579–582, Mar. 1998.

125. Zergioti, I., Mailis, S., Vainos, N.A., Fotakis, C., Chen, S., Grigoropoulos, C.P., Microdeposition of metals by femtosecond excimer laser. *Appl. Surf. Sci.*, 127-129, 601–605, Feb. 1998.
126. Tóth, Z., Szörényi, T., Toth, A.L., Ar+ laser-induced forward transfer (LIFT): A novel method for micrometer-size surface patterning. *Appl. Surf. Sci.*, 69, 1-4, 317–320, Mar. 1993.
127. Mogyorósi, P., Szörényi, T., Bali, K., Tóth, Z., Hevesi, I., Pulsed laser ablative deposition of thin metal films. *Appl. Surf. Sci.*, 36, 1-4, 157–163, Jul. 1989.
128. Kántor, Z., Toth, Z., Szörényi, T., Tóth, A.L., Deposition of micrometer-sized tungsten patterns by laser transfer technique. *Appl. Phys. Lett.*, 64, 25, 3506–3508, Jun. 1994.
129. Kántor, Z. and Szörényi, T., Dynamics of long-pulse laser transfer of micrometer-sized metal patterns as followed by time-resolved measurements of reflectivity and transmittance. *J. Appl. Phys.*, 78, 4, 2775–2781, Aug. 1995.
130. Zergioti, I., Mailis, S., Vainos, N.A., Papakonstantinou, P., Kalpouzos, C., Grigoropoulos, C.P., Fotakis, C., Microdeposition of metal and oxide structures using ultrashort laser pulses. *Appl. Phys. A Mater. Sci. Process.*, 66, 579–582, 1998.
131. Banks, D.P., Grivas, C., Zergioti, I., Robert, W.E., Ballistic laser-assisted solid transfer (BLAST) from a thin film precursor. *Opt. Express*, 16, 5, 3249–3254, 2008.
132. Klini, A., Loukakos, P.A., Gray, D., Manousaki, A., Fotakis, C., Laser induced forward transfer of metals by temporally shaped femtosecond laser pulses. *Opt. Express*, 16, 15, 11300–11309, 2008.
133. Willis, D.A. and Grosu, V., Microdroplet deposition by laser-induced forward transfer. *Appl. Phys. Lett.*, 86, 24, 244103, 2005.
134. Zergioti, I., Mailis, S., Vainos, N.A., Ikiades, A., Grigoropoulos, C.P., Fotakis, C., Microprinting and microetching of diffractive structures using ultrashort laser pulses. *Appl. Surf. Sci.*, 138, 82–86, 1999.
135. Papakonstantinou, P., Vainos, N.A., Fotakis, C., Microfabrication by UV femtosecond laser ablation of Pt, Cr and indium oxide thin films. *Appl. Surf. Sci.*, 151, 3-4, 159–170, 1999.
136. Feinaeugle, M., Alloncle, A.P., Delaporte, P., Sones, C.L., Eason, R.W., Time-resolved shadowgraph imaging of femtosecond laser-induced forward transfer of solid materials. *Appl. Surf. Sci.*, 258, 22, 8475–8483, 2012.
137. Pohl, R., Visser, C.W., Römer, G.R.B.E., Sun, C., Lohse, D., Imaging of the ejection process of nanosecond laser-induced forward transfer of gold. *J. Laser Micro Nanoeng.*, 10, 2, 154–158, 2015.
138. Pohl, R., Visser, C.W., Römer, G.W., Lohse, D., Sun, C., Huis, B., Ejection regimes in picosecond laser-induced forward transfer of metals. *Phys. Rev. Appl.*, 3, 2, 024001, 2015.

139. Zenou, M., Sa'ar, A., Kotler, Z., Supersonic laser-induced jetting of aluminum micro-droplets. *Appl. Phys. Lett.*, 106, 18, 181905, 2015.
140. Zenou, M., Sa'ar, A., Kotler, Z., Digital laser printing of aluminum micro-structure on thermally sensitive substrates. *J. Phys. D Appl. Phys.*, 48, 20, 205303, May 2015.
141. Yan, J., Huang, Y., Chrisey, D.B., Laser-assisted printing of alginate long tubes and annular constructs. *Biofabrication*, 5, 1, 015002, Mar. 2012.
142. Lin, Y., Huang, Y., Wang, G., Tzeng, T.R.J., Chrisey, D.B., Effect of laser fluence on yeast cell viability in laser-assisted cell transfer. *J. Appl. Phys.*, 106, 4, 043106, Aug. 2009.
143. Lin, Y., Huang, G., Huang, Y., Tzeng, T.R.J., Chrisey, D., Effect of laser fluence in laser-assisted direct writing of human colon cancer cell. *Rapid Prototyp. J.*, 2010.
144. Zergioti, I., Karaiskou, A., Papazoglou, D.G., Fotakis, C., Kapsetaki, M., Kafetzopoulos, D., Time resolved schlieren study of sub-pecosecond and nanosecond laser transfer of biomaterials. *Appl. Surf. Sci.*, 247, 1–4, 584–589, Sep. 2005.
145. Serra, P., Fernández-Pradas, J.M., Colina, M., Duocastella, M., Dominguez, J., Morenza, J.L., Laser-induced forward transfer: A direct-writing technique for biosensors preparation. *J. Laser Micro Nanoeng.*, 1, 3, 236–242, 2006.
146. Fernández-Pradas, J.M., Duocastella, M., Colina, M., Serra, P., Morenza, J.L., Production of miniaturized biosensors through laser-induced forward transfer, in: *Bioengineered and Bioinspired Systems III*, vol. 6592, pp. 281–288, SPIE, Microtechnologies for the New Millennium, 2007, Maspalomas, Gran Canaria, Spain, May 2007.
147. Duocastella, M., Fernández-Pradas, J.M., Morenza, J.L., Zafra, D., Serra, P., Novel laser printing technique for miniaturized biosensors preparation. *Sens. Actuators B Chem.*, 145, 1, 596–600, Jan. 2010.
148. Touloupakis, E., Boutopoulos, C., Buonasera, K., Zergioti, I., Giardi, M.T., A photosynthetic biosensor with enhanced electron transfer generation realized by laser printing technology. *Anal. Bioanal. Chem.*, 402, 10, 3237–3244, Jul. 2012.
149. Tsekenis, G., Chatzipetrou, M., Tanner, J., Chatzandroulis, S., Thanos, D., Tsoukalas, D., Zergioti, I., Surface functionalization studies and direct laser printing of oligonucleotides toward the fabrication of a micromembrane DNA capacitive biosensor. *Sens. Actuators B Chem.*, 175, 123–131, 2012.
150. Di Pietrantonio, F., Benetti, M., Cannatà, D., Varriale, A., D'Auria, S., Palla-Papavlu, A., Serra, P., Verona, E., Surface acoustic wave biosensor based on odorant binding proteins deposited by laser induced forward transfer, in: *2013 IEEE International Ultrasonics Symposium (IUS)*, pp. 2144–2147, 2013.
151. Di Pietrantonio, F., Cannatà, D., Benetti, M., Verona, E., Varriale, A., Staiano, M., D'Auria, S., Detection of odorant molecules via surface acoustic wave biosensor array based on odorant-binding proteins. *Biosens. Bioelectron.*, 41, 328–334, 2013.

152. Chatzipetrou, M., Tsekenis, G., Tsouti, V., Chatzandroulis, S., Zergioti, I., Biosensors by means of the laser induced forward transfer technique. *Appl. Surf. Sci.*, 278, 250–254, 2013.
153. Touloupakis, E., Chatzipetrou, M., Boutopoulos, C., Gkouzou, A., Zergioti, I., A polyphenol biosensor realized by laser printing technology. *Sens. Actuators B Chem.*, 193, 301–305, 2014.
154. Tsekenis, G., Filippidou, M.K., Chatzipetrou, M., Tsouti, V., Zergioti, I., Chatzandroulis, S., Heavy metal ion detection using a capacitive micromechanical biosensor array for environmental monitoring. *Sens. Actuators B Chem.*, 208, 628–635, 2015.
155. Skotadis, E., Voutyras, K., Chatzipetrou, M., Tsekenis, G., Patsiouras, L., Madianos, L., Chatzandroulis, S., Zergioti, I., Tsoukalas, D., Label-free DNA biosensor based on resistance change of platinum nanoparticles assemblies. *Biosens. Bioelectron.*, 81, 388–394, 2016.

Part II
BIOMATERIALS

5

The Effect of the Nanostructured Surface Modification on the Morphology and Biocompatibility of Ultrafine-Grained Titanium Alloy for Medical Application

Dragana Mihajlović[1*], Marko Rakin[1†], Anton Hohenwarter[2], Djordje Veljović[1], Vesna Kojić[3] and Veljko Djokić[1,4‡]

[1]University of Belgrade, Faculty of Technology and Metallurgy, Belgrade, Serbia
[2]Department of Materials Science, Chair of Materials Physics, Montanuniversität Leoben, Leoben, Austria
[3]Oncology Institute of Vojvodina, Sremska Kamenica, Serbia
[4]Innovation Centre of the Faculty of Technology and Metallurgy, University of Belgrade, Belgrade, Serbia

Abstract

Primary implant stability after implantation is associated with its good mechanical contact with the touching tissue. Adequate integration of an implant material with the bone is required to provide safety and efficiency of the implant over its life. Generally, two surface properties are the most important facts for tissue replay to the implant material: the surface morphology and chemical structure. Compared with a smooth implant surface, a controlled rough surface provides more surface area for integration with the surrounding tissues and allows successful implant ingrowth into the tissues. It was found that the nanostructured modification of the titanium surface on the level of nano-sized pores influences the adhesion, spreading, and growth of osteoblastic cells. There are many methods for nanostructure modification of biomedical alloy surfaces, but one of the common techniques is electrochemical anodization (anodic oxidation). Electrochemical anodization is a method by which a nanostructured thin film of a mixture of oxides is obtained

*Corresponding author: dbarjaktarevic@tmf.bg.ac.rs
†Corresponding author: marko@tmf.bg.ac.rs
‡Corresponding author: vdjokic@tmf.bg.ac.rs

Jay Prakash Srivastava, Dražan Kozak, Vinayak Ranjan, Pankaj Kumar, Ranjan Kumar and Shubham Tayal (eds.) Mechanical Engineering in Biomedical Applications: Bio-3D Printing, Biofluid Mechanics, Implant Design, Biomaterials, Computational Biomechanics, Tissue Mechanics, (121–150) © 2024 Scrivener Publishing LLC

on the material surface area. The advantage of anodic oxidation is the possibility of controlling the nanostructured morphology of the surface and dimensions of nanotubes (diameter, length, and wall thickness of nanotubes) and their shape through variation of the solution, pH value, potential, or duration of anodic oxidation.

Surface modification of Ti–13Nb–13Zr alloy in the received state and the ultrafine-grained state was conducted by an anodic oxidation process in a 1 M H_3PO_4 + NaF solution, for a period of anodizing time from 30 to 120 min. Atomic force microscopy (AFM) was utilized to characterize the topography of the modified surface, while scanning electron microscopy (SEM) was used to analyze the modified surface morphology. The results showed that a homogeneous nanotubular oxide film consisting of nanotubes could be obtained using the electrochemical anodization treatment, while the roughness of the nanostructured surface increased in comparison to the non-anodized surface. The studies given in this chapter aim to examine the morphology of the nanostructured surface and estimate *in vitro* biocompatibility of the abovementioned titanium alloy after the creation of the nanotubular oxide film. *In vitro* examinations were performed on mouse fibroblast (L929) and human fibroblast (MRC-5) cell lines. The results showed that the nanotubular oxide film obtained on the coarse-grained Ti–13Nb–13Zr (CG TNZ) and the ultrafine-grained Ti–13Nb–13Zr (UFG TNZ) increased the fractions of surviving cells compared with their counterpart alloy, while the cells had better spreading and adhesion on the nanostructured and bare surfaces of the UFG titanium alloy.

Keywords: High-pressure torsion process, ultrafine-grained Ti–13Nb–13Zr alloy, electrochemical anodization process, nanotubular oxide layer, biocompatibility

5.1 Introduction

5.1.1 Titanium-Based Materials for Biomedical Application

Biomaterials can be defined as materials from which components and whole implants are made, intended for utilization in the human body in order to substitute a certain piece or purpose of the part of the body in an economical, physiologically desirable way and at the same time not expose a person to danger and risk [1]. Materials in touch with the environment in the human body (cells, tissues, and fluids) must not have negative effects on the surrounding tissues and can be natural or synthetic [2]. Biomaterials include several classes of materials for use in medicine: metallic biomaterials, ceramic and glassy biomaterials, economical and more expensive non-degradable synthetic polymers, composite biomaterials, biodegradable polymers, bioderivative polymers and tissues, and passive and bioactive coatings [3]. Most of the materials listed here are used today to make implants for application in dental, orthopedic, cardiovascular, and ophthalmological surgery.

However, some of the materials, which are used to make implants, create complications as a result of negative interactions with the environment in the human body. The usual complications that appear are inflammatory processes, infections in the form of allergic or toxic reactions, and the cessation of function or operation of the implanted device, which can cause harmful consequences for the patient's health.

Biocompatible metal materials must meet certain criteria and possess certain properties in order to be used for the production of medical implants (e.g., the appropriate mechanical properties, corrosion resistance and non-toxicity, biocompatibility, osseointegration, acceptability, and costs of implant materials).

Titanium-based materials show the best properties among metal materials, including excellent protection from corrosion, biocompatibility, the adequate values of Young's modulus compared with other groups of biomaterials [4, 5]. Titanium-based materials are very reactive materials, and when exposed to air, they spontaneously react forming a thin oxide film (mostly made of titanium dioxide) on the surface [6]. The creation of such a layer leads to the passivation of the surface because the oxide film simulates a barrier, which protects the surface against chemical reactions [7]. However, titanium-based materials can react in contact with different materials, whereby corrosion leads to the delivery of metallic ions from the materials into the surrounding organs and tissues, which can lead to problems regarding the acceptance of the implant by the human organism, biocompatibility, and the emergence of local inflammatory processes.

The application of titanium in medicine takes place owing to the ability of good osseointegration with the surrounding bone tissue, which was proven in earlier works [8, 9]. Modern research has shown that surface topography, electrochemical impedance, and the chemical structure of the material notably influence the success of implant osseointegration.

The frequently used titanium-based material for making implants was Ti–6Al–4V until it was determined that Al and V could cause allergic reactions and could be cytotoxic [10]. Today, scientific work focuses on the development of metal materials with less poisonous alloying elements for the environment of the human body than Al and V. Since the alloying elements are substantial for lowering the value of the Young's modulus and reducing the toxic action of ions to a minimum, today, more and more titanium alloys are being developed. One of these newly developed materials is based on the Ti–Nb–Zr alloy with different amounts of alloying elements. Zr and Nb, as the alloying elements, lead to better biocompatibility, and they form a microstructure with α' and β phases or a near β phase leading to higher corrosion stability compared with other alloy types [11]. However, recently developed titanium alloys have limitations in their use in medicine because they have not achieved the appropriate level of biocompatibility and bioactivity. It shows that different nanostructured modifications of metallic biomaterials surface led to better corrosion protection, and if surface-modified biomaterials have appropriate molecules, materials led to better bioactivity. Further development of the material is aimed at making progress in biocompatibility and corrosion resistance by creating certain coatings on the surface or nanostructuring the surface.

5.1.2 Ultrafine-Grained Titanium-Based Materials Obtained by Severe Plastic Deformation (SPD)

Nanostructuring of bulk metal materials is one of the accepted processes for obtaining materials not only with better mechanical properties [12] but also with better corrosion resistance and biocompatibility [13]. A common method for the nanostructuring of metal materials as well as for titanium-based materials is SPD [14, 15]. The microstructure grains are refined to the submicron level, so the materials with grain values ranging from 100 to 1,000 nm, produced using SPD methods, are parted within UFG, while materials with grain size less than 100 nm are parted into nanostructured (NS) [16]. In terms of generally accepted definitions, a microstructure can be considered ultrafine-grained when the realized UFG structure has high-angle grain boundaries [17]. It is the grain boundary angle that is the difference between ultrafine-grained materials obtained by SPD processes and standard materials with grains obtained by usual metal processing and deformation methods.

Previous tests carried out in the field of manufacturing metal materials using SPD procedures indicate that by refining the grains of the microstructure, the characteristics of titanium-based materials, such as resistance to wear and harm and tensile strength, may be upgraded. Also, the mechanical and physical characteristics of titanium and titanium alloy are significantly improved, refining the grains of the microstructure [17–19]. By controlling the parameters of the SPD procedures of titanium-based materials, there is a possibility to get three dissimilar microstructures: lamellar, equiaxed, or bimodal. An equiaxed microstructure shows high values of tensile strength and elasticity and relatively low toughness of the material. In contrast, the lamellar structure shows good toughness, while the values of tensile and strength are slightly lower compared with the previous structure. All three structures showed an increase in tensile strength and yield stress by at least 20%, compared with the material that was not produced by the SPD procedures, while the tensile elongation that can be achieved in this way is advantageous for the application of the material in medicine [17–19]. It has been proven that by the application of SPD procedures, an increase in microhardness can be achieved as well as a decline in the values of elastic modulus of titanium-based materials. There is also an improvement in biocompatibility.

High-pressure torsion (HPT) is one of the SPD methods [20]. The HPT apparatus is composed of an upper and lower anvil (Figure 5.1(a)). The metal is placed between the plunges and then a pressure of several GPa is exerted on the sample. Finally, one plunge is revolved opposing to the other causing large shear deformations in the material. As a result of the HPT process, metal materials obtain ultrafine-grained or nanostructured microstructures [21, 22].

It has been shown that the thickness of the disk during the high-pressure torsion procedure notably affects the pressure distribution in the volume of the sample [23]. When the disk has a small thickness, only a narrow area around the edges of the disk is not under hydrostatic pressure. The significance of a low ratio between the thickness and the diameter of the disk, which leads to a linear increase in deformation from the center to the edges of the disk, is indicated in numerous research studies. It was also proposed that the size rate of disk thickness and diameter should not be greater than 1/10 so that the strain at the edge of the specimen is the

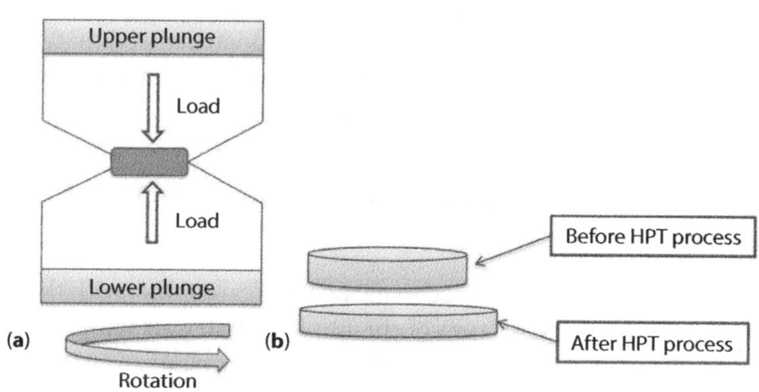

Figure 5.1 Scheme of (a) the HPT apparatus and (b) the difference in the dimensions of the sample before and after the HPT process.

highest and decreases linearly along the diameter of the specimen (strain is null in the center). On the other hand, the number of rotations during the high-pressure torsion process has a significant effect on the mechanical and physical characteristics of the titanium-based material. It has been shown that the rise in the number of revolutions notably contributes to the rise in the microhardness of the material [24]. Different modifications of the metal implant structure affect the surface adhesion of the cells, as well as the cellular response to the presence of implant materials in the human organism. The surface adhesion of osteoblast cells on the UFG Ti–13Nb–13Zr was studied using *in vitro* methods, and the results were comparatively analyzed with the results obtained for CG Ti–13Nb–13Zr [25]. It was shown that there was no important distinction in the percentage of merged cells on the UFG and CG materials; however, osteoblast cells had better adhesion to the UFG material than to the CG material. The increase in the surface adhesion of osteoblast cells on the UFG material occurs mainly due to better surface hydrophilicity. Cell adhesion on the surface of UFG cpTi obtained by the SPD process has been discussed and compared with that on the surface of the conventional cpTi [28]. It was shown that the number of cells on the surface of UFG cpTi was almost three times larger than that on the surface of CG cpTi. In addition, it was shown that the ultra-fine-grained structure of the material leads to the binding and spreading of osteoblasts in the early phases of development due to the observation of a large number of cytoplasmic extensions [26]. The refined microstructure brought a great number of surface-active sites and, thus, more intense cell growth [27]. Examination of the adhesion of osteoblasts and fibroblasts on the surface of UFG cpTi obtained by the HPT process indicated the presence of initial growth and expansion of osteoblasts [28]. The explanation for better bone cell adhesion and spreading in the case of UFG materials is based on the fact that the human bone is a nanostructured material. The logical consequence of nanostructuring of an implant material is adequate proliferation of the osteoblast and fibroblast cells.

5.1.3 Electrochemical Anodization of Titanium-Based Materials

Primary implant stability is related to its mechanical contact with the touching tissue [29]. Good integration of the implant materials with the bone is required to provide safety and efficiency of the implant over its life. The osseointegration process is presented as a direct structural and functional interconnection of the surrounding bone tissue and the biomaterial surface area. Osseointegration is a mechanism implying that the implant is reliably and adequately incorporated into the bone of the human body. Implant surface characteristics also influence the primary stability. Generally, two surface properties have been proposed as the most important facts for the tissue response to the implant material: surface morphology and chemical structure [30]. Compared with a smooth implant surface, a controlled rough surface provides more surface area for integration with the surrounding tissues [31] and allows successful implant growth into the tissues [32, 33]. Reports show that the roughness of the titanium implant surface enhances the rate of osseointegration and biomechanical fixation [34, 35]. The nanostructured implant surface is important for the osseointegration rate, protein adsorption, and fibroblast and osteoblastic adhesion [36, 37]. It was found that the nanostructured modification of the titanium surface on the level of nano-sized pores affects the adhesion, spreading, and growth of osteoblastic cells. The creation of TiO_2 nanoporous surface on titanium affects the osteoblastic proliferation rate and initiates fibroblast growth. Better clinical results were obtained for titanium implants with a nanostructured oxide film on the surface compared with bare titanium surfaces for similar shapes of implants [38]. Comparison of the adhesion, elongation, and growth of osteoblastic and fibroblast cells for various topographies of nanostructured modified surfaces and various nanotube dimensions has been analyzed in many papers. It was shown that outstanding elongation of the cells is obtained on titanium-based nanotubes with diameters ranging between 70 and 100 nm. Diameter values of the nanotubes ranging from 30 to 50 nm are enough for the human cells to improve their elongation [39]. Moreover, a larger diameter has a higher influence on human body cell elongation. On the contrary, a larger dimension of the nanotube produces deterioration of cell adhesion. This happens due to the enlarged expanses between the nanotubes, as more expanses between nanotubes can improve cell spreading [39, 40]. *In vitro* examinations showed that extraordinary adhesion of cells occurs when the diameter of titanium-based nanotubes had a lower value than 100 nm [41].

There are many methods for the nanostructure modification of biomedical alloy surfaces, but the chemical technique is utilized in order to improve osseointegration, biocompatibility, wear resistance, contamination removal, and corrosion resistance [42]. In most chemical methods, a reaction occurs between the surface of the sample and the solution. Materials are treated with an acid solution, alkaline solution, or hydrogen peroxide, as well as heat or passivation treatments. For this reason, chemical methods of surface modification are utilized in order to remove surface oxides and contamination of the surface of the material, which affects the improvement of biocompatibility.

Electrochemical anodization is a method that can form nanotubular oxide thin film on the surface [43]. The nanoporous oxide thin layer on the surface of titanium-based materials can be formed in an acidic solution without F^- ions, while the creation of the nanotubular oxide film consisting of nanotubes requires the existence of F^- ions in the used solution [44]. The creation of the nanotubular oxide film during the process of anodic oxidation consists of two steps: first, the contact of the electrolyte and metal materials brings to the creation of a barrier oxide film:

$$Ti^{4+} + 2H_2O \rightarrow TiO_2 + 4H^+$$

The next step is the creation of a crack on the surface, starting due to the higher separation of the oxide film. The diffusion of the solution into the crack produces a higher rate of oxide film dissolution:

$$TiO_2 + 6F + 4H^+ \rightarrow [TiF_6]^{2-} + 2H_2$$

The creation of nanotubes on the oxide layer is conditioned by the diffusion of F^- ions and the making of $[TiF_6]^{2-}$ [45, 46]. The electrochemical anodization process improves the adhesion and bonding, increases the corrosion resistance, and decreases the ion release rate [47]. The precedence of electrochemical anodization is the possibility of controlling the nanotubular morphology of the surface and dimension of nanotubes (diameter, wall thickness, and length) as well as the shape of nanotubes through variation of the solution, pH value, potential, or duration of anodization [48]. The duration of the electrochemical anodization procedure also influences the thickness of the nanotubular thin layer, as the extension of the duration of the electrochemical anodization increases the thickness of the created nanostructured film [49]. Nevertheless, if the same surface is subjected to electrochemical anodization for a long time, the walls of the nanotubes become extremely thin [50].

The primary aim of this study is to reveal the morphology and properties of the nanostructured surface and to assess its influence on the biocompatibility of the UFG Ti–13Nb–13Zr. Other studies examined the biocompatibility of titanium and its alloys after electrochemical anodization of their surface [51–53], but there are not enough scientific analyses that evaluated the biocompatibility of the nanostructured surface of the UFG Ti–13Nb–13Zr.

5.2 Materials and Methods

5.2.1 High-Pressure Torsion Process

For the investigations, a near β-type Ti–13Nb–13Zr alloy made for medical usage (Standard ASTM F1713) was used and made as a bar with a diameter of 28 mm in an as-casted condition. Disks with a thickness of ≈2.3 mm were cut as the starting samples. To produce ultrafine-grained Ti–13Nb–13Zr (UFG TNZ), the HPT procedure was used. The HPT procedure was performed by rotating one of the anvils with a uniform speed of 0.2 rpm at a pressure of 4.1 GPa. After five revolutions, the disk thickness was reduced by ≈23%, while the diameter was raised by ≈18% (Figure 5.1(b)). To analyze the differences in the structure of the TNZ before and after the high-pressure torsion procedure, scanning (Zeiss Leo 1525) and transmission electron microscopy (TEM) with a Philips CM12 were used. A standard Ar-ion milling process was used for the preparation of the TEM samples. TEM was used in combination with selected area diffraction (SAD).

5.2.2 Electrochemical Anodization

The platinum, as the counter electrode (cathode), and a sample of TNZ in the basic state or after the high-pressure torsion process as a working electrode (anode) were used as the system for the electrochemical anodization (anodic oxidation) process. Both electrodes had an area of 1 cm^2 and were placed one against the other. The solution was 1 M H_3PO_4 + 0.5 wt.% NaF and the duration of anodic oxidation was varied from 30 to 120 min, with a constant potential of 25 V. A *PEQLAB EV 231* was used as the power source device. To analyze the structure of the created nanostructured thin film, obtained after the anodic oxidation, a scanning electron microscope (SEM) MIRA3 XMU was used. The SEM (JEOL 5800) with energy-dispersive spectroscopy (EDS) was used to define the chemical structure of the modified surface on top of the specimen.

5.2.3 Characterization of the Surface Topography by Atomic Force Microscopy (AFM)

In order to analyze the surface topography of the abovementioned alloys, as well as the surface topography for both alloys after electrochemical anodization for 30, 60, 90, and 120 min, atomic force microscopy (AFM, Veeco-Nanoscope III d) in the tapping mode was utilized. The AFM fields were analyzed at a scan speed of 1 Hz, while all the measurements were conducted under ambient conditions.

5.2.4 Biocompatibility Examination

In vitro examination was performed on the animal (L929) and human (MRC-5) fibroblast cell lines, which were grown hooked on the surface of the vessel in Dulbecco's adapted Eagle's agent (DMEM) (*Gibco BRL, England*), containing 100 IU m/L penicillin and 100 μg m/L streptomycin. An agent of 4.5 g/L glucose and 10% fetal calf serum (FCS) (*Sigma-Aldrich, Germany*) was added. Cells were conserved under regular conditions, in an atmosphere of 5% CO_2 and 95% air with an increased moisture level at 37°C. To provoke the growth and development of the cells, the population was transplanted and put in a new agent two times a week. In the experiment, just cells in the phase of growth were utilized, and their viability was >90%. The viability and number of cells were estimated using the color assay rejection of 0.1% Trypan Blue. Cells were planted into Petri dishes on the sterilized surface of alloys at a density of 2.0×10^5 cells ml^{-1}. A sample without cells was used as a control sample. The cytotoxicity of the alloys was evaluated as a percent of cell growth inhibition by two kinds of *in vitro* tests: colorimetric methyl-thiazol-tetrazolium (MTT) test and agar diffusion test (ADT). The growth of the cells was controlled after 96 h for the MTT test, and the experiment was repeated three times to ensure high statistical results. The procedure of the MTT test was described earlier [13, 54].

To analyze the morphology and adhesion of L929 and MRC-5 cells, the alloys were put in Petri dishes and the cells were planted on their surfaces at a concentration of 1.0×10^5 cells/6 ml DMEM with 10% FCS. The Petri dishes were thermostated for 48 h at 37°C in the existence of 5% CO_2. After the incubation time, the cells were fixed using 2.5% glutaraldehyde (*Sigma-Aldrich, Germany*) for 48 h and were dehydrated using 3% acetic acid, 3% acetic acid and 25% ethanol at a ratio of 1:1, 3% acetic acid and 50% ethanol at a ratio of 1:1, or 70% ethanol solution. To analyze the morphology and adhesion of L929 and MRC-5 cells, an SEM MIRA3 XMU guided at 20 keV was used. To provide a comparison of the statistical results, standard deviation (SD) was established.

5.3 Results and Discussion

5.3.1 The Microstructure of the Ultrafine-Grained Two-Phase Ti–13Nb–13Zr Alloy

To get access to the microstructure before and after the HPT procedure, TEM analysis in combination with selected area diffraction (SAD) and SEM was performed on the disk samples. The characteristic starting structure is shown in Figure 5.2(a) and a TEM bright field image of the UFG microstructure is presented in Figure 5.2(b).

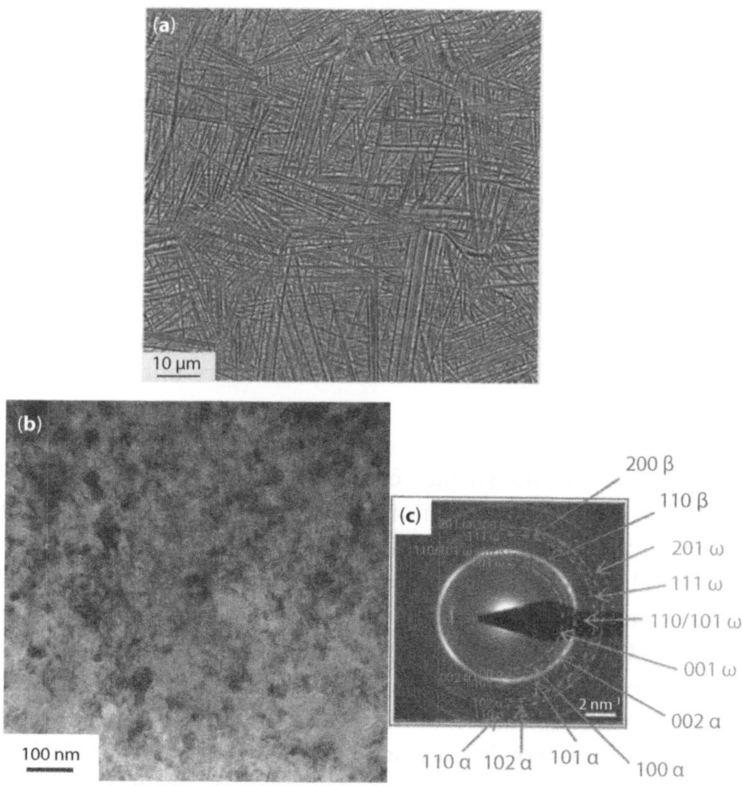

Figure 5.2 The microstructure of the undeformed (a) and high-pressure torsion-deformed TNZ alloy (b) and representative diffraction patterns of the UFG TNZ alloy (c).

The Ti–13Nb–13Zr in the basic state is compounded of α' and β phases (Figure 5.2(a)), while these phases are not easily noticeable after the HPT process [55]. In addition, it was shown that the HPT procedure brought a notable reduction in grain size [55]. As could be seen in Figure 5.2(b), the structure of Ti–13Nb–13Zr after HPT had a grain size in the range of several hundred nanometers, which indicates that the alloy is an ultrafine-grained material. Moreover, in Figure 5.2(b), a blurred perspective can be seen, which is the effect of internal stresses that are often associated with the unbalanced condition of grain boundaries in the materials subjected to SPD. The SAD analysis (inset image of Figure 5.2(b)) confirms the subsistence of the α phase and the β phase for the UFG TNZ but also probably the creation of a ω phase, which would be formed at the positions as presented.

5.3.2 Morphology of Nanostructured Surfaces of the Materials

The appearance of the nanostructured modified surfaces created on the CG and UFG TNZ after the duration of anodic oxidation of 30, 60, 90, and 120 min is shown in Figure 5.3. The nanotubular film was created after 90 and 120 min, while a random film and a nanoporous oxide film were created after 30 and 60 min. During the shorter duration of electrochemical anodization, the nanotubes were connected, while for the longer duration of electrochemical anodization, the nanotubes were unconnected due to dissolution.

Moreover, the anodic oxidation after 30 min did not bring to the creation of nanotubes on the surface of the CG TNZ but led to surface erosion, while a nanoporous oxide film with an average pore diameter of 46.0 nm was produced on the UFG TNZ alloy surface (Figure 5.3(a, e)). After 60 min of the electrochemical anodization process, a random porous oxide layer was formed, as nanotubes, which formed in certain places on the surface of the CG TNZ (Figure 5.3(b)). Nanoporous oxide film with an average nanopore diameter of 54.86 nm was created on the UFG TNZ surface, under the same condition of the electrochemical anodization process (Figure 5.3(f)). These results indicate that the alloy after the HPT procedure requires a shorter duration of electrochemical anodization to form a homogeneous nanoporous structure on the surface.

With time, the nanopores grew, and finally, a nanotubular oxide layer was observed. As presented in Figure 5.3(c, d, g, h), the electrochemical anodization process after 90 and 120 min led to the creation of a nanotubular oxide film with clearly discernible nanotubes. They are homogeneously distributed and the average diameter increases with increasing

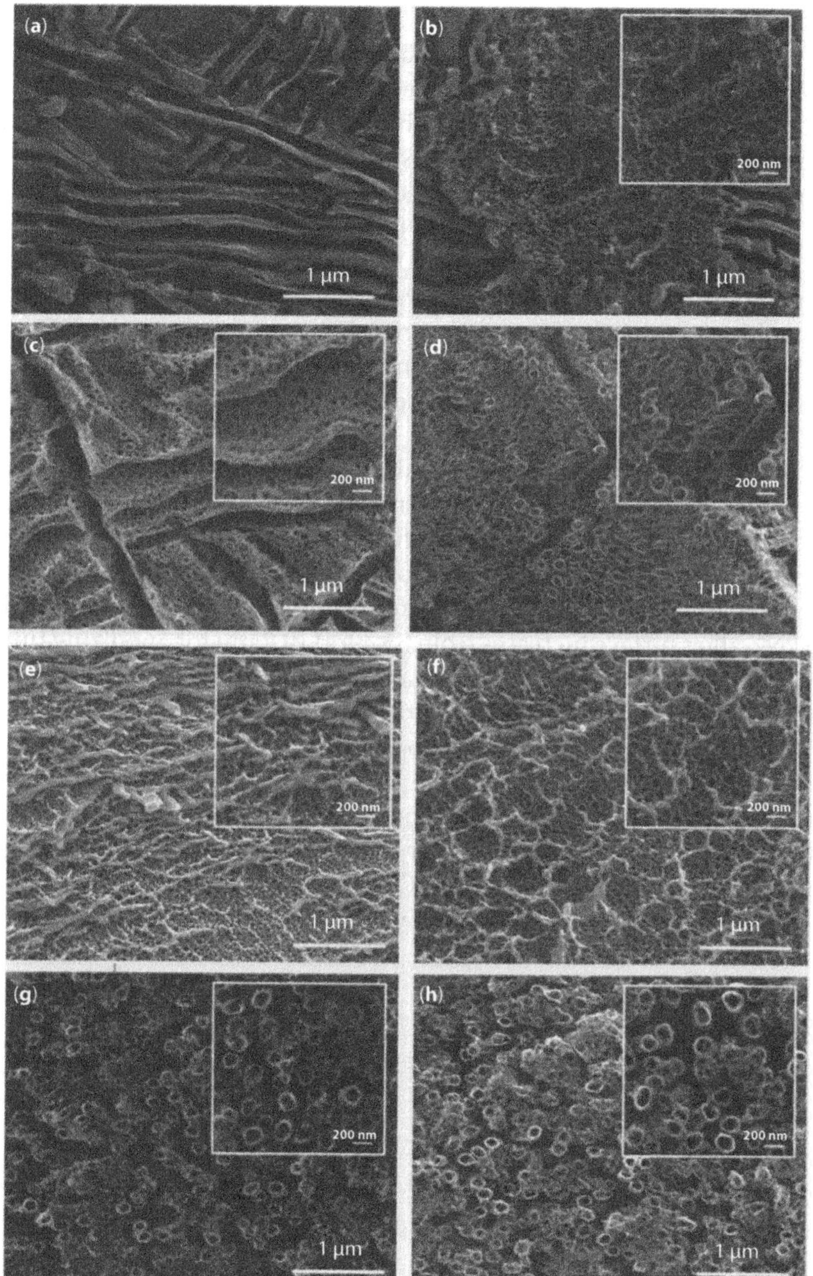

Figure 5.3 The morphology of the nanostructured surface of CG TNZ (a–d) and UFG TNZ alloys (e–h), formed in 1 M H_3PO_4 + 0.5 wt.% NaF, after a period of time from 30 to 120 min of electrochemical anodization, respectively, for both alloys.

anodizing time. This is in accordance with Taveira *et al.* [56], who examined the formation of a titanium oxide surface during anodizing times from 0.5 to 120 min. They concluded that the electrochemical anodization process had a few stages. In the first stage, a barrier oxide film was created, after which porous and random oxide layers were formed, and finally, a nanotubular film was formed. The homogeneous morphology of the nanotubular film was created on the surface of the UFG TNZ with a time modification of 90 and 120 min (Figure 5.3(g, h)). On the contrary, the inhomogeneous morphology of the nanotubular film was obtained on the CG TNZ surface, as could be seen by the formed channels between the nanotubes (Figure 5.3(c, d)). This inhomogeneous morphology of the created oxide film can be induced by two phases (α+β phase) in the microstructure of the CG TNZ [57, 58].

The average nanotube diameter with the standard deviation (SD) of 30 measurements as a function of anodizing time is presented in Figure 5.4(a).

For both coarse-grained and ultrafine-grained TNZ alloys, the results indicate that the average diameter becomes greater with a longer anodizing time. The average value of the diameter of the nanotubes formed on the CG TNZ alloy surface increases from 55.4 to 110.2 nm for anodizing times of 60 to 120 min, respectively. On the other hand, the average nanotube diameter produced on the surface of the UFG TNZ increases from 46.0 to 117.5 nm for anodizing times of 30 to 120 min, respectively. This is consequent with the analyses of Feng *et al.* [59], who showed that the diameter increased with increasing time of modification. As can be seen in Figure 5.4(a), the average nanotube diameter produced on the UFG TNZ is larger compared with that created on the CG TNZ alloy, after the duration of anodic oxidation of 90 and 120 min, while it is slightly higher for the CG

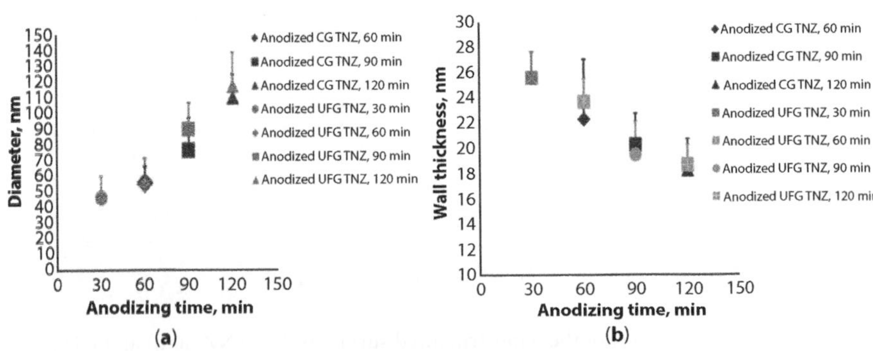

Figure 5.4 Variation of the average value of (a) the diameter of the nanotubes and (b) the wall thickness of the nanotubes for different anodizing times.

TNZ alloy after an anodizing time of 60 min. The results also show that the nanostructured surface consisted of larger and smaller nanotubes. This is confirmed by the large difference in the dimensions of the diameter of the nanotubes, ranging from 35 to 130 nm, for both alloys. The SD values indicate that a notable difference between the sizes of the nanotube diameters formed on the alloys under the same condition as the electrochemical anodization process does not exist. The maximum uniform distribution of the diameter is achieved in the nanoporous structure produced on the UFG TNZ surface at 60 min, while the minimum uniformity is obtained in a nanotubular film produced on the UFG TNZ surface at 120 min. Moreover, by comparing the diameters of the nanotubular layers for both alloys, it can be concluded that the nanotubes created on the surface of the UFG TNZ after 90 min had a diameter of maximum uniformity. The average thickness of the nanotube walls with the SD of 30 measurements as a function of anodizing time is shown in Figure 5.4(b). The wall thickness ranges between 20.3 and 24.3 nm for the CG TNZ, while the wall thickness ranges between 18.1 and 25.8 nm for the UFG TNZ. With decreasing anodizing time, the thickness of the nanotube wall increases [60].

During the electrochemical anodization process, the current density changed, as given in Figure 5.5(a, b). The changing current densities with anodizing time are shown in Figure 5.5 for all alloys. According to many authors, higher current density brings to the creation of a more homogeneous distribution of the nanotubes [61, 62]. As presented in Figure 5.5, the electrochemical anodization process of the UFG alloys shows a higher current density, which indicates that these alloys have thicker and more

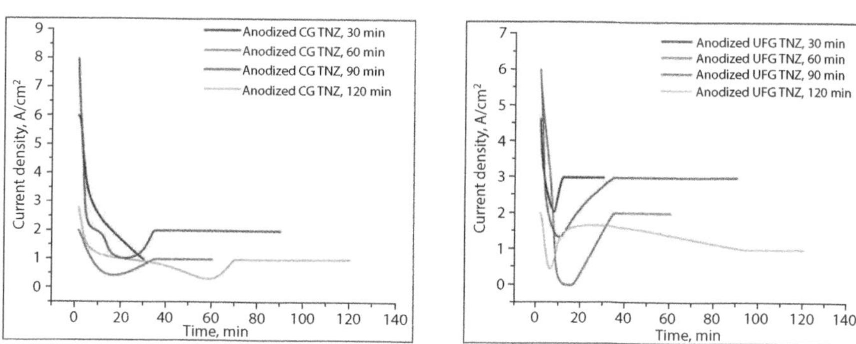

Figure 5.5 The dependence of current density on the anodic oxidation time at a potential of 25 V for (a) CG TNZ and (b) UFG TNZ.

homogeneous oxide layers in comparison with the CG TNZ alloys for the same processing time.

The chemical structure of the obtained nanostructured oxide film is in accordance with the structure of the alloy and the alloying element ratio. It was shown that the electrochemical anodization process when applied to conventional biomedical alloys, such as titanium alloys, with the existence of Al, V, Nb, and Zr elements, brings a significant inhomogeneity in the morphology and chemical structure of the nanostructured surface [57, 63]. This happens due to the different dissolutions of one of these alloying elements and conditioned considerably with the microstructure composition of the phases in these two-phase materials. The characteristic spectra of the elements obtained by EDS are presented in Figure 5.6.

EDS detected all the alloying elements and, additionally, oxygen on the surface. For example, for the UFG TNZ alloy after anodic oxidation for 90 min, the obtained results indicate the existence of O, Ti, Zr, and Nb with values of 40.1%, 38.0%, 11.4%, and 7.8%, respectively. Similarly, UFG TNZ anodized during 60 min shows the existence of O, Ti, Zr, and Nb with values of 42.7%, 40.5%, 11.3%, and 7.6%, respectively. Chemical characterization of the thin oxide films shows no large dissimilarity in the structure of the surfaces of both alloys after various durations of electrochemical anodization. The average values of the chemical composition of three measurements on top of the nanostructured surface are presented in Table 5.1. The detected elements may indicate that a mixture of titanium-, niobium-, and zirconium-based oxides is formed on the surfaces. A smaller amount of Nb after electrochemical anodization may be associated with the higher chemical stability of Nb_2O_5 oxide in a solution with fluoride ions, while Zr and Ti in such solutions are more rapidly dissolved [60, 64]. Moreover, Nb has been shown to decrease the modulus of elasticity, while Zr is responsible

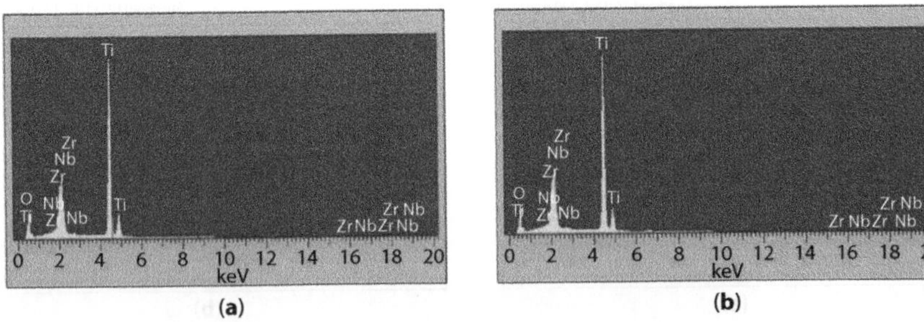

Figure 5.6 The elements of the spectra of the modified surface of (a) CG TNZ and (b) UFG TNZ formed during 90 min.

Table 5.1 EDS analysis of the nanostructured surface of the CG and UFG TNZ formed during various anodizing times.

Material	Anodizing time, min	Elements detected on the surface area, wt.%			
		Ti	Nb	Zr	O
CG TNZ	30	43.7	7.5	11.3	37.2
	60	41,4	7.3	11.4	41.4
	90	39.0	7.2	11.6	42.1
	120	36.5	8.0	11.8	42.4
UFG TNZ	30	43.3	7.6	11.4	37.1
	60	40.1	7.0	11.4	41.4
	90	38.6	8.8	12.0	40.5
	120	42.7	7.8	11.3	38.0

for a higher level of biocompatibility and better corrosion stability because of the steady creation of an oxide layer on the surface [60]. Moreover, Zr has excellent characteristics in physiological fluids (thermal and chemical stability) [64, 65].

5.3.3 Characterization of the Surface Topography

Figure 5.7 presents the AFM characterization of the surface topography of each tested sample. The AFM testing presents that the anodic oxidation procedure resulted in the creation of a rough topography surface of the CG and UFG TNZ. The UFG TNZ alloys anodized for 30, 60, 90, and 120 min (Figure 5.7(f–i)) show a typical rough topography, while the CG TNZ alloys anodized for 60, 90, and 120 min (Figure 5.7(b–d) show a rise in the surface roughness value in comparison with bare alloy with a wavy topography. Electrochemical anodization of the surfaces of titanium alloys at 25 V produces roughness and porosity. Nevertheless, most surfaces showed a rough surface because of dissimilarity in the dismissal rate between the two-phase attendance in titanium alloy. The AFM detected that the surfaces are granular in appearance with a granule value of a few nanometers. The usual value of surface roughness for anodized titanium, which is approximately 10–100 nm, is formed by creating a nanoporous

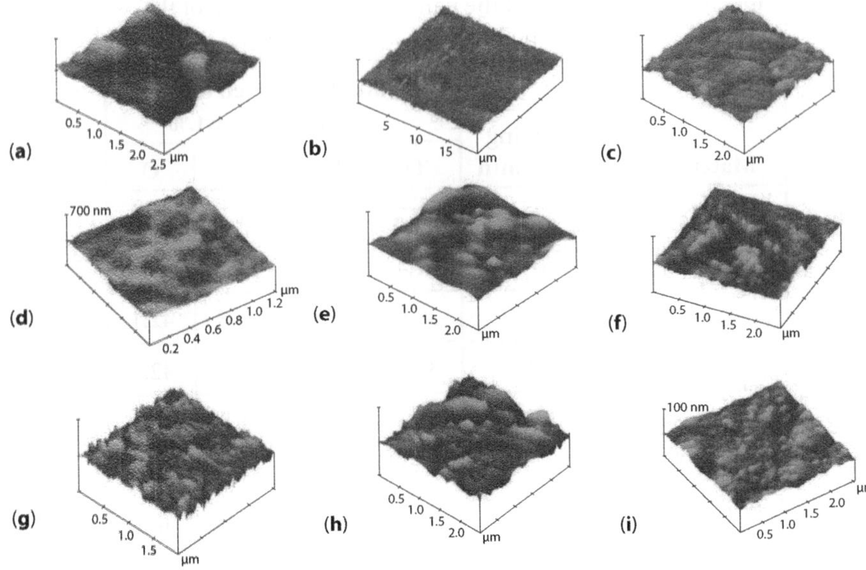

Figure 5.7 AFM surface images for the (a) initial CG TNZ, (b) CG TNZ surface anodized for 60 min, (c) CG TNZ surface anodized for 90 min, (d) CG TNZ surface anodized for 120 min, (e) initial UFG TNZ, (f) UFG TNZ surface anodized for 30 min, (g) UFG TNZ surface anodized for 60 min, (h) UFG TNZ surface anodized for 90 min, and (i) UFG TNZ surface anodized for 120 min.

surface [66]. It was shown that an oxide film with nanotubes of diameters of 120 nm forms a surface roughness of 46.5 nm [67]. Table 5.2 gives the roughness root mean square (R_{rms}/nm) of the CG and UFG TNZ surface during the anodic oxidation process for 30, 60, 90, and 120 min.

From Table 5.2, it can be noticed that the anodized UFG materials have a roughness almost twice as large as anodized CG materials for all anodizing times. As can be seen, the roughness of bare alloys is 6.6 nm for CG TNZ, while the ultrafine-grained alloy has a roughness of 3.6 nm, which shows that the HPT procedure slightly reduces the roughness of the surface. On the other hand, anodized alloys have one order of magnitude greater roughness than the bare alloy. The results show that by extending the anodizing time, the roughness of the alloys also increases as a consequence of the rise in the dimensionality and amount of nanopores on the surface [68]. For example, it was observed that nanotubular oxide layers produced on the surface of both alloys during 90 min had a slightly higher roughness than the surface of alloys with nanoporous oxide layers formed for 60 min. The general conclusion is that the formation of the

Table 5.2 Roughness values of the bare and nanostructured surfaces of CG and UFG TNZ.

Characteristic	Material	Anodizing time, min				
		/	30	60	90	120
R_{rms}/nm	CG TNZ	6.6	/	16.6	19.1	23.5
	UFG TNZ	3.6	20.5	33.3	34.1	35.0

nanostructured surface leads to an increased roughness, but by increasing the average value of the nanotube diameter by about 54 nm for the CG TNZ and from 46.0 to 117.5 nm for the UFG TNZ, no significant changes are made in the surface roughness.

Some studies report that the nanostructural modified surface of the material has an effect on the adsorption and adhesion of proteins, adhesion of the cell, and proliferation [69, 70]. Additionally, it was shown that surface roughness and surface topography are strictly related to the biological response of the cells to the implant surface [71]. Manjaiah *et al.* [68] showed that an increase of the surface roughness from 16.3 nm for unanodized commercially pure titanium to approximately 37.1 nm for titanium anodized at 20 V had a positive effect on cytocompatibility. Sowa *et al.* [72] investigated the bioactivity of an anodized TNZ alloy and showed that surface topography had a strong influence on the viability, appearance, dimensions, and position of the cells. According to the obtained homogeneous nanotubular oxide film with a larger nanotube diameter and higher roughness of the surface, the coarse-grained and ultrafine-grained Ti–13Nb–1Zr alloys anodized for 90 min were chosen for studying biocompatibility and estimating the impact of the surface morphology on it.

5.3.4 Biocompatibility Examination

The viability of the mouse fibroblast (L929) and human fibroblast (MRC-5) cells in touch with a bare surface of the CG and UFG TNZ and with an anodized surface of the CG and UFG TNZ, after 90 min, was estimated using two *in vitro* tests: the colorimetric methyl-thiazol-tetrazolium (MTT) test and the agar diffusion test (ADT). Figure 5.8 presents the valuations of the MTT test and demonstrates that the mean value of the viability of both cells notably increases after the creation of the nanotubular oxide film.

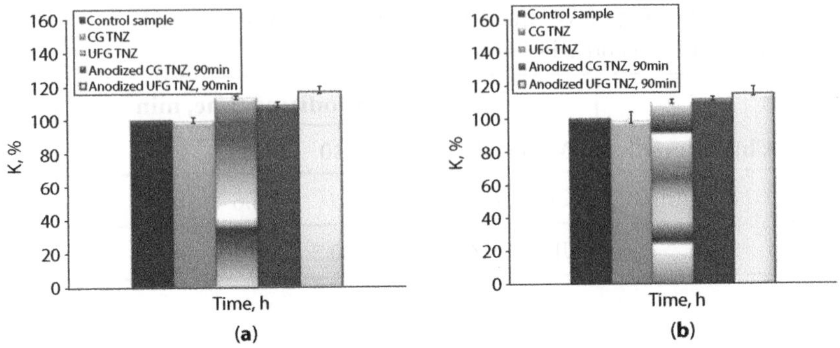

Figure 5.8 Fractions of surviving (*K*) for (a) MRC-5 and (b) L929 cells during 96 h in the MTT test for all examined alloys.

The outcomes of this testing are consistent with other research works, showing that a suitable roughness and nanostructured surface could provide better cell viability [73–76]. After 96 h of the MTT test, the L929 and MRC-5 cells in touch with the bare surface of the CG TNZ alloy had an approximately similar mean value of viability as the fibroblast cells in touch with a control sample. Furthermore, the mean value of the viability of the cells progressively increases in touch with a bare surface of the UFG TNZ and anodized surfaces of both alloys. These facts indicate that none of the examined alloys exhibited a cytotoxic effect. As can be seen in Figure 5.8, the fraction of surviving (*K*) of L929 and MRC-5 cells in touch with the bare surface of the UFG TNZ is higher compared with cells in contact with a bare surface of the CG TNZ, after 96 h. As can be seen in Figure 5.8, both alloys anodized for 90 min show a greater fraction of surviving cells compared with their counterpart alloy, while the UFG TNZ alloy after anodic oxidation for 90 min has the highest fraction of surviving cells (*K*), which indicates that the anodized UFG TNZ alloy could be an excellent material for biomedical applications. This is in accordance with Frandsen *et al.* [77], who showed that the creation of a nanotubular oxide film led to an increasing number of cells and cell elongation, independent of the diameter of the nanotubes. Li *et al.* [78] showed that anode-oxidized titanium had better cell proliferation than polished titanium, after 96 h of the MTT assay. CG TNZ alloy anodized for 90 min had almost the same *K* as the CG TNZ alloy after the HPT process, but it was higher than that in the initial CG TNZ alloy. The *K* for all samples is larger than that of the control sample, except for the CG TNZ alloy. So, we could conclude that the biocompatibility of all the tested alloys, regardless of whether having or not having the anodized surface, is very good. Additionally, the examined alloys were not cytotoxic according to ADT because decolorization was not detected. Figures 5.9 and 5.10 show the appearance of the mouse fibroblast and human fibroblast cells on the tested bare and nanostructured alloy surfaces.

The SEM images show that the cells in touch with the anodized surface of the UFG TNZ are interconnected, spindly elongated, and triangular in shape (Figures 5.9(e) and 5.10(e)), while the cells in touch with the nanostructured anodized surface of the CG TNZ are regular triangular (Figures 5.9(b) and 5.10(b)). On the other hand, the cells in touch with the bare surface of the UFG TNZ are mostly oval and triangular in shape (Figures 5.9(d) and 5.10(d)), while the cells in contact with the bare surface of the CG TNZ are mostly oval (Figures 5.9(a) and 5.10(a)). The shape of the cells on the bare and nanostructured anodized surface of the UFG TNZ indicates their excellent spreading over the surface of the material (Figures 5.9(d, e) and 5.10(d, e)). As can be noticed in Figures 5.9(e) and 5.10(e), the cells are well connected, showing a high tendency to integrate and form a continual cell stratum on the anodized surface of the UFG TNZ, which is a confirmation of good metabolic activity. Good fibroblast cell morphology and suitable attachment in contact with the UFG TNZ surface are evident, which indicates an excellent interaction between fibroblast cells on the bare and anodized surface of the ultrafine-grained alloys. In contact with bare and anodized UFG TNZ alloys, the cells are connected, which is favorable for the proliferation potential of the cells. The position of cytoplasm extensions demonstrated the tendency of the cells to realize a connection with

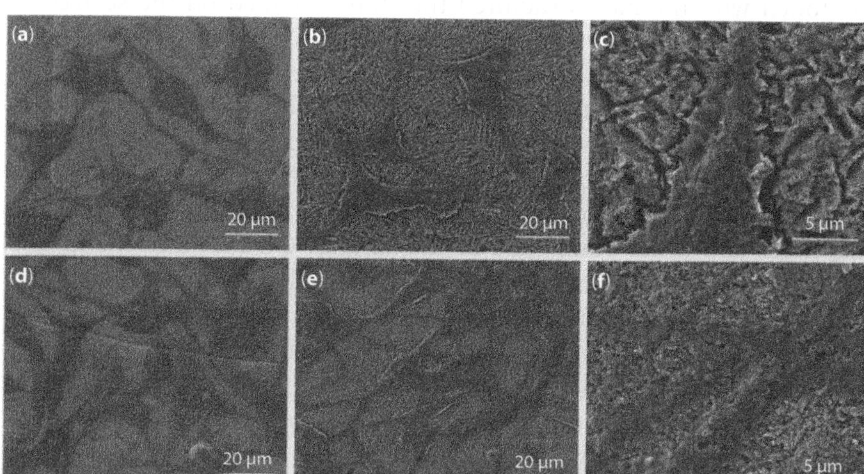

Figure 5.9 The MRC-5 cells on the surface of the (a) initial CG TNZ, (b, c) anodized CG TNZ for 90 min, (d) initial UFG TNZ, and (e, f) anodized UFG TNZ for 90 min.

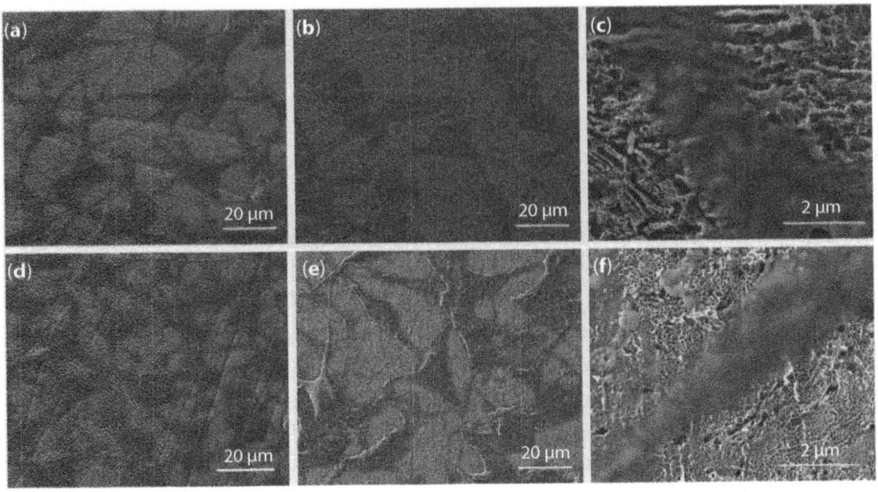

Figure 5.10 The L929 cells on the surface of the (a) initial CG TNZ, (b, c) anodized CG TNZ for 90 min, (d) initial UFG TNZ, and (e, f) anodized UFG TNZ for 90 min.

neighboring fibroblast cells, for the surfaces of all alloys. Furthermore, the cells have a greater number of visible cytoplasm extensions on the UFG alloy surfaces, regardless of whether they have the nanostructured surface, which showed that the cells had more desirable spreading on these alloy surfaces than on CG alloys.

Figures 5.9(f) and 5.10(f) show that cytoplasm extensions had a very fine touch with the nanostructured thin film obtained on the surface of the UFG TNZ without visible intervening spaces between the cytoplasm extensions and the surface of nanotubes. On the contrary, Figure 5.9(c) shows that cytoplasm extensions do not have such good contact with the nanostructured thin film on the CG TNZ, in the case of MRC-5 cells. The reason is that the anodic oxidation of the CG TNZ after 90 min creates a nanotubular oxide film with a less uniform surface morphology of the nanotubes and lower roughness of the surface. On the other hand, the electrochemical anodization of the UFG TNZ alloy for 90 min creates a nanotubular film with a homogeneous morphology and higher roughness, which allows better contact and spreading of the fibroblast cells along the nanotubular surface.

It was shown that nanotube size and surface nanotopography had an impact on biocompatibility and cell morphology and that an increasing nanotube diameter led to an increase in cell elongation [79]. The anodic oxidation process leads to an increasing number of attached cells and their viability, irrespective of the diameter of the nanotubes, while the diameter value of the nanotubes to about 100 nm leads to a notable increase in the elongation of the human cells [80]. Oh *et al.* [81] showed that the measure of cell adhesion on a titanium surface after electrochemical anodization for 30 min, when nanotubes with a diameter greater than 60 nm were formed, was higher than on the titanium before the surface nanostructural modification, i.e., the electrochemical anodization process. On the contrary, Huang *et al.* [82] determined that the evenly formed nanotubes with an average diameter value of 8 nm on the surface of the Ti–25Nb–25Zr alloy have a beneficial behavior for the cells.

This study showed that nanotubes with an average diameter of 76.4 nm created on the CG TNZ alloy and nanotubes with an average diameter of 90.1 nm produced on the UFG TNZ alloy lead to an increasing fraction of cell survival, while elongation and adhesion of the cells are better with a nanostructured surface with a larger diameter of the nanotubes, for the anodized UFG TNZ.

Conclusions

The nanostructured thin films are successfully created on the surfaces of the CG and UFG TNZ using electrochemical anodization in a solution of orthophosphoric acid with F^- ions during different times. The electrochemical anodization process for 90 and 120 min led to the creation of a regular nanotubular film, while randomly oriented oxide films and nanoporous oxide films were created after 30 and 60 min of anodizing. The results indicated that oxide layers exhibited various structures and surface morphologies regarding the phase type, thickness, size, and shape of the nanotubes. The diameters of the obtained nanotubes increased with the increasing duration of electrochemical anodization, while the thickness of the nanotube walls decreased. Nanostructured surfaces of titanium alloys resulted in the creation of a rough topography. By creating a nanoporous oxide film and a nanotubular oxide film, the roughness of the surface also increased. However, increasing the average size of the diameter of the nanotubes did not lead to important changes in the value of surface roughness. Biocompatibility examinations showed that the existence of a nanotubular film on both alloys enhanced the fractions of surviving cells compared with the untreated alloys. The UFG TNZ with a nanotubular film on the surface had the highest fraction of surviving cells. The nanostructured and bare surface of the UFG alloy had more cytoplasm extensions with a better contact, indicating that the cells had better spreading and adherence to the nanostructured surface of the UFG alloys compared with the nanostructured surface of the CG alloys. This fact was more noticeable in the experiments with MRC-5 cells than with L929. Here, the presented results together with the studies published in [55, 83–85] make the entirety.

Acknowledgments

This work was supported by the Ministry of Science, Technological Development and Innovations of the Republic of Serbia (Contract No. 451-03-47/2023-01/200135 and 451-03-47/2023-01/200287). The authors gratefully acknowledge Dr. Sanja Stevanović from the Institute of Chemistry, Technology and Metallurgy, University of Belgrade, Serbia, for performing the AFM measurements. The authors also thank Dr. Christoph Gammer and Dr. Huaping Sheng for the TEM analyses, from Erich Schmid Institute of Materials Science, Austrian Academy of Sciences.

References

1. Park, J. and Lakes, R., *Biomaterials: An Introduction*, Springer Science & Business Media LLC, New York, NY, USA, 2007.
2. Hollinger, J., *An Introduction to Biomaterials*, Taylor & Francis Group, Boca Raton, FL, USA, 2012.
3. Raković, D. and Uskoković, D., *Biomaterials*, Institute of Technical Sciences of SASA, Materials Research Society Serbia, Belgrade, Serbia, 2010.
4. Raković, D., *Osnovi Biofizike*, IASC & IEFPG, Beograd, Serbia, 2008.
5. Ratner, B., Brunette, D., Tengvall, P., Textor, M., Thomsen, P., *Titanium in Medicine*, Berlin, Germany, Springer, 2001.
6. Bai, Y., Deng, Y., Zheng, Y., Li, Y., Zhang, R., Lv, Y., Zhao, Q., Wei, S., Characterization corrosion behavior, cellular response and *in vivo* bone tissue compatibility of titanium–niobium alloy with low young's modulus. *Mater. Sci. Eng. C*, 59, 565–576, 2016.
7. Mohammed, M., Khan, Z., Siddiquee, A., Beta titanium alloys: The lowest elastic modulus for biomedical applications: A review. *Int. J. Chemical, Nuclear, Metallurgical Mater. Eng.*, 8, 822–827, 2014.
8. Rack, H. and Qazi, J., Titanium alloys for biomedical applications. *Mater. Sci. Eng. C*, 26, 1269–1277, 2006.
9. Brånemark, P., *Introduction to Osseointegration*, Quintessence, Chicago, IL, USA, 1985.
10. Chaturvedi, T., Allergy related to the dental implant and its clinical significance. *Clin. Cosmet. Investig. Dent.*, 5, 57–61, 2013.

11. Kopova, I., Stracky, J., Harcuba, P., Landa, M., Janeček, M., Bačakova, L., Newly development Ti-Nb-Zr-Ta-Si-Fe biomedical beta titanium alloys with increased strength and enhanced biocompatibility. *Mater. Sci. Eng. C*, 60, 230–238, 2016.
12. Sharman, K., Bazarnik, P., Brynk, T., Bulutsuz, A.G., Lewandowska, M., Huang, Y., Langdon, T., Enhancement in mechanical properties of α-titanium alloy by high-pressure torsion. *J. Mater. Res. Technol.*, 4, 79–83, 2015.
13. Dimić, I., Cvijović-Alagić, I., Hohenwarter, A., Pippan, R., Kojić, V., Bajat, J., Rakin, M., Electrochemical and biocompatibility examinations of high-pressure torsion processed titanium and Ti–13Nb–13Zr alloy. *J. Biomed. Mater. Res. Part B*, 103, 1097–1107, 2017.
14. Shirooyeh, M., Xu, J., Langdon, T., Microhardness evolution and mechanical characteristics of commercial purity titanium processed by high-pressure torsion. *Mater. Sci. Eng. A*, 614, 223–231, 2014.
15. Valiev, R., Semenova, I., Jakushina, E., Latysh, V., Rack, H., Lowe, T., Petruželka, J., Dluhoš, L., Hrušák, D., Sochová, J., Nanostructured SPD processed titanium for medical implants, in: *Materials Science Forum*, vol. 584–586, pp. 49–54, 2008.
16. Sankar, M., Gopal, V., Alexander, R., Manivasagam, G., Ramalingam, M., *Nanobiomaterials-Classification, Fabrication and Biomedical Applications*, Wiley-VCH Verlag GmbH & Co, 2018. https://doi.org/10.1002/9783527698646.ch2
17. Yilmazer, H., Niinomi, M., Nakai, M., Cho, K., Hieda, J., Todaka, Y., Miyazaki, T., Mechanical properties of a medical β-type titanium alloy with specific microstructural evolution through high-pressure torsion. *Mater. Sci. Eng. C*, 33, 2499–2507, 2013.
18. Valiev, R., Islamgaliev, R., Alexandrov, I., Bulk nanostructured materials from severe plastic deformation. *Prog. Mater. Sci.*, 45, 103–189, 2000.
19. Kim, H., Sasaki, T., Okutsu, K., Kim, I., Inamura, T., Hosoda, H., Miyazaki, S., Texture and shape memory behavior of Ti–22Nb–6Ta alloy. *Acta Mater.*, 54, 423–433, 2006.
20. Edalati, K. and Horita, Z., A review on high-pressure torsion (HPT) from 1935 to 1988. *Mater. Sci. Eng. A*, 652, 325–352, 2016.
21. Wang, C., Gao, N., Gee, M., Wood, R., Langdon, T., Effect of grain size on the micro-tribological behavior of pure titanium processed by high-pressure torsion. *Wear*, 280-281, 28–35, 2012.
22. Dimić, I., Cvijović-Alagić, I., Völker, B., Hohenwarter, A., Pippan, R., Veljović, Đ., Rakin, M., Bugarski, B., Microstructure and metallic ion release of pure titanium and Ti–13Nb–13Zr alloy processed by high pressure torsion. *Mater. Des.*, 91, 340–347, 2016.
23. Pippan, R., Scheriau, S., Taylor, A., Hafok, M., Hohenwarter, A., Bachmaier, A., Saturation of fragmentation during severe plastic deformation. *Ann. Rev. Mater. Res.*, 40, 319–343, 2010.

24. Ashida, M., Chen, P., Doi, H., Tsutsumi, Y., Hanawa, T., Horita, Z., Microstructures and mechanical properties of Ti-6Al-7Nb processed by high-pressure torsion. *Proc. Eng.*, 81, 1523–1528, 2014.
25. Park, H., Lee, C., Kim, Y., Jang, J., Suh, J., Park, W., Improved pre-osteoblast response and mechanical compatibility of ultrafine-grained Ti–13Nb–13Zr alloy. *Clin. Oral. Implants Res.*, 22, 735–742, 2011.
26. Estrin, Y. and Ivanova, E., Accelerated stem cell attachment to ultrafine grained titanium. *Acta Biomater.*, 7, 900–906, 2011.
27. Awang Shri, D., Tsuchiya, K., Yamamoto, A., Effect of HPT deformation on surface properties and biocompatibility. *Biointerphases*, 9, evd. br. 029007, 2014.
28. Faghihi, S. and Zhilyaev, A., Nanostructuring of a titanium material by high-pressure torsion improves pre-osteoblast attachment. *Adv. Mater.*, 19, 1069–1073, 2007.
29. Javed, F., Bashirahmed, H., Crespi, R., Romanos, G., Role of primary stability for successful osseointegration of dental implants: Factors of influence and evaluation. *Int. Med. Appl. Sci.*, 5, 162–167, 2013.
30. Turkyilmaz, I., *Implant Dentistry-A Rapidly Evolving Practice*, pp. 57–82, InTech, 2011. https://doi.org/10.5772/16936
31. Branemark, R., Rydevik, B., Myers, R., Osseointegration in skeletal reconstruction and rehabilitation: A review. *J. Rehabil. Res. Dev.*, 38, 175–181, 2001.
32. Jungner, M., Lundqvist, P., Lundgren, S., A retrospective comparison of oxidized and turned implants with respect to implant survival, marginal bone level and peri-implant soft tissue conditions after at least 5 years in function. *Clin. Implant Dent. Relat. Res.*, 16, 230–237, 2014.
33. Yi, Y., Park, Y., Choi, H., Lee, K., Kim, S., Kim, K., Oh, S., Shim, J., The evaluation of osseointegration of dental implant surface with different size of TiO2 nanotube in rats. *J. Nanomater.*, 2015, 1–11, 2015.
34. Lin, L., Wang, H., Ni, M., Rui, Y., Cheng, T., Cheng, C., Pan, X., Li, G., Lin, C., Enhanced osteointegration of medical titanium implant with surface modifications in micro/nanoscale structures. *J. Orthop. Transl.*, 2, 35–42, 2014.
35. Lacefield, R., Materials characteristics of uncoated/ceramic-coated implant materials. *Adv. Dent. Res.*, 13, 21–26, 2007.
36. Brammer, K., Oh, S., Frandsen, C., Jin, S., Biomaterials and biotechnology schemes utilizing TiO2 nanotube arrays, in: *Biomaterials Science and Engineering*, R. Pignatello (Ed.), pp. 193–211, InTech, 2011. https://doi.org/10.5772/23320
37. Mendonça, G., Mendonça, D., Aragao, F., Cooper, L., Advancing dental implant surface technology, from micron-tonanotopography. *Biomaterials*, 29, 3822–3835, 2008.
38. Jungner, M., Lundqvist, P., Lundgren, S., Oxidized titanium implants (Nobel Biocare Ti Unite) compared with turned titanium implants (Nobel Biocare mark III) with respect to implant failure in a group of consecutive patients

treated with early functional loading and two-stage protocol. *Clin. Oral. Implants Res.*, 16, 308–320, 2005.
39. Oh, S., Brammer, K., Li, Y., Teng, D., Engler, A., Chien, S., Jin, S., Stem cell fate dictated solely by altered nanotube dimension. *Proc. Natl. Acad. Sci. U.S.A.*, 106, 2130–2135, 2009.
40. Khudhair, D., Bhatti, A., Li, Y., Hamedani, H., Garmestani, H., Hodgson, P., Nahavandi, S., Anodization parameters influencing the morphology and electrical properties of TiO2 nanotubes for living cell interfacing and investigations. *Mater. Sci. Eng. C*, 59, 1125–1142, 2016.
41. Minagar, S., Wang, J., Berndt, C., Ivanova, E., Wen, C., Cell response of anodized nanotubes on titanium and titanium alloys. *J. Biomed. Mater. Res. A*, 101, 2726–2739, 2013.
42. Kulkarni, M., Mazare, A., Schmuki, P., Iglič, A., Biomaterial surface modification of titanium and titanium alloys for medical applications, in: *Nanomedicine*, One Central Press, 2014. http://www.onecentralpress.com/chapter-05-biomaterial-surface-modification-of-titanium-and-titanium-alloys-for-medical-applications/
43. Webster, T. and Yao, C., A Promising nano modification technique of titanium-based implants for orthopedic applications, in: *Surgical Tools and Medical Devices*, pp. 55–79, 2016.
44. Sobieszczyk, S., Self-organized nanotubular oxide layers on Ti and Ti alloys-Review paper. *Adv. Mater. Sci.*, 9, 25–41, 2009.
45. Galstyan, V., Comini, E., Faglia, G., Sberveglieri, G., TiO2 nanotubes: Recent advances in synthesis and gas sensing properties: Review. *Sensors*, 13, 14814–14838, 2013.
46. Roy, P., Berger, S., Schmuki, P., TiO2 nanotubes: Synthesis and applications: Reviews. *Chem. Int. Ed.*, 50, 2904–2939, 2011.
47. Liu, X., Chu, P., Ding, C., Surface modification of titanium, titanium alloys, and related materials for biomedical applications. *Mater. Sci. Eng. R*, 47, 49–121, 2004.
48. Ossowska, A., Sobieszczyk, S., Supernak, M., Zielinski, A., Morphology and properties of nanotubular oxide layer on the Ti–13Zr–13Nb alloy. *Surf. Coat. Technol.*, 258, 1239–1248, 2014.
49. Minagar, S., Berndt, C., Wang, J., Ivanova, E., Wen, C., A review of the application of anodization for the fabrication of nanotubes on metal implant surfaces. *Acta Biomater.*, 8, 2875–2888, 2012.
50. Wang, L., Jin, M., Zheng, Y., Guan, Y., Lu, X., Luo, J., Nanotubular surface modification of metallic implants via electrochemical anodization technique. *Int. J. Nanomedicine*, 9, 4421–4435, 2014.
51. Lewandowska, Z., Piszczek, P., Radtke, A., Jedrzejewski, T., Kozak, W., Sadowska, B., The evaluation of the impact of titania nanotube covers morphology and crystal phase on their biological properties. *J. Mater. Sci. Mater. Med.*, 26, 162–174, 2015.

52. Echeverry-Rendón, M., Galvis, O., Aguirre, R., Robledo, S., Castaño, J., Echeverría, F., Modification of titanium alloys surface properties by plasma-electrolytic oxidation (PEO) and influence on biological response. *J. Mater. Sci. Mater. Med.*, 28, 169–182, 2017.
53. Fan, X., Feng, B., Liu, Z., Tan, J., Zhi, W., Lu, X., Wang, J., Weng, J., Fabrication of TiO2 nanotubes on porous titanium scaffold and biocompatibility evaluation *in vitro* and *in vivo*. *J. Biomed. Mater. Res. A*, 100A, 3422–3427, 2012.
54. Aguirre, R., Echeverry-Rendon, M., Quintero, D., Castano, J., Harmsen, M., Robledo, S., Echeverrıa, F., Formation of nanotubular TiO2 structures with varied surface characteristics for biomaterial applications. *J. Biomed. Mater. Res. A*, 106A, 1341–1354, 2018.
55. Barjaktarević, D., Djokić, V., Bajat, J., Dimić, I., Cvijović-Alagić, I., Rakin, M., The influence of the surface nanostructured modification on the corrosion resistance of the ultrafine-grained Ti-13Nb-13Zr alloy in artificial saliva. *Theor. Appl. Fract. Mech.*, 103, 102307, 2019.
56. Taveira, L., Macák, J., Tsuchiya, H., Dick, L., Schmuki, P., Initiation and growth of self-organized TiO2 nanotubes anodically formed in NH4F/(NH4)2SO4 electrolytes. *J. Electrochem. Soc.*, 152, B405–B410, 2005.
57. Saji, V., Choe, H., Brantley, W., Nanotubular oxide layer formation on Ti-13Nb-13Zr alloy as a function of applied potential. *J. Mater. Sci.*, 44, 3975–3982, 2009.
58. Tsuchiya, H., Akaki, T., Nakata, J., Terada, D., Tsuji, N., Koizumi, Y., Minamino, Y., Fujimoto, S., Anodic oxide nanotube layers on Ti–Ta alloys: Substrate composition, microstructure and self-organization on two-size scales. *Corros. Sci.*, 51, 1528–1533, 2009.
59. Feng, X., Macak, J., Albu, S., Schmuki, P., Electrochemical formation of self-organized anodic nanotube coating on Ti–28Zr–8Nb biomedical alloy surface. *Acta Biomater.*, 4, 318–323, 2008.
60. Indira, K., Mudali, U., Nishimura, T., Rajendran, N., A review on TiO2 nanotubes: Influence of anodization parameters, formation mechanism, properties, corrosion behavior, and biomedical applications. *J. Bio. Trib. Corros.*, 1, 28–50, 2015.
61. Mello, M., Taipina, M., Rabelo, G., Cremasco, A., Caram, R., Production and characterization of TiO2 nanotubes on Ti-Nb-Mo-Sn system for biomedical applications. *Surf. Coat. Technol.*, 326, 126–133, 2017.
62. Hu, N., Gao, N., Starink, M., The influence of surface roughness and high pressure torsion on the growth of anodic titania nanotubes on pure titanium. *Appl. Surf. Sci.*, 387, 1010–1020, 2016.
63. Macak, J., Tsuchiya, H., Taveira, L., Ghicov, A., Schmuki, P., Self-organized nanotubular oxide layers on Ti-6Al-7Nb and Ti-6Al-4V formed by anodization in NH4F solutions. *J. Biomed. Mater. Res. A*, 75A, 928–933, 2005.
64. Tsuchiya, H., Macak, J., Taveira, L., Schmuki, P., Fabrication and characterization of smooth high aspect ratio zirconia nanotubes. *Chem. Phys. Lett.*, 410, 18–191, 2005.

65. Simka, W., Krzakała, A., Masełbas, M., Dercz, G., Szade, J., Winiarski, A., Michalska, J., Formation of bioactive coatings on Ti–13Nb–13Zr alloy for hard tissue implants. *RSC Adv.*, 3, 11195–11204, 2013.
66. Alla, R., Ginjupalli, K., Upadhya, N., Shammas, M., Ravi, R., Sekhar, R., Surface roughness of implants: A review. *Trends Biomater. Artif. Organs*, 25, 112–118, 2011.
67. Yu, W., Jiang, X., Zhang, F., Xu, L., The effect of anatase TiO2 nanotube layers on MC3T3-E1 preosteoblast adhesion, proliferation, and differentiation. *J. Biomed. Mater. Res. A*, 94, 1012–1022, 2010.
68. Manjaiah, M. and Laubscher, R., Effect of anodizing on surface integrity of Grade 4 titanium for biomedical applications. *Surf. Coat. Technol.*, 310, 263–272, 2017.
69. Durdu, S., Cihan, G., Yalcin, E., Altinkok, A., Characterization and mechanical properties of TiO2 nanotubes formed on titanium by anodic oxidation. *Ceram. Int.*, 47, 10972–10979, 2021.
70. Tan, A., Pingguan-Murphy, B., Ahmad, R., Akbar, S., Review of titania nanotubes: Fabrication and cellular response. *Ceram. Int.*, 38, 4421–4435, 2012.
71. Tan, A., Ismail, R., Chua, K., Ahmad, R., Akbar, S., Pingguan-Murphy, B., Osteogenic potential of *in situ* TiO2 nanowire surfaces formed by thermal oxidation of titanium alloy substrate. *Appl. Surf. Sci.*, 320, 161–170, 2014.
72. Sowa, M., Piotrowska, M., Widziołek, M., Dercz, G., Tylko, G., Gorewoda, T., Osyczka, A., Simka, W., Bioactivity of coatings formed on Ti–13Nb–13Zr alloy using plasma electrolytic oxidation. *Mater. Sci. Eng. C*, 49, 159–173, 2015.
73. Nguyen, T., Park, I., Lee, M., Bae, T., Enhanced biocompatibility of a precalcified nanotubular TiO2 layer on Ti–6Al–7Nb alloy. *Surf. Coat. Technol.*, 236, 127–134, 2013.
74. Li, B., Li, Y., Li, J., Fu, X., Li, H., Wang, H., Sh. Xin, N., Zhou, L., Liang, C., Influence of nanostructures on the biological properties of Ti implants after anodic oxidation. *J. Mater. Sci. Mater. Med.*, 25, 199–205, 2014.
75. Sun, S., Zhang, Y., Zeng, D., Zhang, S., Zhang, F., Yu, W., PLGA film/titanium nanotubues as a sustained growth factor releasing system for dental implants. *J. Mater. Sci. Mater. Med.*, 29, 141–149, 2018.
76. Veljović, D., Čolić, M., Kojić, V., Bogdanović, G., Kojić, Z., Banjac, A., Palcevskis, E., Petrović, R., Janaćković, D., The effect of grain size on the biocompatibility, cell-materials interface, and mechanical properties of microwave-sintered bioceramics. *J. Biomed. Mater. Res. Part A*, 100A, 3059–3070, 2012.
77. Frandsen, C., Brammer, K., Jin, S., Variations to the nanotube surface for bone regeneration. *Int. J. Biomater.*, 2013, 1–10, 2013.
78. Li, B., Li, Y., Li, J., Fu, X., Li, C., Wang, H., Liu, S., Guo, L., Xin, S., Liang, C., Li, H., Improvement of biological properties of titanium by anodic oxidation and ultraviolet irradiation. *Appl. Surf. Sci.*, 307, 202–208, 2014.

79. Huang, H., Wu, C., Sun, Y., Yang, W., Lee, T., Surface nanotopography of an anodized Ti–6Al–7Nb alloy enhances cell growth. *J. Alloys Compd.*, 615, S648–S654, 2014.
80. Brammer, K., Oh, S., Cobb, C., Bjursten, L., Heyde, H., Jin, S., Improved bone-forming functionality on diameter-controlled TiO2 nanotube surface. *Acta Biomater.*, 5, 3215–3223, 2009.
81. Oh, S., Daraio, C., Chen, L., Pisanic, T., Fi˜nones, R., Jin, S., Significantly accelerated osteoblast cell growth on aligned TiO2 nanotubes. *J. Biomed. Mater. Res. A*, 78, 97–103, 2006.
82. Huang, H., Wu, C., Sun, Y., Huang, H., Lee, T., Enhanced corrosion resistance and biocompatibility of β-type Ti–25Nb–25Zr alloy by electrochemical anodization. *Thin Solid Films*, 549, 87–92, 2013.
83. Barjaktarević, D., Medjo, B., Štefane, P., Gubeljak, N., Cvijović-Alagić, I., Djokić, V., Rakin, M., Tensile and corrosion properties of anodized ultrafine-grained Ti–13Nb–13Zr biomedical alloy obtained by high-pressure torsion. *Met. Mater. Int.*, 26, 3325–3341, 2020.
84. Barjaktarević, D., Rakin, M., Đokić, V., Characterisation of the nanotubular oxide layer formed on the ultrafine-grained titanium. *Metall. Mater. Eng.*, 24, 261–270, 2018.
85. Barjaktarević, D., Međo, B., Đokić, V., Rakin, M., Morphology and nanomechanical properties of the ultrafine-grained Ti-13Nb-13Zr alloy surface obtained using electrochemical anodization, in: *Experimental and Numerical Investigations in Materials Science and Engineering*, N. Mitrović, G. Mladenović, A. Mitrović (Eds.), pp. 123–141, Springer, 2022. https://doi.org/10.1007/978-3-030-86009-7_7

6

Powder Metallurgy-Prepared Ti-Based Biomaterials with Enhanced Biocompatibility

Šugár, P.[1*], Antala, R.[1], Šugárová, J.[1] and Kováčik, J.[2]

[1]Slovak University of Technology in Bratislava, Faculty of Materials Science and Technology in Trnava, Institute of Production Technologies, Trnava, Slovakia
[2]Slovak Academy of Sciences, Institute of Materials and Machine Mechanics, Bratislava, Slovakia

Abstract

Low density, lightweight, chemical and corrosion resistance, good strength-to-weight ratio, high fatigue resistance, and biocompatibility are the major reasons behind the frequent use of Ti-based materials in a variety of industrial applications. As a result of their non-toxic behavior and good biocompatibility, these materials present a very suitable option in orthopaedic and dental implantology. They are being used as commercially pure Ti (cpTi) with various degrees of purity which influences its mechanical properties, as multicomponent metallic alloys or as Ti matrix composites (TiMCs) strengthened with different types of particles or whiskers. In comparison to traditional metallurgical procedures, TiMCs represent a cutting-edge generation of biomaterials with exceptional properties and affordable production techniques. The solid attempt to develop this kind of material by powder metallurgy is visible in many studies. The reason is the effort to minimalize the mismatch between Young's modulus of Ti and the human bone by the manufacture of Ti implants of certain porosity.

Surface integrity, together with mechanical properties, is also a very crucial quality aspect of biomedical materials. Highlighted are the properties such as surface roughness, surface topography, surface energy, wettability, surface chemistry, and thickness of the Ti-based oxide or nitride layers.

**Corresponding author*: peter.sugar@stuba.sk

Jay Prakash Srivastava, Dražan Kozak, Vinayak Ranjan, Pankaj Kumar, Ranjan Kumar and Shubham Tayal (eds.) Mechanical Engineering in Biomedical Applications: Bio-3D Printing, Biofluid Mechanics, Implant Design, Biomaterials, Computational Biomechanics, Tissue Mechanics, (151–184) © 2024 Scrivener Publishing LLC

Surface integrity leading to the improved formation of a direct bond between the implant and the bone, called osseointegration, may be achieved through different surface engineering techniques. Laser surface treatment is of great significance in modifying the surface morphology and also surface and near-surface region microstructures.

This chapter reviews the literature on Ti-based materials, prepared by powder metallurgy, dedicated for application in dental and orthopaedic implantology. Also, the topic of laser surface modification of these materials to enhance human cells' osteodifferentiation is presented.

***Keywords*:** Powder metallurgy, biomaterial, dental, orthopaedic, implant, Ti matrix composites

6.1 Introduction

Industries from aviation to biomedicine use Ti-based materials thanks to their outstanding strength, high strength-to-weight and good resistance-to-corrosion ratios, high melting point, low density, high fracture toughness, low thermal and electrical conductivity, and excellent biocompatibility [1, 2]. However, titanium is in general at least 25 times more expensive than carbon steel and approximately six to seven times more expensive than stainless steel. The high cost involved in the fabrication of Ti components is associated with the high affinity of titanium to oxygen and nitrogen; therefore, it is necessary to use a vacuum casting process [3].

The long-established production routes of Ti-based materials follow the same procedures as for steel, including the primary production of metal, alloy ingot melting and pouring, production of mill products via forging and rolling, and finally, manufacturing of components or structures by forming or machining. The material utilization in this case is very low. The buy-to-fly ratio (in aerospace industry terminology) for titanium parts fabricated from mill products (plates or bar stocks) is typically 12:1 and often more than 20:1 [4]. For example, the F-22 fighter jet has a buy-to-fly ratio of 12.2:1, meaning that 82% of the Ti became scrap metal before the final goods are produced.

The reduction of titanium production and manufacturing costs is currently very actual. One of the solutions is the application of powder metallurgy (PM), as a near-net-shape manufacturing technology, which has been applied to Ti research and development for more than four decades [5]. Powder metallurgy consists of the production of powders, their compaction and shaping, and then sintering, as well as post-sintering processes to fabricate ready-to-use components. Because of its capacity to streamline processing and reduce the amount of scrap, the potential of the PM approach for Ti manufacturing is very promising [4]. PM-prepared materials, including metal matrix composites (TiMCs), are going to be very progressive for biomedical applications, where attention is paid to the development of materials with tunable mechanical and functional properties with a reduced mismatch between the implant and natural bone elastic modulus and improved osseointegration [6–8].

This chapter gives a view of the actual trends in the field of PM Ti material development supplemented by the authors' research results of laser surface modification of warm compacted biocompatible Ti–graphite composite.

6.2 Powder Metallurgy of Ti-Based Materials

The first basic cost factor of PM Ti-based materials is the cost of powders and their amount on the market. Powders are prepared from Ti sponge by-products, such as sponge fines, by hydrogenating the Ti sponge, ingot, or mill product, crushing and/or grinding it to the desired powder size, and then dehydrogenating the material to form hydride–dehydride (HDH) Ti powder (Figure 6.1). Additionally, Ti scrap metals can also be hydrogenated and dehydrogenated to produce HDH powder.

A large number of various technologies, including widely known techniques such as the Armstrong [9] and the Fray, Farthing, and Chen (FFC)

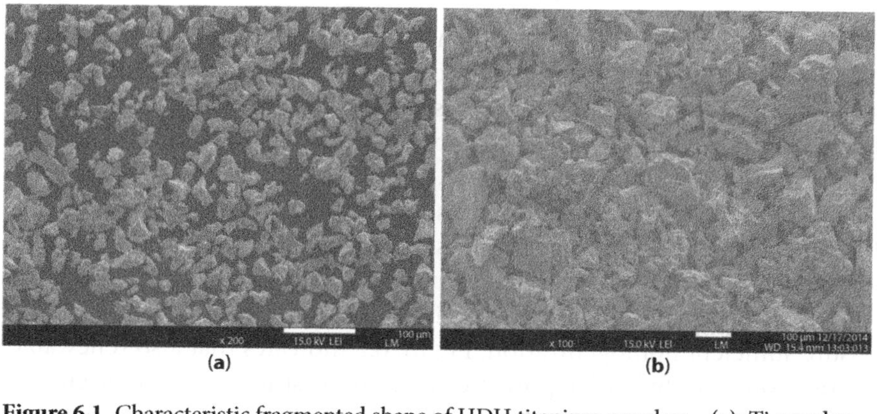

(a) (b)

Figure 6.1 Characteristic fragmented shape of HDH titanium powders—(a): Ti powder below 30 μm; (b): Ti powder below 150 μm (Kimet Special Metal Precision Casting Co., Ltd., China).

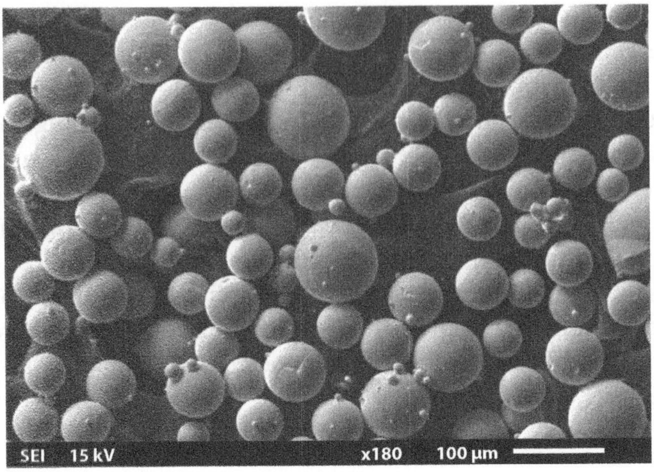

Figure 6.2 Characteristic spherical plasma atomized titanium powder: powder size between 45 and 106 µm (AP&C, Quebec, Canada).

processes [10], have been developed over the recent decades to produce Ti powder. The most important from the point of view of additive manufacturing of Ti and Ti alloy are powders made by a variety of highly advanced atomizing techniques, which consist of first making the metal alloy, melting the material, and then atomizing the molten metal. Atomized alloy powders (Figure 6.2) have significantly higher costs than HDH powder; however, their production undergoes new renaissance and growth.

6.2.1 Powder Metallurgy of Ti and Ti Alloys

Many processes are applied to the consolidation and sintering of Ti and Ti alloy powders to transform them into functional components. They usually include both pressureless and pressure-assisted sintering processes. As there are strict requirements for low oxygen, sintering under a vacuum or argon atmosphere usually takes place.

Ti has reportedly been sintered using both inert gas and vacuum sintering, which are both types of pressureless sintering. However, vacuum sintering is far more frequently employed. Temperatures of 1,200°C or more are often required during sintering to enhance densification and also ensure sufficient homogenization. Figure 6.3 indicates that also renewable energy-concentrated solar power can be successfully used as an energy source for the pressureless sintering of titanium [11]. Very often, pressureless sintering techniques possess residual porosity.

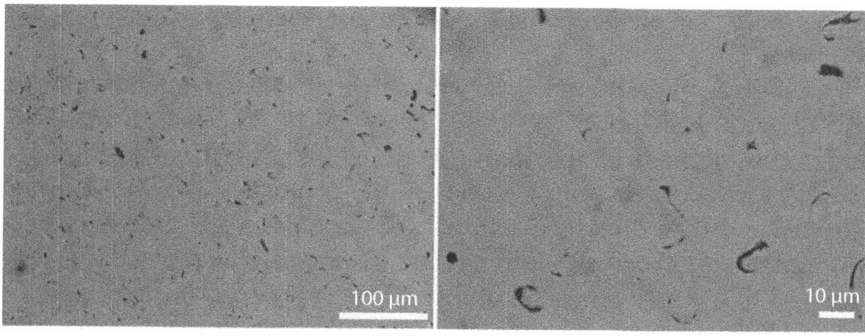

Figure 6.3 The microstructure of solar pressureless sintered Ti in a solar furnace at PSA, Spain (SEM, porosity 9.2%, sintered at 1,237°C for 30 min - HDH Ti powder below 30 µm) [11].

Pressure-assisted techniques lead to full density and good mechanical properties although accompanied by a relatively high cost. The preferred powder for pressure-assisted processes is spherical—usually, plasma-atomized powder (PA).

Hot isostatic pressing (HIP) is the most common method used for consolidating PA Ti powders. HIP generally operates between 850 and 1,200°C. The isostatic nature of the deformation densifies the material at lower temperatures than pressureless sintering. After the HIP, the titanium product may be processed with another form of transformation technique to refine the microstructure and improve the mechanical properties [12].

Another technology is vacuum hot pressing (VHP). VHP is usually done between 900 and 1,100°C under various external pressures. There is no apparent benefit of using VHP versus conventional pressureless sintering. Higher temperatures are still required to homogenize the alloy, concerning the microstructure.

The CIP-sinter-HIP (CHIP) process is a multistep PM process in which a green body/preform is produced via cold isostatic pressing (CIP), conventionally sintered under vacuum, and then undergoes HIP to close the remaining porosity [12].

Warm compaction of titanium patented by Kondoh *et al.* [13] has been used to compact commercially pure titanium (cpTi), Ti–6Al–4V, and Ti–6Al–4V–6TiB_2 powders at 150°C. A compaction pressure ranging from 392 to 1,568 MPa was used. After warm compaction with a pressure of 588 MPa, the relative density of the cpTi powder compact ranges from 82.5% to 97.5% of the theoretical density (TD). Luo *et al.* [14] applied the warm die compaction to HDH Ti powder at 200°C and achieved a green density increment of 65–97% TD under the compaction pressure from 200 to 1,000 MPa.

The slightly higher temperature variation of warm compaction of Ti and Ti composites up to 500°C was developed in the last 10 years for dental and biomedical applications [4, 13–20]. The first experiments were performed with simply CIPed HDH Ti powders at a pressure of 200 MPa and then directly extruded (DE) at 500°C. The fine (below 30 μm) and coarse (below 150 μm) HDH Ti powders were tested. It was observed that direct extrusion of fine HDH Ti powder was not possible due to high friction and shear forces during extrusion. Moreover, the concentration of oxygen, nitrogen, and hydrogen (ONH) analysis of CIPed HDH Ti powders below 150 μm indicated an increased concentration of oxygen and nitrogen in the final good (Figures 6.4 and 6.5). Note that commercial cpTi grade 1 titanium has the following concentration limits: 0.18 wt.% oxygen, 0.03 wt.% nitrogen, and 0.015 wt.% hydrogen.

For this reason, the technology was changed in such a way, that CIP blanks were put into sleeves—containers made from Cu or Al with a wall thickness of 3 mm and an outer diameter of 30 mm. They were further instantly pressed by uniaxial vacuum hot pressing (VHP) at a pressure of 500 MPa, vacuum of 8 Pa, and temperature of 470°C. CIPs and also sleeves were fully consolidated and sheared by direct extrusion into rod bars with a diameter of 6 mm using a flat face die at a reduction ratio of 16:1 and a ram speed of 10 mm s^{-1}. The extrusion temperature was in the range of 450–500°C.

This approach enabled to limit the oxygen concentration increase in warm compacted Ti goods (Figure 6.4a). The nitrogen concentration (Figure 6.4b) was the best when an aluminum container was used. Anyway, the final concentration of impurities is ruled primarily by their concentration in the HDH Ti powder used. It must be noted that the material of the container rests on the surface of the extruded good and needs to be machined out before further exploitation of titanium good. The impurities and porosity affect significantly the mechanical properties of titanium. Figure 6.4c shows that the achieved theoretical density of two different warm compacted HDH Ti powders was above 99% of TD.

To investigate their mechanical properties, the samples from the warm compacted titanium materials were machined for tensile tests with a gauge diameter of 3 mm and a gauge length of 32 mm. They were tested in tension on a Zwick testing machine at a crosshead speed of 1 mm min^{-1} (Figure 6.5).

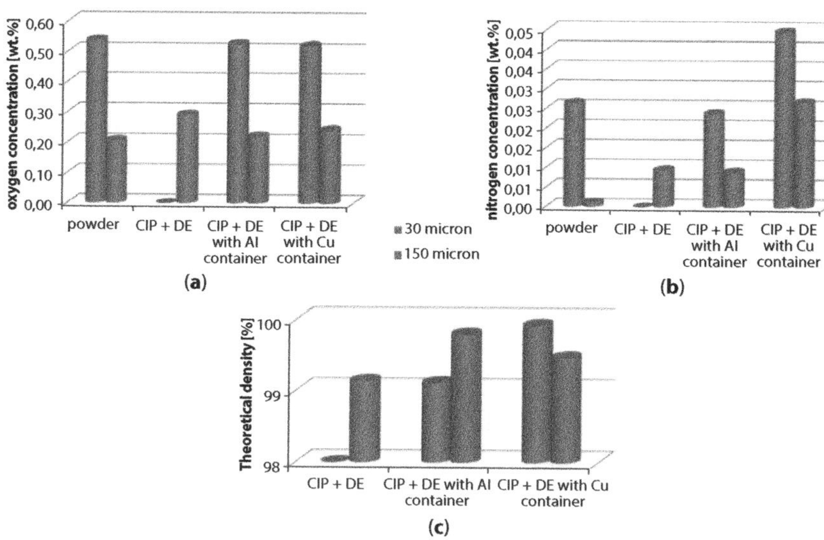

Figure 6.4 Comparison of (a) oxygen and (b) nitrogen concentration in powders and warm compacted materials and achieved theoretical density (c) for warm compacted Ti materials.

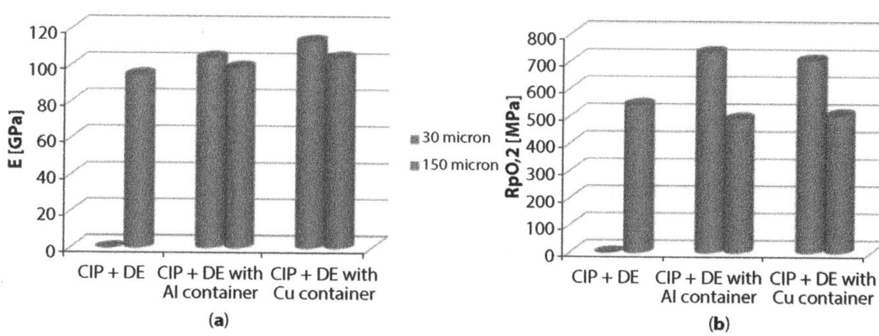

Figure 6.5 Comparison of (a) modulus of elasticity and (b) yield stress for various warm compacted Ti materials.

The modulus of elasticity of cpTi grade 1 is 105 GPa with a yield stress of 170–310 MPa, a tensile strength of 240 MPa, and an elongation to fracture of 24% [21]. When comparing the mechanical properties of warm compacted materials from the fine and coarse HDH Ti powders the warm compacts from fine HDH Ti powder are stronger (yield stress, tensile strength), with a comparable modulus of elasticity, but their elongation to fracture is significantly smaller as the corresponding properties of warm compacted goods from coarse HDH Ti powder. This is the result of the fact that the starting Ti powder with a powder size below 30 μm has a significantly higher concentration of oxygen and nitrogen than coarse powder below 150 μm. After warm compaction, this ratio remains in the final goods (Figures 6.5 and 6.6).

Finally, other special hot consolidation techniques for titanium powders are also developed such as spark plasma sintering (Figure 6.7), microwave sintering, powder rolling, forging, gaseous isostatic forging, metal injection molding, and additive manufacturing [4].

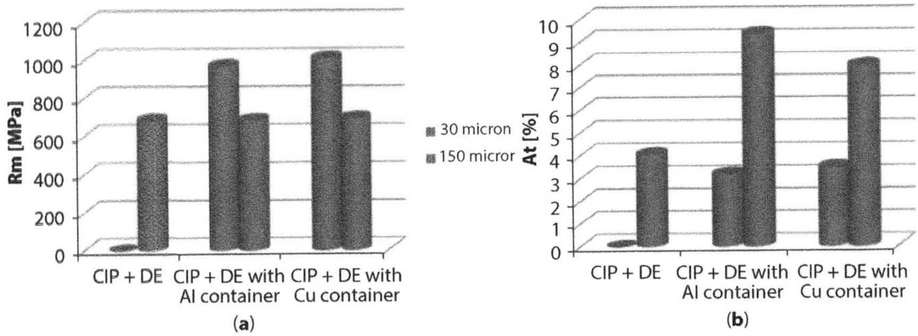

Figure 6.6 Comparison of (a) tensile strength and (b) elongation to fracture for various warm compacted Ti materials.

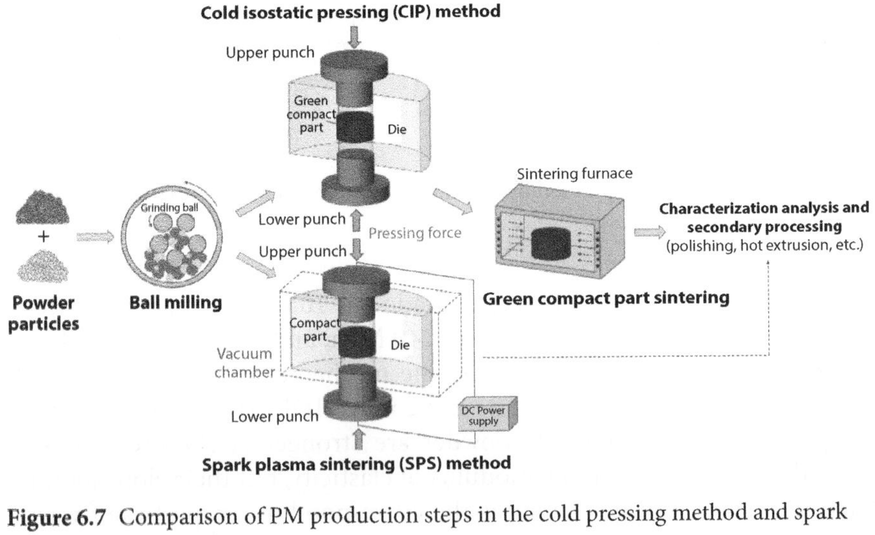

Figure 6.7 Comparison of PM production steps in the cold pressing method and spark plasma sintering (SPS).

6.2.2 Powder Metallurgy of Ti-Based Composites

Uniform and homogeneous dispersion of reinforcement and strong adhesion to the matrix are crucial properties to ensure stable composite performance. Reinforcing particles can be incorporated into TiMCs either through *ex situ* or *in situ* production techniques. While *in situ* fabrication is aided by high temperature, the solid-state reaction of dispersed reinforcing particles, typically based on C, B, and N phases (due to the high reactivity of Ti), results in a uniformly distributed structure with stronger bonding, low number of impurities, and low contamination. In contrast, the *ex situ* technique utilizes premade particles, often thermodynamically stable ceramic phases, e.g., TiB, SiC, TiC, and ZrC, dispersed into the composite matrix [22–24]. The *in situ* technique generally results in improved tribological and mechanical performance (strength, hardness, elastic modulus), whereas *ex situ* methods do not affect particle size or morphology and also result in extraordinarily good mechanical properties [25, 26]. While surface contaminants can cause a significant decrease in wettability and interface reaction between the *ex situ* reinforcement and matrix, the interface between the small intermetallic *in situ* reinforcing particles and the matrix is usually clear and stable [25, 27].

Nowadays, mostly due to the combination of their superior isotropic properties, high modulus, and strength with improved thermal stability, discontinuously reinforced TiMCs [28] are given special attention over continuously reinforced composites [29]. Powder metallurgy with discontinuous fillers (short fibers or particles) presents cost-effective technology that helps to prevent liquid-state processing limitations (high reactivity and melting point of Ti) by the solid processing route below the titanium matrix melting point. In general, both Ti alloys and TiMCs suffer from poor tribological properties; thus, reinforcing with stiff and strong ceramic particles or fibers helps to significantly improve their strength, hardness, and wear resistance at both low and high temperatures [24, 30]. According to studies [8, 22, 26, 29, 31–35], the most commonly used ceramic reinforcing particles include TiC, TiN, TiB, TiB_2, B_4C, WC, SiC, ZrC, Cr_3C_2, and Ti_5Si_3.

The PM route enables the production of complex-shaped and dimensionally accurate Ti parts with improved mechanical and structural properties and enhanced microstructural and chemical homogeneity; however, the residual porosity of manufactured parts, as well as the price of alloy powders due to the complicated preparation process, may be relatively high [36–38].

6.2.2.1 Porosity of PM Ti-Based Materials

The surface porosity of PM-processed biomaterials might be one of the key solutions for managing the problems with stress shielding while creating favorable conditions for the promotion of tissue into the implant surface. The porous structure of the body implant possesses low density and high surface area with good penetrability and, thus, may act as a strong foundation for the tissue and also improve the flow of body fluid. For instance, the distribution, number, size, and shape of individual pores significantly affect the biomechanical functionalities of the implant material [17, 37, 39–41]. According to Wang *et al.* [40], higher porosity that mimics the environment of the human bone helps to achieve better osteoconduction with lower mechanical properties. On the other hand, lower porosity often inhibits bone ingrowth but ensures excellent strength. From the point of view of human body anatomy, the density of cortical bone is roughly 1.85 g cm^{-3} with porosity from 5% to 30%, while the density of spongy trabecular human bone ranges between 0.1 and 0.60 g cm^{-3} (porosity in the range from 70% to 95%) [42, 43].

Enhanced osteoconductivity requires a connection with pore openings, while porous titanium implants, used in load-bearing applications, should have, in terms of mechanical safety, a strength comparable to that of human bone [40, 44]. Human bone, which naturally consists of inorganic hydroxyapatite crystals (roughly 65 wt.%), organic collagenous and non-collagenous proteins (25 wt.%), and water (10 wt.%), has an elastic modulus of approximately 19.6 GPa [43, 45]. Young's modulus is higher than that of the human bone causing stress distribution on the bone to be altered; thus, the phenomenon of stress shielding is induced, which weakens the bone next to the metallic dental or orthopaedic implant. Therefore, attention is paid to the reduction of mismatch between implants and natural bone elastic modulus, which helps to improve the osseointegration process, avoid bone atrophy, and prevent the stress-shielding effect, which impedes bone remodeling and leads to bone resorption, connection failure, and loosening of body implant [6, 7, 46, 47].

Pore interconnections in the structure of porous materials compared with bulk materials help to increase vascularization and bone ingrowth, by creating adequate room for the flow of body fluid and migration of cells through the surrounding pores [36, 44] (Figure 6.8).

According to various authors, enhancement of implant osseointegration can be achieved by a porosity ratio ranging from 40% to 90% [36] with the presence of well-interconnected pores (approximately 51% to 52%), with a pore diameter ranging from 100 to 400 μm [37, 44]. While the presence of

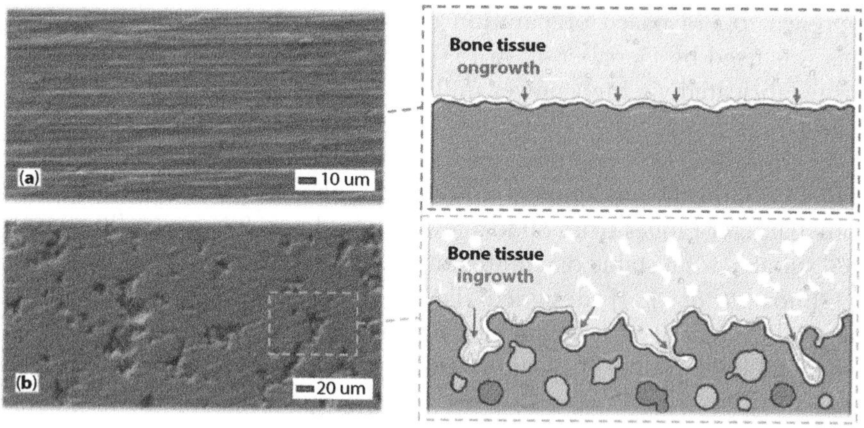

Figure 6.8 Bone tissue behavior on the PM-prepared Ti compact with (a) low surface porosity and (b) bone tissue fixation on the highly porous PM-prepared Ti-C composite [20, 48].

small (shallow) and narrow pores can inhibit cell proliferation and migration, the presence of too wide and deep pores, on the other hand, generally impede living cell habitation [43]. For instance, Wubneh *et al.* [49] claimed that a pore diameter larger than 75 μm is adequate for the creation of both unmineralized and completely mineralized bone tissues, while a pore diameter less than 75 μm is often correlated with unfavorable fibrous tissue formation. For instance, the minimal pore diameter for a stable bone formation process can be determined by osteoblast size (roughly 100 μm), and thus, pore size in between 100 and 150 μm might seem to be sufficient for optimal osteoblast adhesion and proliferation [49, 50]. According to the study of Wang *et al.* [51], a porous structure with 40% porosity and 0.4 mm pore diameter resulted in strong MG63 osteoblast cell adhesion and rapid proliferation. In contrast, pores wider than 0.5 mm led to the formation of less-dense ingrowing bone tissue.

Nowadays, many researchers try to examine the effect of porosity on the behavior of potential biomedical composite materials prepared via powder metallurgy techniques. Xu *et al.* in 2019 [52] designed innovative porous Ti35Zr28Nb scaffolds, prepared via a space holder using the PM technique, that exhibited mechanical properties similar to the frequently used biomedical Ti6Al4V alloy. The porous structure with pore size in the range from roughly 230 to 430 μm (considered as positive for bone tissue ingrowth) and porosity range of 50% to 65% resulted in no cytotoxicity effect on *in vitro*-tested MC3T3-E1 osteoblast cell's proliferation and also excellent *in vivo* hemocompatibility tested on rat distal femurs. A different approach to PM-based preparation of biomedical porous implant materials was used by Dercz *et al.* in [53] who studied porous Ti–50Ta composite fabricated via high-power milling, isostatic pressing, and sintering. Researchers found that a longer sintering time (72 h at 1,000°C) resulted in the formation of a homogeneous structure with a porosity rate of more than 14%, and the presence of a higher number of pores within Ferret's pore diameter ranges from 50 to a few hundred μm. It is generally considered suitable for a stable osseointegration process.

Olmos *et al.* [54] designed Ti6Al4V-based biomedical composite material potential for dental application manufactured via the powder metallurgy route. They intended to mimic the properties of human bone with a porous core and dense surface layer. Therefore, the investigated implant prototype was composed of three different layers, i.e., hard Ti6Al4V/20CoCrMo outer shell, dense layer of Ti6Al4V/40 with relatively small pores in the middle, and Ti6Al4V base metal in the core with numerous large pores. The results showed that the studied material composition had mechanical properties similar to human bone (slightly higher than trabecular bone) with a wide range of pore diameter (from 2 to 600 μm) that can help to enhance cell adhesion and bone regeneration.

However, despite the enhanced biocompatible behavior of porous biomedical composites with often reduced Young's modulus, porous structures seem to have significant drawbacks, such as the fact that increased porosity usually results in poor wear resistance and poor machinability, which makes it challenging to obtain specific shapes frequently used in dental or orthopaedic implantology [54]. In addition, the highly porous structure of the body implant might be insufficient during load-bearing applications, thus is often determined by low mechanical strength [36, 51, 54].

6.2.2.2 *Effect of Reinforcing Particles on the Biological Behavior of Ti-Based Composites*

Biological compatibility, corrosion resistance, surface integrity, and chemical and mechanical properties all represent the most essential factors that researchers take into account when developing components for biomedical applications [55]. In terms of the implant's biomechanical properties, changing the implant's composition via the incorporation of reinforcing particles seems like a promising strategy [30]. TiMCs can be generally classified as continuously (CRTCs) and discontinuously reinforced (DRTCs), while in terms of biomedical applications, DRTCs are prominent due to their combination of good mechanical properties (high wear resistance, stiffness, or strength), thermal durability, and stability [56]. Therefore, numerous studies have been conducted to look at the effect of alloying elements on composite material resultant properties after the incorporation of different metallic or ceramic reinforcing particles via the powder metallurgy method [33, 57, 58].

Incorporation of Metallic-Based Reinforcing Particles

Despite the prominent properties of Ti-based alloys and composites, the addition of some of the metallic-based reinforcing particles can help improve either the mechanical properties, corrosion resistance, or biocompatibility of the body implant, without any allergic or inflammatory reaction. According to various authors, metals such as Ni, Cr, Al, V, Mn, Co, Fe, and Ag can be classified as hazardous to the human body environment, while the presence of different metals, e.g., Mo, Pd, Ta, Pt, Au, W, Sn, and, Mg can help enhance the biological compatibility of biomedical composites [59, 60]. In addition, the reduction of implant-associated bacterial infection and formation of bacterial biofilms, often caused by the absence of any antibacterial behavior of pure Ti, can be accomplished by incorporation of antibacterial metallic agents, e.g., Ag, Zn, or Cu ions, into the titanium matrix during the PM process, in the form of nanoparticles or through deposition of antibacterial coating [39, 61–63].

One of the promising metallic elements to improve bone healing efficiency during orthopaedic applications is bioactive and also biodegradable Mg, which is one of the naturally occurring metallic elements in the human body with good biocompatibility to osteoblast, low elastic modulus (approx. 40 GPa), and its biologically safe daily intake dosage released during degradation (roughly 240 to 420 mg) [17, 47, 64–66].

Ouyang *et al.* [65] manufactured spark plasma-sintered Ti–Mg composite that exhibited during *in vitro* testing improved SaOS-2 cell cytocompatibility and osseointegration compared with the cpTi-based specimen. Li *et al.* [64] studied the innovative biomedical Ti–35Nb–xMg (x = 3 to 5 wt.%) composite prepared via mechanical alloying and spark plasma sintering, and they found that increasing Mg content from 3 to 5 wt.% resulted in fracture strains and strength deterioration, while the Ti–35Nb–5Mg alloy prepared under a 20:1 ball-to-powder ratio showed good MC3T3-E1 cell spreading after Mg degradation. Li *et al.* [47] investigated bioactive Ti–13Nb–13Zr–1.25Mg composite material, prepared via mechanical alloying and the spark plasma sintering method, that exhibited an increase in corrosion resistance, elastic modulus, density, yield, and compressive strength as the sintering temperature rises. The studied samples sintered at 800°C or 900°C all exhibited improved SaOS-2 cell adhesion and proliferation; however, a sintering temperature of 700°C resulted in a higher burst release of Mg^{2+} inhibiting cell proliferation compared with samples sintered at higher temperatures (release of Mg^{2+} in appropriate dosage can stimulate initial attachment and growth of bone tissues).

Balog *et al.* [67] reported a study of bioactive Ti–12 vol.% Mg metal composite named BIACOM for dental implant applications, manufactured via low-temperature extrusion. The composite consisting of biodegradable Mg distributed homogeneously within the bioinert titanium matrix exhibited low density (4.12 g cm^{-3}) and significantly reduced elastic modulus (92.1 GPa) while maintaining fatigue performance and mechanical strength of the reference Ti6Al4V alloy. The corrosion of Mg filaments leaves pores within the titanium matrix, which caused a further decrease of Young's elastic modulus and improvement of both micro- and macrosurface roughness. In addition, Balog and his team [17] successfully produced a new type of bimetallic Ti with 12, 17, and 24 vol.% Mg composite via a low-temperature powder metallurgy process, which proved its application potential as a load-bearing dental implant material. The addition of biodegradable Mg components resulted in a reduction of all mechanical properties, while *in vitro* testing in HBSS revealed pore formation as well as micro- and macrosurface roughness alternation that provides favorable conditions for a stable osseointegration process.

Incorporation of Ceramic-Based Reinforcing Particles

The application of one of the most promising ceramic reinforcing particles of TiC in TiMC was studied by Pan *et al.* [30]. They observed a positive impact of the ceramic reinforcing particles on increasing elastic modulus, hardness, and oxidation resistance. Otte *et al.* [24], on the other hand, studied TiB nanowhisker-reinforced TiMC, which exhibited improvement in strength, wear resistance, corrosion resistance, and biocompatibility, which are all crucial properties for long-lasting biomedical implant application. The addition of TiB particles, which were formed by BN and Ti powder reaction *in situ*, caused a 304% increase in hardness and 170% increase in Young's modulus compared with pure Ti while retaining strong osseointegration behavior. Angel *et al.* [68] developed an innovative hybrid nanocrystalline Ti and TiB whisker-reinforced titanium nanocomposite via the powder metallurgy route, where the presence of TiB mono boride that was formed by thermal decomposition of TiB_2 resulted in improved hardness and yield strength compared with the non-hybrid alternative. Prakash *et al.* in a research paper from 2018 [69] successfully eliminated the low wear resistance of titanium alloys by reinforcing atomized Ti–6Al–4V with

B_4C ceramic particles through a powder metallurgy process. The addition of B_4C also helped to significantly decrease wear rate and density, while hardness and corrosion resistance had been increased.

Nowadays, in addition to the metallic and ceramic reinforcing particles, nanofiller materials, e.g., metal nitrides or oxides (Si_3N_4, ZrO_2, Al_2O_3, Y_2O_3, La_2O_3, MnO_2), bioglasses, carbon nanostructures (nanodiamonds, graphene, graphene oxides, or carbon nanotubes), calcium phosphate, and hydroxyapatite, are extensively used to improve the mechanical properties and also enhance the biocompatibility of TiMCs by mimicking bone tissue behavior [70–80].

In 2022, Massooth *et al.* [71] investigated the properties of PM-prepared Ti6Al4V alloy reinforced by biocompatible ceramic phase ZrO_2, frequently used in biomedical applications, that helped to increase compression strength, hardness, and shrinkage rate, with the best results at 6 wt.% of ZrO_2. Liu *et al.* [81] studied the mechanical properties and microstructure of graphene oxide-reinforced TiMCs prepared via hot-pressed sintering in an Ar atmosphere using different manufacturing parameters. The titanium matrix was reinforced by TiC particles that were formed *in situ* after titanium and graphene oxide reactions during the sintering process. The presence of graphene oxide particles caused an increase in mechanical properties, while the amount of TiC was affected primarily by graphene oxide content and sintering temperature.

Šugár *et al.* [18, 20, 48] investigated the properties of novel PM-prepared, highly porous Ti–graphite composite material, designed specifically for dental implant application. The authors found that the mixture of powdered components did not react with each other during the low-temperature powder metallurgy process and thus prevented the formation of brittle and hard titanium carbides. The resultant lightweight material exhibited lower mechanical properties, e.g., strength and Young's modulus of 96.83 ± 25.74 GPa, compared with the frequently used Ti grade 5 alloy (113.8 GPa).

One of the most popular ceramic reinforcements is highly biocompatible hydroxyapatite (HA) $Ca10(PO_4)_6(OH)_2$, i.e., calcium and phosphorous-based nanocrystal phase, naturally occurring in human teeth and bone tissue, that can be used as either dispersive phase and bioactive coating layer to improve bone tissue orientation, accelerate bone in-growth, and promote adhesion and proliferation of osteoblast cells resulting into higher healing rate and prolonged life of the body implant [37, 55, 82].

However, using hydroxyapatite as a bulk load-bearing implant material, e.g., for dental implant applications, is limited due to its low strength and high brittleness, and thus, HA is frequently used as graft materials in low load implant fabrication [42, 79, 83]. Krishna *et al.* [82] used the powder metallurgy route to prepare bioactive Ti–HA composite potential for orthopaedic and dental applications, which is incorporated with nanohydroxyapatite powder through high-energy ball milling and sintering. The presence of nano-HA grains resulted in increased wettability, crucial for initial protein adhesion, and also promoted higher deposition of the Ca/P phase from simulated body fluid solution in comparison with the cpTi reference sample. The same results were achieved by Farrahnoor *et al.* [76], who focused their research on the bioactivity of HA-enriched Ti–Nb alloy matrix composite manufactured via mechanical alloying and powder metallurgy. All the studied Ti–Nb–HA composite samples with HA content from 0 to 15 wt.% exhibited accelerated apatite formation and high surface wettability, while 15 wt.% HA composite indicated the best apatite coverage among all the investigated specimens. On the other hand, Mohammadtaheri *et al.* [78], who studied the properties of Ti–Al_2O_3–HA composite material prepared via the spark plasma sintering method, found that the rate of dead cells on Ti–Al_2O_3 was higher than on pure Ti sample; however, the addition of HA resulted in enhanced cell viability and cytocompatibility, compared with both the studied Ti–Al_2O_3 and cPTi.

Some authors also attempted to synergize the impact of the already mentioned advances of structural porosity with HA phase addition. Prakash *et al.* [37] studied the porous Ti–Nb(50-x)–HA(x) (x = 0, 10, 20 wt.%) composite prepared via PM and found that HA reinforcement helped to increase compressive strength and improve corrosion resistance, while structural porosity (roughly 55%, with pore size in the range of 105 to 265 μm) resulted into low elastic modulus of 15 to 29 GPa, which is similar to the human bone. Based on *in vitro* biological testing with MG-63 cells, it can be stated that the investigated composite material is non-cytotoxic, while its good cell adhesion and growth can be correlated with an increase in resultant porosity. The obtained results also showed the presence of biocompatible $Ca_3(PO_4)_2$, CaO, and Ti_5P_3 phases that support the proliferation and differentiation of human cells.

Zhang *et al.* [80] prepared a novel Ti–Nb–Zr/Ti–HA composite via the spark plasma sintering method, with a dense Ti–13Nb–13Zr core (porosity 92%) that provides superior mechanical properties, enclosed by external bioactive Ti–HA coating (porosity roughly from 35% to 40%) with low elastic modulus and the presence of micro- and macropores that help to induce apatite phase deposition. The authors stated that the increase in HA concentration resulted in higher porosity and thickness of the homogeneous apatite layer. The elastic modulus of the composite (from 10.7 to 18.9 GPa) decreased due to the porosity increase. The BMSC cell adhesion tests showed that the presence of pore structure with HA content enhanced the three-dimensional proliferation of mesenchymal cells on the porous Ti–HA coating layer, compared with only two-dimensional movement, adhesion, and proliferation of cells on the dense Ti–13Nb–13Zr core.

Yilmaz *et al.* [42] studied porous bone-like Ti/Ti–HA functionally graded composite material that could mimic the structural, biological, and mechanical properties of cortical bone. An outer shell-like layer of this novel implant material was composed of bioactive Ti–HA composite coating prepared via the PM method (10 wt.%, porosity 28.88%), while the inner layer was based on porous Ti with superior mechanical properties (porosity 58.44%). A promising behavior for biomedical implant employment was confirmed when the contact angle of the shell layer incorporated with HA (32°) decreased significantly compared with pure Ti inner layer (71°), cell viability on the HA layer improved from 54% to 84%, while the Ca/P ratio exhibited similar values to natural bone apatite.

Aside from the predominant advantages of HA, numerous studies have conducted other alternatives, such as equine heterogeneous bones [84, 85], to further enhance biomechanical contact between the implant and human tissue. For example, Jeong *et al.* [85] paid attention to the development of PM-manufactured and heat-treated Ti6Al4V–equine bone (EB) biomedical nanocomposite with excellent bioactivity and corrosion resistance. A test at normal body temperature in 0.9% NaCl solution showed that dispersion of EB particles containing natural HA phase caused enhancement of MC3T3-E1 cell adhesion, proliferation, spreading, and corrosion resistance compared with conventional Ti6Al4V (Ti grade 5). In addition, the authors in one of their previous studies [84] stated that the hardness of spark plasma-sintered Ti–6Al–4V/EB composite increased with the content of equine bone phase, thanks to the uniform distribution of EB particles in the titanium matrix.

6.3 Laser Surface Treatment of Materials for Enhanced Human Cell Osteodifferentiation

Along with mechanical properties, surface integrity represents a very important factor in biomedical material quality. Underlined are the properties such as surface topography, roughness, surface chemistry, surface free energy, wettability, the thickness of the oxide and nitride layers, and also the presence of undesirable impurities [86]. These properties can be achieved using surface modification methods such as sandblasting, plasma spraying, acid etching, laser beam micromachining, anodization, electropolishing, inorganic and organic coating, and ion implantation [87, 88]. These production methods can help shorten healing time by enhancing cell attachment as well as the osseointegration process, which is represented by the structural and functional connection between the bone and the implanted material [86, 88].

By laser beam micromachining, a material is removed from the surface by a pulsed laser beam of high energy density, the cavities are formed in the irradiated area, and the roughness of the machined surface is increasing. Both the properties of the machined material and the input process parameters have an impact on the interaction between the laser beam and the processed material.

Many studies [89–97] have investigated the proper surface roughness of implants. The surface roughness at the micrometer scale ought to enhance cell attachment and differentiation, and the roughness at the nanometer scale increases surface energy [98–100]. The suitable surface roughness depends on the cell size, e.g., the adequate roughness Ra value for the osteoblasts is in the range from 1 to 2 μm [97, 101, 102]. From a topographic point of view, a surface with grooves or depressions, having the same size as the cells, i.e., about 30 μm or larger, is preferred for its colonization by cells [103].

6.3.1 Laser-Treated Surfaces of PM Ti-Based Materials

The laser surface modification on the biological and mechanochemical behavior of potential biomedical Ti–Nb–Mo alloys with 15.5% Mo_{eq} prepared via powder metallurgy was studied by Tendero *et al.* [104]. For surface modification, a high-power CW Nd:YAG laser was used with 1,064 nm wavelength and energy densities of 50, 75, and 100 J mm^{-2} and helium shielding gas at 20 L min^{-1} flow rate. Laser fusion caused an improvement of maximum bending resistance (max. 1,500 MPa for the Ti–27Nb–8Mo alloy), an increase of microhardness in both the irradiated zone and base material (max. 350 HV in the FZ of the Ti–20Nb–10Mo alloy), and a decrease of elastic modulus with either laser modification (by around 10%) and Nb content. In Ti–35Nb–6Mo and Ti–27Nb–8Mo alloys, laser surface modification also lowered the corrosion rate, while Nb release did not reach the physiological limit.

Laser surface modification of beta Ti–30Nb–4Sn (TNS) and Ti–35Nb–10Ta (TNT) alloys prepared via the powder metallurgy process was studied by Rossi *et al.* [105]. Two combinations of laser power and laser beam scanning speed with a specific energy of 75 J mm^{-2} were used (laser power of 1,000 and 1,500 W and scanning speed of 6.7 and 10 mm s^{-1}). A CW Nd:YAG laser system operating at 1,064 nm with helium as the shielding gas (flow rate 20 L min^{-1}) helped to improve the surface quality and cause changes in the mechanical properties in both irradiated samples. For the TNS alloy, lower values of laser power and scanning speed caused an increase in hardness and a decrease in elastic modulus, and higher parameters caused a decrease in both hardness and elastic modulus. On the other hand, lower values of laser power and scanning speed caused a decrease in hardness and an increase in elastic modulus for the TNT alloy, while higher values of process parameters did not influence the mechanical properties compared with the untreated specimen. Laser irradiation in both cases resulted in the formation of a dense, approximately 200-mm-thick dendritic surface layer, with a structure consisting of martensitic phases α″, α, and β. In addition to a slight increase in hardness and a decrease in elastic modulus for the TNS alloy, no significant differences were noted in terms of the laser irradiation parameters used.

Šugár *et al.* [19] paid attention to the investigation of the laser micromachining effect on novel Ti-based material prepared via low-temperature hydride–dehydride powder compaction by cold isostatic pressing at 200 MPa, followed by preheating (30 min at 450°C in the air) and direct extrusion at 500°C. The laser modification process was carried out on an industrial-grade Lasertec 80 Shape machining center equipped with a nanosecond fiber Yb-doped laser, operating at 20 kHz pulse frequency, 100 mm s^{-1} scanning speed, 50 μm spot diameter, 90% lateral and 50% transverse pulse overlap, and 0.2 to 1 mJ pulse energies (E_p), while the composite material was ablated under ambient air in one layer. As a general trend, higher pulse energy resulted in surface roughness alternation (investigation revealed the surface roughness Ra in the range of 2.18–11.7 μm), and loss of uniformity of the surface profile is observed. The higher values of energies correlate to a rougher machined surface. Laser scanning tracks separated by 25 μm were observed when the low laser pulse energy was applied. At higher pulse energies, tracks vanished and no surface texture was seen. During laser machining, not only significant changes in surface morphology and roughness were identified. Also, significant changes were observed in the surface chemistry. Considering that the studies were conducted in ambient air, the increase in pulse energy encouraged the increase of the treated surface's oxygen content.

The SEM analysis of the experimental surfaces showed an increase in ablation depth (Rz up to 70.16 μm) resulting from the increased laser pulse energy. The resultant surface topography was formed by the ridges of the molten redeposited and solidified titanium voids, craters, and globules.

It is formed by the solidified titanium particles agglomerated with irregularly shaped macro- and micropores, with sizes ranging from 10 to 25 μm and from 1 to 3 μm, respectively (Figure 6.9). The characteristic microscale irregularities caused by the melting and partial melting of the irradiated material may be seen. These occurrences are advantageous because they increase the adhesion between the bone tissue and the implant while also enhancing cellular bioactivity and overall implantation performance. The amount of O substantially increased as the laser beam pulse intensity increased. On the other hand, X-ray diffraction analysis only identified the TiO phase, while the oxide layer was hardly noticeable due to its very low thickness. Interestingly, despite the maximum input of laser energy, no TiO_2 was found on the laser-treated surfaces.

An industrial-grade fiber nanosecond laser working with a wavelength of 1,064 nm was employed for the surface modification of the Ti–graphite composite prepared by vacuum low-temperature extrusion of hydrogenated–dehydrogenated titanium powder of particle size below 32 μm mixed with 15.0 vol.% of graphite flakes of an average particle with an aspect ratio

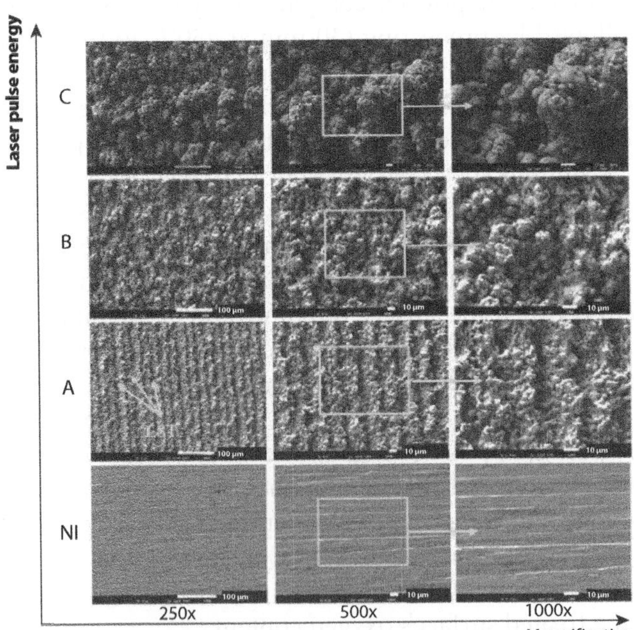

Figure 6.9 SEM images of the irradiated surfaces demonstrating how laser fluence influences surface morphology: NI—non-irradiated surface (Ra 0.25 μm); surface A (E_p = 0.2 mJ, Ra 2.95 μm); surface B (E_p = 0.6 mJ, Ra 6.38 μm); surface C (E_p = 1 mJ, Ra 11.67 μm); LST—laser scanning tracks [18].

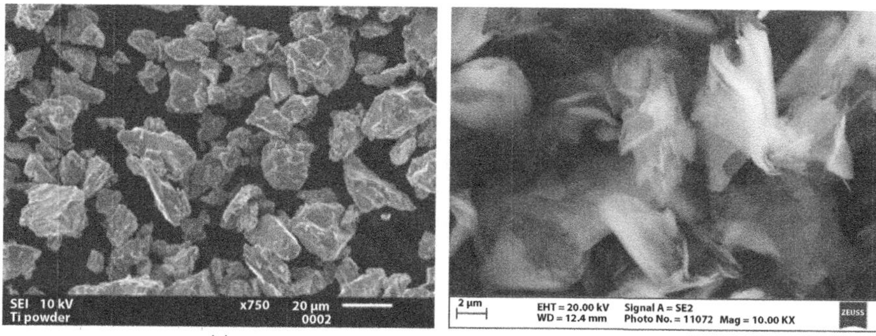

Figure 6.10 Characteristic HDH titanium powder and graphite particle shapes: (a) HDH Ti powder particles at the magnification of 750× and (b) graphite flakes at the magnification of 10.00k× [20].

of the flakes of 10 (diameter to thickness ratio), density of 99.9%, and size of 16 µm (Figure 6.10). The density of the samples was in the range of 4.1–4.15 g cm^{-3} with porosity of the resultant compact of 2.44% ± 0.15% [20].

The experiment was conducted with a laser spot diameter of 50 µm and a constant pulse length of 120 ns applied. Variable process parameters, pulse energy, and lateral pulse (pulse-to-pulse) distances were used. A cross-hatching strategy of laser beam movement was applied with a transverse track distance of the pulses of 10 µm. It means a transverse pulse overlap of 80%. Laser treatment under ambient air and with an Ar shielding atmosphere with a gas flow rate of 10 L min^{-1} was investigated.

Surface machining conducted by laser beam resulted in surface morphology consisting of the ascended ridges of the molten redeposited and solidified globules, cavities, and craters. We noticed several surface types that were affected by the applied energy. The first one was a rough surface with a porous structure formed by the highest fluences (Figure 6.11), the second one was formed by the protrusions forming a small square-shaped texture, and the third surface type was formed by visible laser beam paths that created squares as a result of the bi-directional traces layout (Figure 6.12). Higher energies transferred into the irradiated material at the same place resulted in deep valley formation as a result of the material evaporation. Microdefects (cavities with sizes ranging between 1 and 5 µm) were identified in the laser-treated surfaces, probably as a result of the melted material interaction with shielding gas flow. The final porosity was a combination of the original material porosity and the porosity induced by laser irradiation.

The treated surface oxidation is affected by the energy of the laser beam interacting with the material. The oxygen content increases along with

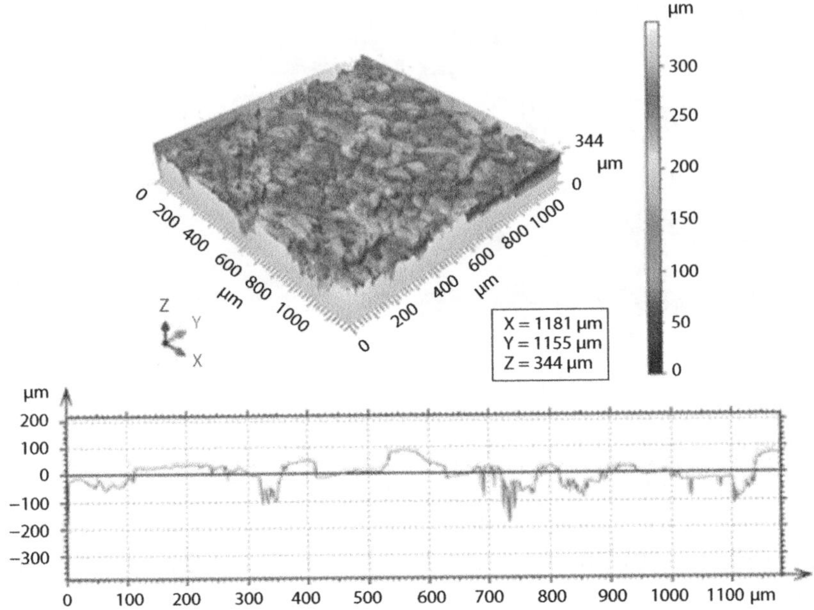

Figure 6.11 3D surface maps of the laser-modified PM Ti–graphite surface (E_p = 1 mJ, fluence of 50.92 J cm^{-2}) [18].

energy density and decreases when the lateral pulse distance decreases. The presence of oxygen was documented on the surfaces after laser ablation in the argon shielding atmosphere, but the amount of oxygen was lower compared with the treatment without a shielding atmosphere. After laser ablation in the argon shielding atmosphere, only two types of oxides, namely, TiO and TiO_2, were identified.

The *in vitro* analysis of the biocompatibility and osteoinductivity of the laser-treated PM Ti–graphite composite surfaces using human mesenchymal stem cells confirmed that the surface profile with platykurtic distribution curve, positive kurtosis, and higher value of peaks spacing exhibits better bioactivity compared with the surface profile with negative kurtosis parameter values. When compared with a surface that was machined using higher energy, the surface produced by employing a lower incident energy level encourages cell proliferation and osteodifferentiation.

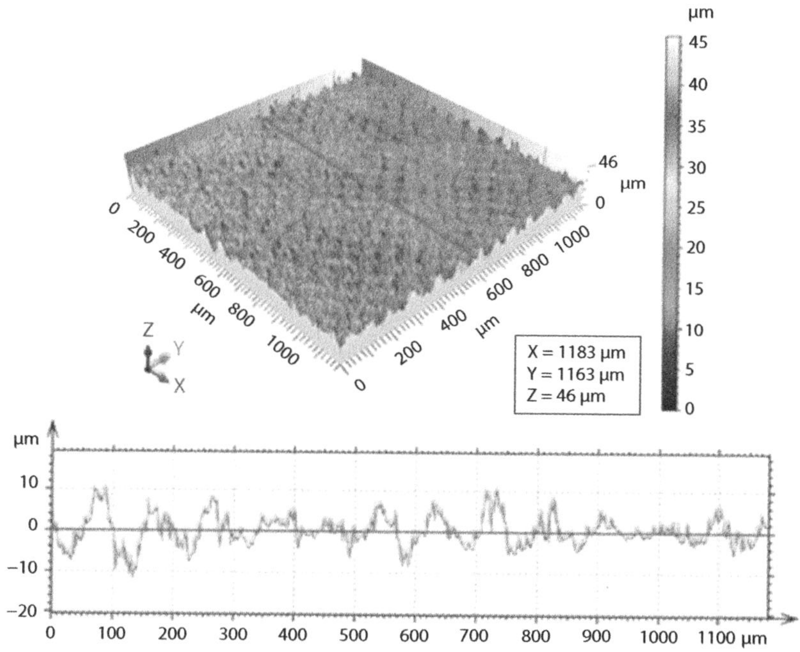

Figure 6.12 3D surface maps of the laser-modified PM Ti–graphite surface (E_p = 0.2 mJ, fluence of 10.18 J cm^{-2}) [18].

Conclusion

There are many questions related to the development of biocompatible materials with specific mechanical and surface layer properties, considering a high degree of osseointegration while maintaining the conditions of availability and economic acceptability, which have not been answered yet. The manufacture and application of PM Ti-based materials that meet not only technical but also health and biocompatibility criteria are still in the experimental research stage. Systematic studies on the optimal production processes of powders and their compaction and sintering need to be undertaken. The process of the surface treatments to create conditions for ideal cell attachment and simultaneously the least bacterial biofilm attachment is another open area of investigation where attention is needed and expected in the future to speed up the clinical application of these materials.

Acknowledgments

This research was supported by the VEGA 2/0054/23 and the KEGA 026STU-4/2023 research projects. The authors express their sincere thanks to the Ministry of Education, Science, Research and Sport of the Slovak Republic for the financial contribution. The authors also thank for a financial contribution from the STU Grand Scheme for Support of Young Researchers, project: KOMPLAS.

References

1. Cui, C., Hu, B.M., Zhao, L., Liu, S., Titanium alloy production technology, market prospects and industry development. *Mater Design*, 32, 3, 1684–91, 2011. Available from: https://doi.org/10.1016/j.matdes.2010.09.011.
2. Selvan, J.S., Subramanian, K., Nath, A.K., Gogia, A.K., Balamurugan, A.K., Rajagopal, S., Hardness, microstructure and surface characterization of laser gas nitrided commercially pure titanium using high power CO2 laser. *J. Mater. Eng. Perform.*, 7, 5, 647–55, 1998.
3. Qian, M. and Froes, F.H., *Titanium Powder Metallurgy: Science, Technology and Applications*, Waltham, MA: Elsevier, Batterworth-Heinemann, 2015.
4. Fang, Z.Z., Paramore, J.D., Sun, P., Chandran, K.S.R., Zhang, Y., Xia, Y. *et al.*, Powder metallurgy of titanium–past, present, and future. *Int. Mater Rev.*, 63, 7, 407–59, 2018. Available from: https://doi.org/10.1080/09506608.2017.1366003.
5. Caron, R.N. and Staley, J.T., Titanium and titanium alloys, in: *ASM Handbook Volume 20- Materials Selection and Design*, G.E. Dieter (Ed.), pp. 399–403, ASM International, Materials Park, 1997.
6. Brizuela, A., Herrero-Climent, M., Rios-Carrasco, E., Rios-Santos, J.V., Pérez, R.A., Manero, J.M. *et al.*, Influence of the elastic modulus on the osseointegration of dental implants. *Materials*, 12, 6, 1–7, 2019. Available from: https://doi.org/10.3390/ma12060980.
7. Gao, X., Zhao, Y., Wang, M., Liu, Z., Liu, C., Parametric design of hip implant with gradient porous structure. *Front. Bioeng. Biotechnol.*, 10, 1–15, 2022. Available from: https://doi.org/10.3389/fbioe.2022.850184.
8. Yoganandam, K., Mohanavel, V., Vairamuthu, J., Kannadhasan, V., Mechanical properties of titanium matrix composites fabricated via powder metallurgy method. *Mater Today Proc.*, 33, 3243–7, 2020.
9. D.R. Armstrong, R.P. Anderson, L.E. Jacobsen, Preparation of alloys by the armstrong method. US Patent 7,041,150 B2, 2006.
10. Chen, G.Z., Fray, D.J., Farthing, T.W., Direct electrochemical reduction of titanium dioxide to titanium in molten calcium chloride. *Nature*, 407, 6802, 361–364, 2000. Available from: https://doi.org/10.1038/35030069.

11. Kováčik, J., Emmer, S., Rodríguez, J., Cañadas, I., Sintering of HDH Ti powder in a solar furnace at plataforma solar de almeria. *J. Alloys Compd.*, 695, 52–9, 2017. Available from: https://doi.org/10.1016/j.jallcom.2016.10.147.
12. Abkowitz, S., Abkowitz, S., Fisher, H., Titanium alloy components manufacture from blended elemental powder and the qualification process, in: *Titanium Powder Metallurgy*, M. Qian, (Eds.), pp. 299–312, Butterworth-Heinemann, Oxford UK, 2015.
13. M. Kondoh, T. Saito, H. Takamiya, Green compact and process for compaction the same, metallic sintered body and process for producing the same, worked component part and method of working. US Patent, US20040013558A1.
14. Luo, S.D., Yang, Y.F., Schaffer, G.B., Qian, M., Warm die compaction and sintering of titanium and titanium alloy powders. *J. Mater. Process. Technol.*, 214, 3, 660–6, 2014. Available from: https://doi.org/10.1016/j.jmatprotec.2013.10.010.
15. Ibrahim, A.M.H., Takacova, M., Jelenska, L., Csaderova, L., Balog, M., Kopacek, J. *et al.*, The effect of surface modification of TiMg composite on the *in-vitro* degradation response, cell survival, adhesion, and proliferation. *Mater. Sci. Eng. C*, 127, 112259, 2021. Available from: https://doi.org/10.1016/j.msec.2021.112259.
16. Balog, M., Viskic, J., Krizik, P., Schauperl, Z., Snajdar, M., Stanec, Z. *et al.*, CP Ti fabricated by low temperature extrusion of HDH powder: Application in dentistry. *Key Eng. Mater.*, 704, 351–9, 2016. Available from: https://doi.org/10.4028/www.scientific.net/KEM.704.351.
17. Balog, M., Ibrahim, A.M.H., Krizik, P., Bajana, O., Klimova, A., Catic, A. *et al.*, Bioactive Ti + Mg composites fabricated by powder metallurgy: The relation between the microstructure and mechanical properties. *J. Mech. Behav. Biomed. Mater.*, 90, 45–53, 2019. Available from: https://doi.org/10.1016/j.jmbbm.2018.10.008.
18. Šugár, P., Ludrovcová, B., Kováčik, J., Sahul, M., Šugárová, J., Laser-based ablation of titanium-graphite composite for dental application. *Materials*, 13, 10, 2312, 2020. Available from: https://doi.org/10.3390/ma13102312.
19. Šugár, P., Kováčik, J., Šugárová, J., Ludrovcová, B., A study of laser micromachining of PM processed Ti compact for dental implants applications. *Materials*, 12, 14, 2246, 2019. Available from: https://doi.org/10.3390/ma12142246.
20. Šugár, P., Ludrovcová, B., Kalbáčová, M.H., Šugárová, J., Sahul, M., Kováčik, J., Laser surface modification of powder metallurgy-processed Ti-graphite composite which can enhance cells' osteo-differentiation. *Materials*, 14, 20, 6067, 2021. Available from: https://doi.org/0.3390/ma14206067.
21. Boyer, R., Welsch, G., Collings, E.W., *Materials Properties Handbook : Titanium Alloys*, ASM International, Materials Park, OH, 1994.
22. Jeje, S.O., Shongwe, M.B., Rominiyi, A.L., Olubambi, P.A., Spark plasma sintering of titanium matrix composite-a review. *Int. J. Adv. Manuf.*

Technol., 117, 2529–2544, 2021. Available from: https://doi.org/10.1007/s00170-021-07840-7.

23. Liu, Y., Pan, Y., Sun, J., Kuang, F., Yu, A., Yang, S. *et al.*, High strength and high wear-resistant Ti composites fabricated by powder metallurgy pressureless sintering. *Adv. Mater. Technol.*, 7, 11, 2200219, 2022. Available from: https://doi.org/10.1002/admt.202200219.
24. Otte, J.A., Zou, J., Patel, R., Lu, M., Dargusch, M.S., Tib nanowhisker reinforced titanium matrix composite with improved hardness for biomedical applications. *Nanomaterials*, 10, 12, 1–11, 2020. Available from: https://doi.org/10.3390/nano10122480.
25. Wan, Q., Li, F., Wang, W., Hou, J., Cui, W., Li, Y., Study on *in-situ* synthesis process of Ti-Al intermetallic compound-reinforced Al matrix composites, *Materials*, 12, 12, 1967, 2019. Available from: https://doi.org/10.3390/ma12121967.
26. Zglobicka, I., Zybala, R., Kaszyca, K., Molak, R., Wieczorek, M., Recko, K. *et al.*, Titanium matrix composites reinforced with biogenic filler. *Sci. Rep.*, 12, 1, 1–10, 2022. Available from: https://doi.org/10.1038/s41598-022-12855-5.
27. Attar, H., Ehtemam-Haghighi, S., Kent, D., Dargusch, M.S., Recent developments and opportunities in additive manufacturing of titanium-based matrix composites: A review. *Int. J. Mach. Tools Manuf.*, 133, 85–102, 2018. Available from: https://doi.org/10.1016/j.ijmachtools.2018.06.003.
28. Jiao, Y., Huang, L., Geng, L., Progress on discontinuously reinforced titanium matrix composites. *J. Alloys Compd.*, 767, 1196–215, 2018. Available from: https://doi.org/10.1016/j.jallcom.2018.07.100.
29. Jeje, S.O., Shongwe, M.B., Maledi, N., Rominiyi, A.L., Adesina, O.S., Olubambi, P.A., Synthesis and characterization of TiN nanoceramic reinforced Ti–7Al–1Mo composite produced by spark plasma sintering. *Mater. Sci. Eng. A*, 807, 140904, 2021. Available from: https://doi.org/10.1016/j.msea.2021.140904.
30. Pan, Y., Li, W., Lu, X., Hayat, M.D., Yin, L., Song, W. *et al.*, Microstructure and tribological properties of titanium matrix composites reinforced with *in situ* synthesized TiC particles. *Mater. Charact.*, 170, 110633, 2020. Available from: https://doi.org/10.1016/j.matchar.2020.110633.
31. Myalska-Głowacka, H., Chmiela, B., Godzierz, M., Sozańska, M., Residual stress induced by addition of nanosized TiC in titanium matrix composite. *Materials*, 15, 7, 1–15, 2022. Available from: https://doi.org/10.3390/ma15072517.
32. Rominiyi, A.L., Shongwe, M.B., Jeje, S.O., Olubambi, P.A., Microstructure, tribological and oxidation behaviour of spark plasma sintered Ti-Ni-xTiCN composites. *J Alloys Compd.*, 890, 161857, 2022. Available from: https://doi.org/10.1016/j.jallcom.2021.161857.
33. Fattahi, M., Delbari, S.A., Namini, A.S., Ahmadi, Z., Azadbeh, M., Asl, M.S., Characterization of triplet Ti–TiB–TiC composites: Comparison of *in-situ*

formation and *ex-situ* addition of TiC. *Ceram. Int.*, 46, 8, 11726–34, 2020. Available from: https://doi.org/10.1016/j.ceramint.2020.01.204.
34. Zhou, Y., Yang, F., Chen, C., Shao, Y., Lu, B., Sui, Y. *et al.*, Mechanical property and microstructure of *in-situ* TiB/Ti composites via vacuum sintering and hot rolling. *J. Alloys Compd.*, 911, 165042, 2022. Available from: https://doi.org/10.1016/j.jallcom.2022.165042.
35. Namini, A.S., Dilawary, S.A.A., Motallebzadeh, A., Asl, M.S., Effect of TiB2 addition on the elevated temperature tribological behavior of spark plasma sintered Ti matrix composite. *Compos. B Eng.*, 172, 271–80, 2019. Available from: https://doi.org/10.1016/j.compositesb.2019.05.073.
36. Yamanoglu, R., Bahador, A., Kondoh, K., Fabrication methods of porous titanium implants by powder metallurgy. *Trans. Indian Inst. Met.*, 74, 11, 2555–67, 2021. Available from: https://doi.org/10.1007/s12666-021-02332-4.
37. Prakash, C., Singh, S., Ramakrishna, S., Królczyk, G., Le, C.H., Microwave sintering of porous Ti–Nb-HA composite with high strength and enhanced bioactivity for implant applications. *J. Alloys Compd.*, 824, 153774, 2020. Available from: https://doi.org/10.1016/j.jallcom.2020.153774.
38. Sharma, D.K., Mahant, D., Upadhyay, G., Manufacturing of metal matrix composites: A state of review. *Mater. Today Proc.*, 26, 506–19, 2019. Available from: https://doi.org/10.1016/j.matpr.2019.12.128.
39. Adamek, G., Kozlowski, M., Junka, A., Siwak, P., Jakubowicz, J., Preparation and properties of bulk and porous Ti-Ta-Ag biomedical alloys. *Materials*, 15, 12, 4332, 2022. Available from: https://doi.org/10.3390/ma15124332.
40. Wang, Q., Zhou, P., Liu, S., Attarilar, S., Ma, R.L.W., Zhon.g, Y. *et al.*, Multiscale surface treatments of titanium implants for rapid osseointegration: A review. *Nanomaterials*, 10, 6, 1–27, 2020. Available from: https:/doi.org/10.3390/nano10061244.
41. Kayacan, M.C., Baykal, Y.B., Karaaslan, T., Özsoy, K., Alaca, İ., Duman, B. *et al.*, Monitoring the osseointegration process in porous Ti6Al4V implants produced by additive manufacturing: an experimental study in sheep. *J. Appl. Biomater. Funct. Mater.*, 16, 2, 68–75, 2018. Available from: https://doi.org/10.5301/jabfm.5000385.
42. Yılmaz, E., Kabataş, F., Gökçe, A., Fındık, F., Production and characterization of a bone-like porous Ti/Ti-hydroxyapatite functionally graded material. *J. Mater. Eng. Perform.*, 29, 10, 6455–67, 2020. Available from: https://doi.org/10.1007/s11665-020-05165-2.
43. Chen, X., Fan, H., Deng, X., Wu, L., Yi, T., Gu, L. *et al.*, Scaffold structural microenvironmental cues to guide tissue regeneration in bone tissue applications. *Nanomaterials*, 8, 11, 1–15, 2018. Available from: https://doi.org/10.3390/nano8110960.
44. Pałka, K. and Pokrowiecki, R., Porous titanium implants: A review. *Adv. Eng. Mater.*, 20, 5, 1700648, 2018. Available from: https://doi.org/10.1002/adem.201700648.

45. Kolodziejska, B., Pajchel, L., Zgadzaj, A., Kolmas, J., A new, biomimetic collagen–apatite wound-healing composite with a potential regenerative and anti-hemorrhagic effect in dental surgery. *Materials (Basel)*, 15, 24, 8888, 2022. Available from: https://doi.org/10.3390/ma15248888.
46. Savio, D. and Bagno, A., When the total hip replacement fails: A review on the stress-shielding effect. *Processes*, 10, 1–17, 2022. Available from: https://doi.org/10.3390/pr10030612.
47. Li, G., Li, K., Liang, L., Li, K., Lu, Y., Zhu, W. *et al.*, Microstructure and corrosion resistance of powder metallurgical Ti-Nb-Zr-Mg alloys with low modulus for biomedical application. *SSRN Electron. J.*, 192, 112223, 2022. Available from: https://doi.org/10.1016/j.matchar.2022.112223.
48. Ludrovcová, B., Šugár, P., Sahul, M., Kováčik, J., Czibor, Z., Oxidation of biocompatible graphite-Ti composite after laser ablation in different atmospheres. *IOP Conf. Ser. Mater. Sci. Eng.*, 987, 1, 012032, 2020. Available from: https://doi.org/10.1088/1757-899X/987/1/012032.
49. Wubneh, A., Tsekoura, E.K., Ayranci, C., Uludağ, H., Current state of fabrication technologies and materials for bone tissue engineering. *Acta Biomater.*, 80, 1–30, 2018. Available from: https://doi.org/10.1016/j.actbio.2018.09.031.
50. Dziaduszewska, M. and Zieliński, A., Structural and material determinants influencing the behavior of porous Ti and its alloys made by additive manufacturing techniques for biomedical applications. *Materials*, 14, 4, 1–48, 2021. Available from: https://doi.org/ 10.3390/ma14040712.
51. Wang, D., He, G., Tian, Y., Ren, N., Ni, J., Liu, W. *et al.*, Evaluation of channel-like porous-structured titanium in mechanical properties and osseointegration. *J. Mater. Sci. Technol.*, 44, 160–70, 2020. Available from: https://doi.org/10.1016/j.jmst.2019.10.026.
52. Xu, W., Tian, J., Liu, Z., Lu, X., Hayat, M.D., Yan, Y. *et al.*, Novel porous Ti35Zr28Nb scaffolds fabricated by powder metallurgy with excellent osteointegration ability for bone-tissue engineering applications. *Mater. Sci. Eng. C*, 105, 110015, 2019. Available from: https://doi.org/10.1016/j.msec.2019.110015.
53. Dercz, G., Matuła, I., Zubko, M., Kazek-Kęsik, A., Maszybrocka, J., Simka, W. *et al.*, Synthesis of porous Ti–50Ta alloy by powder metallurgy. *Mater. Charact.*, 142, 124–36, 2018. Available from: https://doi.org/10.1016/j.matchar.2018.05.033.
54. Olmos, L., Mihalcea, E., Vergara-Hernández, H.J., Bouvard, D., Jimenez, O., Chávez, J. *et al.*, Design of architectured Ti6Al4V-based materials for biomedical applications fabricated via powder metallurgy. *Mater. Today Commun.*, 29, 102937, 2021. Available from: https://doi.org/10.1016/j.mtcomm.2021.102937.
55. Topuz, M., Dikici, B., Gavgali, M., Challenges in the production of titanium-based scaffolds bio-functionalized with hydroxyapatite by powder metallurgy technique. *Eur. J. Sci. Technol.*, 28, 46–51, 2021. Available from: https://doi.org/10.31590/ejosat.980402.

56. Hayat, M.D., Singh, H., He, Z., Cao, P., Titanium metal matrix composites: An overview. *Compos. Part A Appl. Sci. Manuf.*, 121, 418–38, 2019. Available from: https://doi.org/10.1016/j.compositesa.2019.04.005.
57. Asl, M.S., Delbari, S.A., Azadbeh, M., Namini, A.S., Mehrabian, M., Nguyen, V.H. *et al.*, Nanoindentational and conventional mechanical properties of spark plasma sintered Ti-Mo alloys. *J. Mater. Res. Technol.*, 9, 5, 10647–58, 2020. Available from: https://doi.org/10.1016/j.jmrt.2020.07.066.
58. Verma, R.P., Titanium based biomaterial for bone implants: A mini review. *Mater. Today Proc.*, 26, 3148–51, 2020. Available from: https://doi.org/10.1016/j.matpr.2020.02.649.
59. Yehia, H.M., El-Tantawy, A., Ghayad, I.M., Eldesoky, A.S., El-kady, O., Effect of zirconia content and sintering temperature on the density, microstructure, corrosion, and biocompatibility of the Ti–12Mo matrix for dental applications. *J. Mater. Res. Technol.*, 9, 4, 8820–33, 2020. Available from: https://doi.org/10.1016/j.jmrt.2020.05.109.
60. Zhang, L.C. and Chen, L.Y., A review on biomedical titanium alloys: Recent progress and prospect. *Adv. Eng. Mater.*, 21, 4, 1801215, 2019. Available from: https://doi.org/10.1002/adem.201801215.
61. Lei, Z., Zhang, H., Zhang, E., You, J., Ma, X., Bai, X., Antibacterial activities and biocompatibilities of Ti-Ag alloys prepared by spark plasma sintering and acid etching. *Mater. Sci. Eng. C*, 92, 121–31, 2018. Available from: https://doi.org/10.1016/j.msec.2018.06.024.
62. Chen, M., Yang, L., Zhang, L., Han, Y., Lu, Z., Qin, G. *et al.*, Effect of nano/micro-Ag compound particles on the bio-corrosion, antibacterial properties and cell biocompatibility of Ti-Ag alloys. *Mater. Sci. Eng. C*, 75, 906–17, 2017. Available from: https://doi.org/10.1016/j.msec.2017.02.142.
63. Xu, N., Fu, J., Zhao, L., Chu, P.K., Huo, K., Xu, N. *et al.*, Biofunctional elements incorporated nano/microstructured coatings on titanium implants with enhanced osteogenic and antibacterial performance. *Adv. Healthc. Mater.*, 9, 23, 2000681, 2021. Available from: https://doi.org/10.1002/adhm.202000681.
64. Li, Z., Xu, H., Dong, A., Qiu, J., He, L., Zhang, T. *et al.*, Characteristics of Ti-Nb-Mg alloy by powder metallurgy for biomedical applications. *Mater. Charact.*, 173, 110953, 2021. Available from: https://doi.org/10.1016/j.matchar.2021.110953.
65. Ouyang, S., Huang, Q., Liu, Y., Ouyang, Z., Liang, L., Powder metallurgical Ti-Mg metal-metal composites facilitate osteoconduction and osseointegration for orthopaedic application. *Bioact. Mater.*, 4, 1, 37–42, 2019. Available from: ttps://doi.org/10.1016/j.bioactmat.2018.12.001.
66. Bommala, V.K., Krishna, M.G., Rao, C.T., Magnesium matrix composites for biomedical applications: A review. *J. Magnes. Alloy.*, 7, 1, 72–9, 2019. Available from: https://doi.org/10.1016/j.jma.2018.11.001.
67. Balog, M., Snajdar, M., Krizik, P., Schauperl, Z., Stanec, Z., Catic, A., Titanium-magnesium composite for dental implants (BIACOM), in: *Minerals, Metals*

and Materials Series Part F6, pp. 271–84, 2017, Available from: https://doi.org/10.1007/978-3-319-51493-2_26.
68. Angel, D.A., Mikó, T., Kristály, F., Benke, M., Gácsi, Z., Development of TiB and nanocrystalline Ti-reinforced novel hybrid Ti nanocomposite produced by powder metallurgy. *J. Mater. Sci.*, 57, 6, 4130–44, 2022. Available from: https://doi.org/10.1007/s10853-021-06833-x.
69. Prakash, K.S., Gopal, P.M., Anburose, D., Kavimani, V., Mechanical, corrosion and wear characteristics of powder metallurgy processed Ti-6Al-4V/B4C metal matrix composites. *Ain Shams Eng. J.*, 9, 4, 1489–96, 2018. Available from: https://doi.org/10.1016/j.asej.2016.11.003.
70. Moriche, R., Beltrán, A.M., Begines, B., Rodríguez-Ortiz, J.A., Alcudia, A., Torres, Y., Influence of the porosity and type of bioglass on the micromechanical and bioactive behavior of coated porous titanium substrates. *J. Non Cryst. Solids*, 551, 120436, 2021. Available from: https://doi.org/10.1016/j.jnoncrysol.2020.120436.
71. Masooth, P.H.S., Bharathiraja, G., Jayakumar, V., Palani, K., Microstructure and mechanical characterisation of ZrO2 reinforced Ti6Al4V metal matrix composites by powder metallurgy method. *Mater. Res. Express*, 9, 2, 020003, 2022. Available from: https://doi.org/10.1088/2053-1591/ac5352.
72. Zhang, F. and Liu, T., Nanodiamonds reinforced titanium matrix nanocomposites with network architecture. *Compos. B Eng.*, 165, 143–54, 2019. Available from: https://doi.org/10.1016/j.compositesb.2018.11.110.
73. Guler, O. and Bagci, N., A short review on mechanical properties of graphene reinforced metal matrix composites. *J. Mater. Res. Technol.*, 9, 3, 6808–33, 2020. Available from: https://doi.org/10.1016/j.jmrt.2020.01.077.
74. Yu, J., Zhao, Y., Zhao, Q., Zhang, W., Huo, W., Zhang, Y., Microstructure and properties of titanium matrix composites synergistically reinforced by graphene oxide and alloying elements. *Adv. Eng. Mater.*, 25, 2101778, 1–8, 2022. Available from: https://doi.org/10.1002/adem.202101778.
75. Okoro, A.M., Awotunde, M., Ajiteru, O.A., Lephuthing, S.S., Olubambi, P.A., Machaka, R., Effects of carbon nanotubes on the mechanical properties of spark plasma sintered titanium matrix composites-a review. *2018 IEEE 9th International Conference on Mechanical and Intelligent Manufacturing Technologies, ICMIMT*, pp. 54–9, 2018, Available from: https://doi.org/10.1109/ICMIMT.2018.8340420.
76. Farrahnoor, A. and Zuhailawati, H., Effects of hydroxyapatite addition on the bioactivity of Ti-Nb alloy matrix composite fabricated via powder metallurgy process. *Mater. Today Commun.*, 27, 102209, 2021. Available from: https://doi.org/10.1016/j.mtcomm.2021.102209.
77. Miranda, G., Araújo, A., Bartolomeu, F., Buciumeanu, M., Carvalho, O., Souza, J.C.M. *et al.*, Design of Ti6Al4V-HA composites produced by hot pressing for biomedical applications. *Mater. Des.*, 108, 488–93, 2016. Available from: http://dx.doi.org/10.1016/j.matdes.2016.07.023.

78. Mohammadtaheri, M., Bozorg, M., Yazdani, A., Salehi, M., Fabrication of Ti–Al2O3–HA composites by spark plasma sintering and its properties for medical applications. *J. Mater. Res.*, 37, 2571–2580, 2022. Available from: https://doi.org/10.1557/s43578-022-00561-x.
79. Peng, Q., Bin, X., Pan, H., Wang, Y., Peng, Z., Tang, Z., Fabrication of boronized Ti6Al4V/HA composites by microwave sintering in mixed gases. *ACS Omega*, 5, 20, 11629–36, 2020. Available from: https://doi.org/10.1021/acsomega.0c00862.
80. Zhang, J., Jiang, Z., Guo, H., Sun, T., Chen, A., Zhou, Y. *et al.*, Function-structure-integrated Ti-HA coatings on TiNbZr with enhanced mechanical properties and bioactivity prepared by spark plasma sintering. *Vacuum*, 184, 109863, 2021. Available from: https://doi.org/10.1016/j.vacuum.2020.109863.
81. Liu, J., Hu, N., Liu, X., Liu, Y., Lv, X., Wei, L. *et al.*, Microstructure and mechanical properties of graphene oxide-reinforced titanium matrix composites synthesized by hot-pressed sintering. *Nanoscale Res. Lett.*, 14, 114, 2019. Available from: https://doi.org/10.1186/s11671-019-2951-9.
82. Krishna, E.S. and Suresh, G., Bioactive titanium-hydroxyapatite composites by powder metallurgy route. *Biointerface Res. Appl. Chem.*, 12, 4, 5375–83, 2022. Available from: https://doi.org/10.33263/BRIAC124.53755383.
83. Arifin, A., Sulong, A.B., Muhamad, N., Syarif, J., Ramli, M.I., Material processing of hydroxyapatite and titanium alloy (HA/Ti) composite as implant materials using powder metallurgy: A review. *Mater. Des.*, 55, 165–75, 2014. Available from: http://dx.doi.org/10.1016/j.matdes.2013.09.045.
84. Jeong, W., Shin, S.E., Choi, H., Microstructure and mechanical properties of titanium–equine bone biocomposites. *Metals (Basel)*, 10, 5, 1–10, 2020. Available from: https://doi.org/10.3390/met10050581.
85. Jeong, W. and Shin, S.E., Biocompatibility and corrosion behavior of heat-treated Ti6Al4V-equine bone nanocomposites. *J. Mater. Sci.*, 57, 18051–18061, 2022. Available from: https://doi.org/10.1007/s10853-022-07370-x.
86. Shahali, H., Jaggessar, A., Yarlagadda, P.K.D.V., Recent advances in manufacturing and surface modification of titanium orthopaedic applications. *Proc. Eng.*, 174, 1067–76, 2017. Available from: http://doi.org/10.1016/j.proeng.2017.01.259.
87. Asri, R.I.M., Harun, W.S.W., Samykano, M., Lah, N.A.C., Ghani, S.A.C., Tarlochan, F. *et al.*, Corrosion and surface modification on biocompatible metals: A review. *Mater. Sci. Eng. C*, 77, 1261–74, 2017. Available from: http://doi.org/10.1016/j.msec.2017.04.102.
88. Alla, R.K., Ginjupalli, K., Upadhya, N., Shammas, M., Krishna Ravi, R., Sekhar, R., Surface roughness of implants: A review. *Trends Biomater. Artif. Organs*, 25, 3, 112–8, 2011.
89. Lario, J., Amigó, A., Segovia, F., Amigó, V., Surface modification of Ti-35Nb-10Ta-1.5Fe by the double acid-etching process. *Materials*, 11, 4, 1–11, 2018. Available from: http://doi.org/ 10.3390/ma11040494.

90. Hara, Y., Komasa, S., Yoshimine, S., Nishizaki, H., Okazaki, J., Effect of nano modified titanium surface on adsorption of rat periodontal ligament cells. *J. Osaka Dent. Univ.*, 52, 1, 37–44, 2018.
91. Ferraris, S., Bobbio, A., Miola, M., Spriano, S., Micro- and nano-textured, hydrophilic and bioactive titanium dental implants. *Surf. Coat. Technol.*, 276, 374–83, 2015. Available from: https://doi.org/10.1016/j.surfcoat.2015.06.042.
92. Gehrke, S.A., de Lima, J.H.C., Rodriguez, F., Calvo-Guirado, J.L., Júnior, J.A., Pérez-Díaz, L. *et al.*, Microgrooves and microrugosities in titanium implant surfaces: An *in vitro* and *in vivo* evaluation. *Materials*, 12, 8, 1–13, 2019. Available from: https://doi.org/10.3390/ma12081287.
93. Marenzi, G., Impero, F., Scherillo, F., Sammartino, J.C., Squillace, A., Spagnuolo, G., Effect of different surface treatments on titanium dental implant micro-morphology. *Materials*, 12, 5, 733, 2019. Available from: https://doi.org/10.3390/ma12050733.
94. Schubert, A., Wassmann, T., Holtappels, M., Kurbad, O., Krohn, S., Bürgers, R., Predictability of microbial adhesion to dental materials by roughness parameters. *Coatings*, 9, 7, 456, 2019. Available from: https://doi.org/10.3390/coatings9070456.
95. Wang, Y., Yu, Z., Guo, X., Hu, J., Surface morphology of modified titanium alloy affects proliferation stability of bone marrow mesenchymal stem cells. *Surf. Coat. Technol.*, 366, 156–63, 2019. Available from: https://doi.org/10.1016/j.surfcoat.2019.02.087.
96. Souza, J.C.M., Sordi, M.B., Kanazawa, M., Ravindran, S., Henriques, B., Silva, F.S. *et al.*, Nano-scale modification of titanium implant surfaces to enhance osseointegration. *Acta Biomater.*, 94, 112–31, 2019. Available from: https://doi.org/10.1016/j.actbio.2019.05.045.
97. Günay-Bulutsuz, A., Berrak, Ö., Yeprem, H.A., Arisan, E.D., Yurci, M.E., Biological responses of ultrafine grained pure titanium and their sand blasted surfaces. *Mater. Sci. Eng. C*, 91, 382–8, 2018. Available from: https://doi.org/10.1016/j.msec.2018.05.056.
98. Rodríguez, Á., Trueba, P., Amado, J.M., Tobar, M.J., Giner, M., Amigó, V. *et al.*, Surface modification of porous titanium discs using femtosecond laser structuring. *Metals (Basel)*, 10, 6, 1–18, 2020. Available from: https://doi.org/10.3390/met10060748.
99. Wennerberg, A. and Albrektsson, T., Effects of titanium surface topography on bone integration: A systematic review. *Clin. Oral. Implants Res.*, 20, SUPPL. 4, 172–84, 2009. Available from: https://doi.org/10.1111/j.1600-0501.2009.01775.x.
100. Kalbacova, M., Rezek, B., Baresova, V., Wolf-Brandstetter, C., Kromka, A., Nanoscale topography of nanocrystalline diamonds promotes differentiation of osteoblasts. *Acta Biomater.*, 5, 8, 3076–85, 2009. Available from: https://doi.org/10.1016/j.actbio.2009.04.020.
101. Nicolas-Silvente, A.I., Velasco-Ortega, E., Ortiz-Garcia, I., Monsalve-Guil, L., Gil, J., Jimenez-Guerra, A., Influence of the titanium implant surface

treatment on the surface roughness and chemical composition. *Materials*, 13, 2, 314, 2020. Available from: https://doi.org/10.3390/ma13020314.

102. Andrukhov, O., Huber, R., Shi, B., Berner, S., Rausch-Fan, X., Moritz, A. *et al.*, Proliferation, behavior, and differentiation of osteoblasts on surfaces of different microroughness. *Dent. Mater.*, 32, 11, 1374–84, 2016. Available from: https://doi.org/10.1016/j.dental.2016.08.217.
103. Gnilitskyi, I., Pogorielov, M., Viter, R., Ferraria, A.M., Carapeto, A.P., Oleshko, O. *et al.*, Cell and tissue response to nanotextured Ti6Al4V and Zr implants using high-speed femtosecond laser-induced periodic surface structures. *Nanomedicine*, 21, 102036, 2019. Available from: https://doi.org/10.1016/j.nano.2019.102036.
104. Tendero, I., Rossi, M.C., Viera, M., Amado, J.M., Tobar, M.J., Vicente, Á. *et al.*, Laser surface modification in Ti-xNb-yMo alloys prepared by powder metallurgy. *Metals (Basel)*, 11, 2, 1–21, 2021. Available from: https://doi.org/10.3390/met11020367.
105. Rossi, M.C., Amado, J.M., Tobar, M.J., Vicente, A., Yañez, A., Amigó, V., Effect of alloying elements on laser surface modification of powder metallurgy to improve surface mechanical properties of beta titanium alloys for biomedical application. *J. Mater. Res. Technol.*, 14, 1222–34, 2021. Available from: https://doi.org/10.1016/j.jmrt.2021.07.037.

7

Total Hip Replacement Response to a Variation of the Radial Clearance Through *In Silico* Models

Alessandro Ruggiero* **and Alessandro Sicilia**

Department of Industrial Engineering, University of Salerno, Fisciano, Italy

Abstract

The *in silico* analysis is a powerful approach able to simulate via calculators a physical phenomenon, in order to, for example, conduct a parametric analysis during the design of a prosthesis. In the biomechanics and biotribology fields, specific attention is paid to the design parameters (i.e., mechanical and geometrical) characterizing the prosthetic implant which will replace an unhealthy synovial articulation: a tribological characterization of the artificial joint is actually needed so that minimum wear and maximum duration are guaranteed.

The aim of this chapter is to analyze the tribological response of an artificial hip joint made of ultrahigh molecular weight polyethylene acetabular cup against a ceramic femoral head to a variation of the implant radial clearance. The task is performed by taking advantage of two numerical models developed by the authors in a Matlab computational environment:

- A multibody model solving the inverse dynamics of musculo-skeletal mechanical systems and elaborating the muscular actions and the joint reactions produced by a selected kinematics input, used for a lower limb apparatus moving following the gait cycle evolution;
- A lubrication model based on the Reynolds equation numerical solution supplied by the hip joint loads and relative angular velocities coming from the multibody simulation, which elaborates the

**Corresponding author*: ruggiero@unisa.it

Jay Prakash Srivastava, Dražan Kozak, Vinayak Ranjan, Pankaj Kumar, Ranjan Kumar and Shubham Tayal (eds.) Mechanical Engineering in Biomedical Applications: Bio-3D Printing, Biofluid Mechanics, Implant Design, Biomaterials, Computational Biomechanics, Tissue Mechanics, (185–230) © 2024 Scrivener Publishing LLC

tribological quantities associated with the mixed elastohydrodynamic lubrication mode (fluid/contact pressure, surfaces' separation, wear penetration depth, etc.).

The whole computational tool obtained by the merging of the two models allowed us to collect information about the behavior of interesting tribological quantities, with respect to a variation of the implant radial clearance, in order to study its role in the framework of the prosthesis wear prediction during its design.

Keywords: Biomechanics, biotribology, prosthetic, implant, Matlab

7.1 Introduction

Human articulations are the natural connections between the bones that allow the relative motion between them. The complex structure realizing the interface between two articular surfaces is essentially composed of cartilage covering the bones' extremities: in case of a wide range of motion permitted by the articulation, it is called a synovial joint.

The synovial articulation is characterized by a cavity separating the linked cartilage surfaces which are filled by the natural human lubricant, i.e., the synovial fluid, supplied through the synovial membrane. The synovial fluid provides for joint mobility with minimum friction and high load capacity through complex lubrication mechanisms which can be divided into full film modes (hydrodynamic and elastohydrodynamic) and boundary modes (mixed and boosted): the coexistence in space and time of the above lubrication mechanisms, together with the mechanical properties of porosity and deformability of the cartilage matrix and the smoothness of its surface, contributes to establish the synovial articulation as an optimal tribological joint able to guarantee minimum friction and wear of the articulated surfaces [1–3].

When the cartilage deteriorates because of aging or articular diseases, the surgical procedure of replacement of the unhealthy joint with an artificial implant could be needed. The tribomechanical design of the prosthesis has to guarantee load capacity, mobility, stability, fixation, biocompatibility, and friction/wear resistance in order to enhance its duration so that the number of surgical revisions is minimized [4, 5].

In particular, the design of the artificial implant has to include an accurate and detailed tribological analysis [6] in order to reach the above requirements, and two main approaches can be followed: the *in vitro* approach, characterized by several experiments aided by the usage of sophisticated time/money-consuming machines, and the *in silico* approach, aimed to simulate the same phenomena in a computational environment, thanks to a deep physical and mathematical definition of the investigated problem. Actually, the scientific research in the biomechanical framework tends to validate computational medicine as a powerful tool during the design and simulation phase due to its versatility and cheaper nature concerning the *in vitro* approach, using software [7–13] or analytical models [14, 15].

This chapter is focused on total hip replacement modeling, since the hip is one of the main synovial joints involved in most daily activities and it is subjected to high load: currently, hip implants could be requested by both old patients, because of aging, and young patients, considering fast and intense sports activities [16–20].

From a biomechanical point of view, the main objective pursued by the authors in their wide research activity in the field was the modeling of:

- The mechanical problem: the musculoskeletal apparatus in relation to the prosthetic patient can be seen as a multibody system (set of bones linked by the articulations, moved by the muscles, and subjected to force fields) moving with a particular kinematics which produces the unknown reaction forces acting as loads on the articulations;
- The tribological problem: the loading configuration of the analyzed hip, together with its motion, produces a lubrication/contact state at the level of the prosthetic surfaces' interface which can be described by lubrication models and contacting surface logics.

Following the above modeling procedure, the objective was to propose a novel multibody inverse dynamics solver able to reproduce the hip loads and rotations related to a musculoskeletal kinematics, in order to supply the successive assembled lubrication/contact algorithm with the elaborated tribological state. Moreover, this approach makes it possible to achieve full control of all the involved variables.

Several authors conducted research studies in the described framework through several numerical approaches. From the mechanical point of view, the first problem is related to the wide choice of the musculoskeletal model in terms of the bodies/joints' topology and muscle recruitment criteria, which results in a data dispersion of the joint loading conditions, with respect to several subjects [21, 22]: the issue does not allow to elaborate hip loadings associated with the gait cycle very close to the ones measured *in vivo* by Bergmann [23].

A high number of papers are available in the scientific literature, in which the authors presented ideas and evaluations that referred to many aspects of the biomechanical inverse dynamics simulations, e.g., the role of the optimization algorithm in the context of muscle recruitment, the effect of the constraint equations on the kinematics of the musculoskeletal system, or the behavior of the muscle–tendon dynamics model [24–26]; moreover, a fundamental role is played by the usage of a musculoskeletal multibody software, e.g., OpenSim or AnyBody, which lay the foundations for understanding the basics of the musculoskeletal models and the classical inverse dynamics pipeline [27–34].

Many other works were published related to the realization of lubrication/contact models associated with hip prosthesis treated as a lubricated spherical joint: the main aspects faced by the researchers are the modeling of the thin full film hydrodynamic lubrication through the Reynolds equation [35–37], the modeling of the elastohydrodynamic mode, when the prosthesis surfaces were considered deformable [38–43] together with the study of the related deformation models [43–46], and the more challenging analysis of the transition between a full lubricated phase to a mixed one, in which direct contact in some areas occurs [47–49].

The aim of the authors is to present in an extended way two of their main works already published [50, 51] as development based on the previous ones [52–54], in which the mechanical and tribological problems were described and modeled in the same computational environment: the realization of an interface between the models could result in a globally effective tool that aims to assist the design of the hip implant requested by a subject characterized by a particular musculoskeletal system moving with particular kinematics, i.e., it can be useful in the context of the realization of customized implants. Moreover, the authors took advantage of the described model coupling to present a sensitivity analysis based on the artificial hip implant tribological response to a variation of its radial clearance.

7.2 The Musculoskeletal Multibody Model

The analysis of the kinematics and the dynamics associated with a set of connected bodies is based on the assumption that the state of each body i along time t can be described by a vector of Lagrangian coordinates $\boldsymbol{q}_i$,

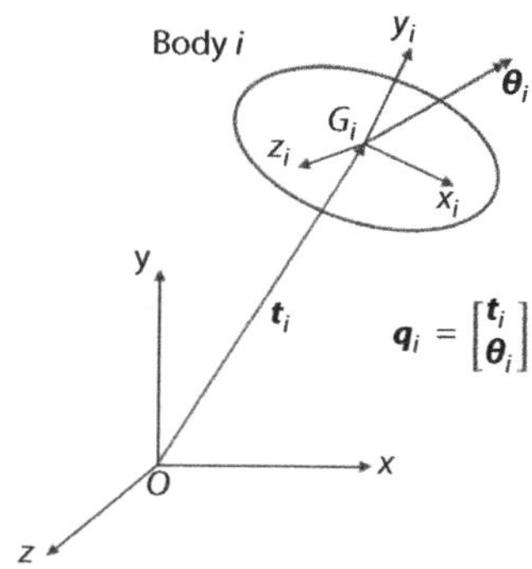

Figure 7.1 The body's Lagrangian coordinates.

which gives information about the body's local reference frame: in our case, the reference frame of the body i is centered in correspondence to its center of mass G_i and it is oriented so that the related inertia tensor is diagonal; then, the Lagrangian coordinate vector $\boldsymbol{q}_i$, Eq. (7.1), is composed by the translation vector $\boldsymbol{t}_i$, locating the body's reference frame $\{G_i x_i y_i z_i\}$ with respect to an inertial one $\{Oxyz\}$ (fixed to the ground), and by a unit quaternion $\boldsymbol{\theta}_i$, describing the body's reference frame orientation with respect to the ground, as shown in Figure 7.1.

$$\boldsymbol{q}_i(t) = \begin{bmatrix} \boldsymbol{t}_i(t) \\ \boldsymbol{\theta}_i(t) \end{bmatrix} \tag{7.1}$$

7.2.1 Kinematical Analysis

In order to solve the inverse dynamics problem associated with the analyzed musculoskeletal system, the kinematical analysis has to be performed first: it consists of assembling and solving with respect to the unknown Lagrangian coordinates a set of constraint equations forcing the bodies to move following the known degrees of freedom $\boldsymbol{q}_{dof}(t)$ and avoiding at the same time the relative movements prevented by the linking joints.

Then, the set of constraint equations $\boldsymbol{C}$ is assembled by analyzing the kinematical behavior of each joint J within the multibody system: the six potential relative motions between bodies i and j linked by joint J are turned into six constraint equations, i.e., $\boldsymbol{C}_J$, dependent on the related Lagrangian coordinates, $\boldsymbol{q}_i$ and $\boldsymbol{q}_j$, and generally on time t; in particular, they are composed by the rheonomic constraint equations $\boldsymbol{C}_{rJ}$, which force the Lagrangian coordinates $\boldsymbol{q}_i$ and $\boldsymbol{q}_j$ to evolve following the related degrees of freedom $\boldsymbol{q}_{dof_J}$, and by the scleronomic constraint equations $\boldsymbol{C}_{s_J}$, which force $\boldsymbol{q}_i$ and $\boldsymbol{q}_j$ to evolve avoiding the related prevented relative movements, as written in Eq. (7.2).

$$\boldsymbol{C}_J(\boldsymbol{q}_i,\boldsymbol{q}_j,t)=\begin{bmatrix}\boldsymbol{C}_{r_J}(\boldsymbol{q}_i,\boldsymbol{q}_j,t)\\ \boldsymbol{C}_{s_J}(\boldsymbol{q}_i,\boldsymbol{q}_j)\end{bmatrix}=\boldsymbol{0} \tag{7.2}$$

Once the joints' list is scanned for each J, the constraint equations forcing the quaternions $\boldsymbol{\theta}_i$ of each body i to evolve keeping the unitary norm, i.e., $C_{b_i}(\boldsymbol{q}_i)$, will be included in order to close the equations set. Concatenating the bodies' Lagrangian coordinates and the constraint equations along single vectors, the kinematical position problem is assembled in (7.3).

$$\boldsymbol{C}(\boldsymbol{q},t)=\begin{bmatrix}\boldsymbol{C}_r(\boldsymbol{q},t)\\ \boldsymbol{C}_s(\boldsymbol{q})\\ \boldsymbol{C}_b(\boldsymbol{q})\end{bmatrix}=\boldsymbol{0} \tag{7.3}$$

The constraint equation (7.3) are generally non-linear, then they are solved through the Newton method by introducing the constraint Jacobian $\boldsymbol{C}_q$, which can be assembled analytically by differentiating each joint constraint equations $\boldsymbol{C}_J$ with respect to the related Lagrangian coordinates $\boldsymbol{q}_i$ and $\boldsymbol{q}_j$, as shown in Eq. (7.4).

$$\boldsymbol{C}_J(\boldsymbol{q}_i,\boldsymbol{q}_j,t)=\boldsymbol{0}\rightarrow$$
$$\frac{\partial \boldsymbol{C}_J}{\partial \boldsymbol{q}}=\boldsymbol{C}_{q_J}=\begin{bmatrix}\cdots\boldsymbol{0}\cdots & \dfrac{\partial \boldsymbol{C}_J}{\partial \boldsymbol{q}_i}=\boldsymbol{C}_{q_{J_i}} & \cdots\boldsymbol{0}\cdots & \dfrac{\partial \boldsymbol{C}_J}{\partial \boldsymbol{q}_j}=\boldsymbol{C}_{q_{J_j}} & \cdots\boldsymbol{0}\cdots\end{bmatrix} \tag{7.4}$$

Then, given a starting Lagrangian coordinate vector guess, the (k) iteration is updated through Eq. (7.5).

$$\boldsymbol{q}^{(k)}=\boldsymbol{q}^{(k-1)}-\boldsymbol{C}_q^{-1}(\boldsymbol{q}^{(k-1)},t)\boldsymbol{C}(\boldsymbol{q}^{(k-1)},t) \tag{7.5}$$

Once the non-linear position problem is solved, the differentiation with respect to time of the constraint equations set leads to the linear problems associated with the velocity and acceleration analysis, in which the Lagrangian velocities $\dot{\boldsymbol{q}}$ and acceleration $\ddot{\boldsymbol{q}}$ are calculated via matrix inversion, as highlighted in Eq. (7.6), in which further derivatives of the constraint equations $\boldsymbol{C}$ (analytically obtainable) are introduced, i.e., $\boldsymbol{C}_t, \dot{\boldsymbol{C}}_q$ and $\dot{\boldsymbol{C}}_t$.

$$\boldsymbol{C}(\boldsymbol{q},t)=\boldsymbol{0} \rightarrow \begin{matrix} \dot{\boldsymbol{C}}(\boldsymbol{q},\dot{\boldsymbol{q}},t)=\boldsymbol{C}_q\dot{\boldsymbol{q}}+\boldsymbol{C}_t=\boldsymbol{0} \\ \ddot{\boldsymbol{C}}(\boldsymbol{q},\dot{\boldsymbol{q}},\ddot{\boldsymbol{q}},t)=\dot{\boldsymbol{C}}_q\dot{\boldsymbol{q}}+\boldsymbol{C}_q\ddot{\boldsymbol{q}}+\dot{\boldsymbol{C}}_t=\boldsymbol{0} \end{matrix} \rightarrow \begin{matrix} \dot{\boldsymbol{q}}=-\boldsymbol{C}_q^{-1}\boldsymbol{C}_t \\ \ddot{\boldsymbol{q}}=-\boldsymbol{C}_q^{-1}(\dot{\boldsymbol{C}}_q\dot{\boldsymbol{q}}+\dot{\boldsymbol{C}}_t) \end{matrix} \tag{7.6}$$

In our multibody model, four types of joints were considered; in order to not overwhelm the text, the constraint equations are reported in the following without their derivatives with respect to the Lagrangian coordinates or to time: they are left to the readers who are redirected to [50].

The fundamental laws governing the determination of the location $\boldsymbol{r}$ with respect to the ground of a point P fixed to a body i through the local position vector $\bar{\boldsymbol{u}}_{P_i}$ and the definition of the unit quaternion $\boldsymbol{\theta}$ are reported in Eq. (7.7): these are the foundations of the constraint mechanics description. In Eq. (7.7), $\boldsymbol{R}$ is the rotation matrix obtained with the usage of the quaternion, while the quaternion is defined as the function of the rotation θ performed by the local reference frame around the unit vector $\hat{\boldsymbol{v}}$ locating the rotation axis defined in the global or, more generally, in the parent reference frame.

$$\boldsymbol{r}_P=\boldsymbol{t}_i+\boldsymbol{R}(\boldsymbol{\theta}_i)\bar{\boldsymbol{u}}_{P_i} \quad \boldsymbol{\theta}(\hat{\boldsymbol{v}},\theta)=\left[\cos\left(\frac{\theta}{2}\right) \quad \hat{\boldsymbol{v}}^T\sin\left(\frac{\theta}{2}\right)\right]^T \tag{7.7}$$

With reference to Figure 7.2, the considered joints are:

- The revolute joint (7.2(a)): it allows one degree of freedom between the linked bodies, which is the rotation θ around the axis oriented by the unit vector $\hat{\boldsymbol{v}}$ defined in the parent body reference frame; hence, it produces one rheonomic and five scleronomic constraint equations reported in (7.8); the rheonomic equation is obtained by imposing the equality between rotation θ and the one associated with the linked bodies' relative quaternion (obtained by the Hamiltonian product matrix $\boldsymbol{H}$ and the quaternion phase operator $\langle\boldsymbol{\theta}\rangle$); the scleronomic equations related to the three locked relative translations are obtained imposing the equality between the global position of the revolute joint J calculated with the Lagrangian coordinates of both linked bodies, while the scleronomic equations related to the two locked relative rotations are obtained imposing the orthogonality of two local vectors $\bar{s}_{i_1}$ and $\bar{s}_{i_2}$ (belonging to the body i, perpendicular to each other and to the rotation axis) with respect to $\bar{s}_j$ (local vector belonging to the body j and parallel to the rotation axis);

$$C_J(q_i,q_j,t) = \begin{bmatrix} C_{r_J}(q_i,q_j,t) = [\langle H(\theta_i)^T \theta_j \rangle - \theta(t)] \\ C_{s_J}(q_i,q_j) = \begin{bmatrix} t_i + R(\theta_i)\bar{u}_{J_i} - t_j - R(\theta_j)\bar{u}_{J_j} \\ \left(R(\theta_i)\bar{s}_{i_1}\right)^T (R(\theta_j)\bar{s}_j) \\ \left(R(\theta_i)\bar{s}_{i_2}\right)^T (R(\theta_j)\bar{s}_j) \end{bmatrix} \end{bmatrix} = 0 \tag{7.8}$$

– The cam joint (7.2(b)): it is kinematically similar to the revolute joint, except for two allowed relative translations dependent on the single degree of freedom θ performed along the two unit vectors $\hat{v}_1$ and $\hat{v}_2$ defined in the parent body reference frame and governed by the known functions $u_1(\theta)$ and $u_2(\theta)$; hence, according to Eq. (7.9), while the rheonomic equation and the two scleronomic equations associated with the two locked relative rotations are equal to the ones presented for the revolute joint, the three scleronomic equations associated with the three relative translations are modified through the definition of the cam joint local position $\bar{u}_{J_i}$ as a function of the allowed rotation θ;

$$\bar{u}_{J_i}(\theta) = u_1(\theta)\hat{v}_1 + u_2(\theta)\hat{v}_2$$

$$C_J(q_i,q_j,t) = \begin{bmatrix} C_{r_J}(q_i,q_j,t) = [\langle H(\theta_i)^T \theta_j \rangle - \theta(t)] \\ C_{s_J}(q_i,q_j,t) = \begin{bmatrix} t_i + R(\theta_i)\bar{u}_{J_i}(\theta(t)) - t_j - R(\theta_j)\bar{u}_{J_j} \\ \left(R(\theta_i)\bar{s}_{i_1}\right)^T (R(\theta_j)\bar{s}_j) \\ \left(R(\theta_i)\bar{s}_{i_2}\right)^T (R(\theta_j)\bar{s}_j) \end{bmatrix} \end{bmatrix} = 0 \tag{7.9}$$

- The spherical joint (7.2(c)): it allows three relative rotations θ_1, θ_2, and θ_3 between the linked bodies performed around three successive rotation axes locally oriented by the unit vectors $\hat{\boldsymbol{v}}_1, \hat{\boldsymbol{v}}_2$, and $\hat{\boldsymbol{v}}_3$, respectively; therefore, according to Eq. (7.10), while the three scleronomic equations associated with the three locked translations are equal to the ones characterizing the revolute joint, the three rheonomic equations are obtained through the same approach presented with the revolute joint in which the Hamiltonian product matrix and the quaternion phase operator are used, by introducing in addition other $\boldsymbol{H}$ matrices related to successive rotations order;

$$\boldsymbol{C}_J(\boldsymbol{q}_i,\boldsymbol{q}_j,t)=\begin{bmatrix} \boldsymbol{C}_{r_J}(\boldsymbol{q}_i,\boldsymbol{q}_j,t)=\begin{bmatrix} \langle \boldsymbol{H}_{\theta_1}^T\boldsymbol{H}(\boldsymbol{\theta}_i)^T\boldsymbol{\theta}_j\rangle-\theta_1(t) \\ \langle \boldsymbol{H}_{\theta_2}^T\boldsymbol{H}(\boldsymbol{\theta}_i)^T\boldsymbol{\theta}_j\rangle-\theta_2(t) \\ \langle \boldsymbol{H}_{\theta_3}^T\boldsymbol{H}(\boldsymbol{\theta}_i)^T\boldsymbol{\theta}_j\rangle-\theta_3(t) \end{bmatrix} \\ \boldsymbol{C}_{s_J}(\boldsymbol{q}_i,\boldsymbol{q}_j)=\left[\boldsymbol{t}_i+\boldsymbol{R}(\boldsymbol{\theta}_i)\bar{\boldsymbol{u}}_{J_i}-\boldsymbol{t}_j-\boldsymbol{R}(\boldsymbol{\theta}_j)\bar{\boldsymbol{u}}_{J_j}\right] \end{bmatrix}=\boldsymbol{0} \tag{7.10}$$

- The free joint (7.2(d)): it permits the full spatial mobility between the linked bodies, by allowing, in addition with respect to the spherical joint, three relative translations t_1, t_2, and t_3, respectively, performed along three unit vectors $\hat{\boldsymbol{u}}_1, \hat{\boldsymbol{u}}_2$, and $\hat{\boldsymbol{u}}_3$ defined in the parent body reference frame; then, according to Eq. (7.11), while the three rheonomic equations related to the allowed rotations are equal to the ones presented for the spherical joint, further three rheonomic equations are introduced in which the free joint local position vector $\bar{\boldsymbol{u}}_{J_i}$ is written as a function of the three translational degrees of freedom.

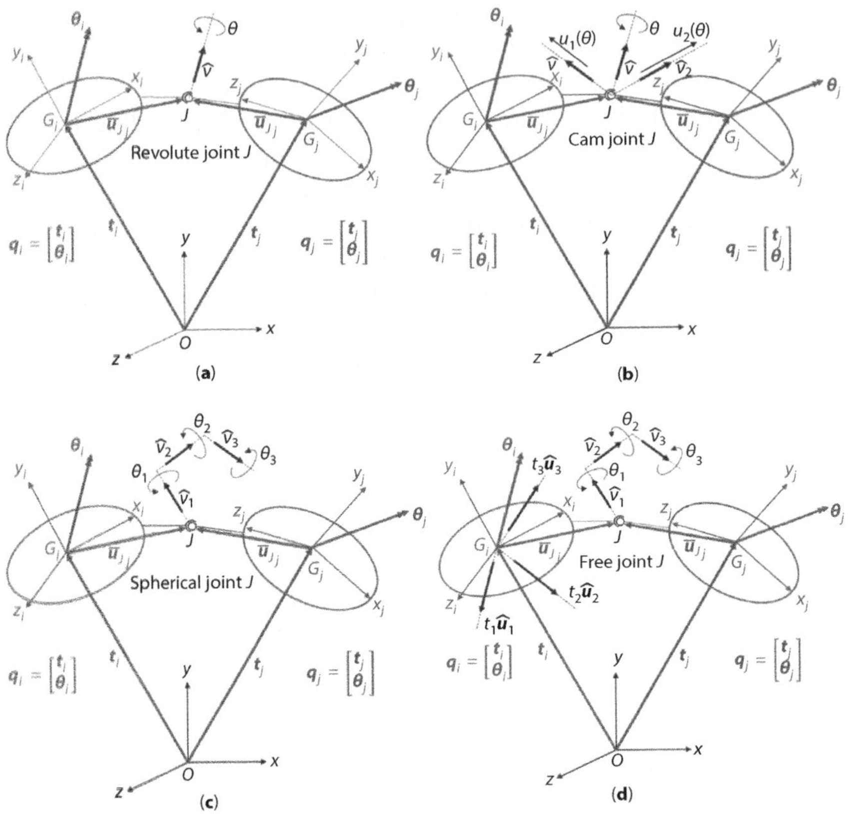

Figure 7.2 Joints considered in the multibody model. (a) Revolute joint scheme, (b) Cam joint scheme, (c) Spherical joint scheme, (d) Free joint scheme.

$$\bar{u}_{J_i}(t) = t_1(t)\hat{u}_1 + t_2(t)\hat{u}_2 + t_3(t)\hat{u}_3 = \begin{bmatrix} \hat{u}_1 & \hat{u}_2 & \hat{u}_3 \end{bmatrix} \begin{bmatrix} t_1(t) \\ t_2(t) \\ t_3(t) \end{bmatrix} = \mathbf{U}\mathbf{t}(t)$$

$$\mathbf{C}_J(\mathbf{q}_i,\mathbf{q}_j,t) = \begin{bmatrix} \mathbf{C}_{r_J}(\mathbf{q}_i,\mathbf{q}_j,t) = \begin{bmatrix} \langle \mathbf{H}_{\theta_1}^T \mathbf{H}(\theta_i)^T \theta_j \rangle - \theta_1(t) \\ \langle \mathbf{H}_{\theta_2}^T \mathbf{H}(\theta_i)^T \theta_j \rangle - \theta_2(t) \\ \langle \mathbf{H}_{\theta_3}^T \mathbf{H}(\theta_i)^T \theta_j \rangle - \theta_3(t) \\ \mathbf{U}^{-1}\mathbf{R}(\theta_i)^T(\mathbf{t}_j + \mathbf{R}(\theta_j)\bar{\mathbf{u}}_{J_j} - \mathbf{t}_i) - \mathbf{t}(t) \end{bmatrix} \end{bmatrix} = \mathbf{0} \tag{7.11}$$

7.2.2 Dynamical Analysis

The kinematical problem (7.6) is completely addressed, so the Lagrangian coordinates $\boldsymbol{q}$, velocities $\dot{\boldsymbol{q}}$, and accelerations $\ddot{\boldsymbol{q}}$ can be calculated along the analyzed time t and can be used to characterize the next step, i.e., the dynamical problem, in which the equation of motion has to be assembled in order to evaluate the driving actions and the consequent joint reaction produced by the known kinematics that moves the analyzed multibody system.

The equation of motion of the analysed $n_B - 1$ rigid bodies (excluding the ground) can be derived from the Lagrange equation (7.12) which states that the global virtual work δW of all the generalized forces for a set of virtual displacements $\delta \boldsymbol{q}_i$ is null and it can be written as a function of the kinetic energy T_i and the external generalized forces $\boldsymbol{Q}_{e_i}$.

$$\delta W = \sum_{i=1}^{n_B-1}\left(\frac{d}{dt}\frac{\partial T_i}{\partial \dot{\boldsymbol{q}}_i} - \frac{\partial T_i}{\partial \boldsymbol{q}_i} - \boldsymbol{Q}_{e_i}^T\right)\delta \boldsymbol{q}_i = 0 \tag{7.12}$$

The kinetic energy of rigid bodies is defined in (7.13) as a function of the bodies' internal point velocity $\dot{\boldsymbol{r}}_i$ and mass density ρ_i within their volume V_i.

$$\boldsymbol{r}_i = \boldsymbol{t}_i + \boldsymbol{R}(\boldsymbol{\theta}_i)\bar{\boldsymbol{u}}_i \quad \rightarrow \quad T_i = \frac{1}{2}\int \rho_i \dot{\boldsymbol{r}}_i^T \dot{\boldsymbol{r}}_i dV_i \tag{7.13}$$

Substituting (7.13) in (7.12), the equation of motion is written in (7.14) as a function of the bodies' mass matrix $\boldsymbol{M}_i$ and of the Coriolis and centrifugal generalized force vector $\boldsymbol{Q}_{v_i}$, dependent on the bodies' mass m_i, local inertia tensor $\bar{\boldsymbol{I}}_i$, and other vector and matrix quantities related to the quaternion θ_i and its derivative $\dot{\boldsymbol{\theta}}_i$, i.e., $\bar{\boldsymbol{G}}_i$ and $\bar{\boldsymbol{\omega}}_i$; moreover, the global virtual work is written as a unique scalar product by concatenating all the introduced vectors and matrices.

$$\boldsymbol{M}_i = \begin{bmatrix} m_i\boldsymbol{I} & \boldsymbol{0} \\ \boldsymbol{0} & \bar{\boldsymbol{G}}_i^T \bar{\boldsymbol{I}}_i \bar{\boldsymbol{G}}_i \end{bmatrix} \quad \boldsymbol{Q}_{v_i} = \begin{bmatrix} \boldsymbol{0} \\ -2\bar{\boldsymbol{G}}_i^T \bar{\boldsymbol{I}}_i \bar{\boldsymbol{\omega}}_i \end{bmatrix}$$

$$\delta W = \sum_{i=1}^{n_B-1} \left(\boldsymbol{M}_i\ddot{\boldsymbol{q}}_i - \boldsymbol{Q}_{v_i} - \boldsymbol{Q}_{e_i}\right)^T \delta\boldsymbol{q}_i = (\boldsymbol{M}\ddot{\boldsymbol{q}} - \boldsymbol{Q}_v - \boldsymbol{Q}_e)^T \delta\boldsymbol{q} = 0 \tag{7.14}$$

In order to consider the internal forces depending on the constraint equations $\boldsymbol{C}$ within the equation of motion, their virtual work (which is null by definition) is included as a function of the Lagrange multipliers vector $\boldsymbol{\lambda}$, turning the energy balance equation into a force balance equation (7.15).

$$\begin{cases} (\boldsymbol{M}\ddot{\boldsymbol{q}} - \boldsymbol{Q}_v - \boldsymbol{Q}_e)^T \delta\boldsymbol{q} = 0 \\ \boldsymbol{C}(\boldsymbol{q},t) = \boldsymbol{0} \end{cases} \rightarrow \boldsymbol{\lambda}^T \boldsymbol{C}_q \delta\boldsymbol{q} = 0 \rightarrow \boldsymbol{M}\ddot{\boldsymbol{q}} - \boldsymbol{Q}_v - \boldsymbol{Q}_e + \boldsymbol{C}_q^T\boldsymbol{\lambda} = \boldsymbol{0} \tag{7.15}$$

The Lagrange multipliers $\boldsymbol{\lambda}$ are the unknowns of the linear problem in Eq. (7.16), since they represent the internal forces acting at the joints' space level in order to satisfy the constraint equations over the analyzed kinematics. In fact, separating the three constraint contributions, the Lagrange multipliers associated with the rheonomic constraints $\boldsymbol{\lambda}_r$ represent the internal driving actions acting on each degree of freedom and forcing the multibody system to evolve following the known kinematics time evolution.

$$\begin{bmatrix} \boldsymbol{C}_{q_r}^T & \boldsymbol{C}_{q_s}^T & \boldsymbol{C}_{q_b}^T \end{bmatrix} \begin{bmatrix} \boldsymbol{\lambda}_r \\ \boldsymbol{\lambda}_s \\ \boldsymbol{\lambda}_b \end{bmatrix} = -(\boldsymbol{M}\ddot{\boldsymbol{q}} - \boldsymbol{Q}_v - \boldsymbol{Q}_e) \tag{7.16}$$

Even if the rheonomic multipliers $\boldsymbol{\lambda}_r$ give information about the net forces and torques that move the degrees of freedom, they are not generated by specific actuators acting solely on the related degrees of freedom: the actuation system of a musculoskeletal set of bones is indeed represented by a redundant group of muscles which have to be modeled in order to calculate the real actuation driving forces and, also, the real joint reactions that cannot be represented by the scleronomic Lagrange multipliers $\boldsymbol{\lambda}_S$ in (7.16).

7.2.3 The Muscle Actuator

The main approach used to model the muscular actuators is to consider a curve delimited by the muscle's attachment point on the related bones: the resulting actuator is a deformable curve able to generate the contractile muscle force transmitted along the path as a function of its length and deformation velocity by way of specific constitutive relations based on the chosen muscle model. The muscle path considered in our model can be composed of both straight lines and curved ones, where the latter occurs when a muscle segment encounters a bony surface around which it wraps.

Then, in addition to the kinematics of the bones, also the kinematics of the wrapping object has to be evaluated through the knowledge of the Lagrangian coordinates and their time derivatives. The more general configuration is reported in Figure 7.3, where the muscle n is fixed to the bodies i and j through the attachment points A and B while it wraps around a surface which is fixed to the body k through the two wrap points P and Q.

While the kinematics of the attachment points A and B is known from the kinematical analysis because they are fixed to the bodies i and j, the one

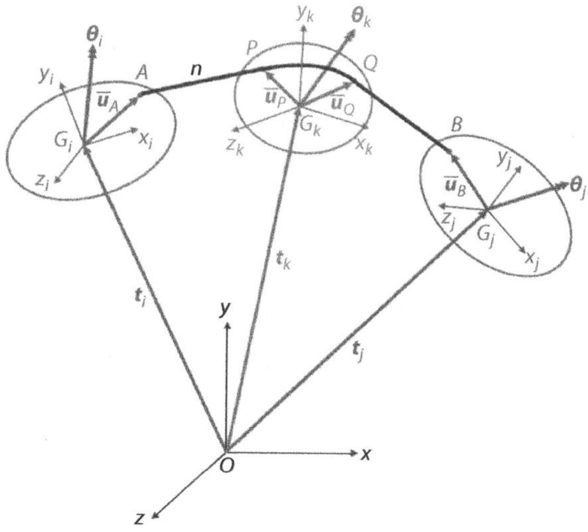

Figure 7.3 The wrapping muscle scheme.

referred to the wrap points P and Q depends on the wrapping algorithm used and it will be discussed below. The virtual work done by the muscle n, i.e., δW_{m_n} in Eq. (7.17), is related only to the muscle's straight lines AP and QB because the curved path PQ will be modeled as a geodesic curve lying on the surface without friction so that the muscle force F_{m_n} is transmitted along the whole muscle path.

$$\begin{cases} \boldsymbol{r}_A = \boldsymbol{t}_i + \boldsymbol{R}(\theta_i)\bar{\boldsymbol{u}}_A \\ \boldsymbol{r}_P = \boldsymbol{t}_k + \boldsymbol{R}(\theta_k)\bar{\boldsymbol{u}}_P \\ \boldsymbol{r}_Q = \boldsymbol{t}_k + \boldsymbol{R}(\theta_k)\bar{\boldsymbol{u}}_Q \\ \boldsymbol{r}_B = \boldsymbol{t}_j + \boldsymbol{R}(\theta_j)\bar{\boldsymbol{u}}_B \end{cases} \rightarrow \begin{cases} \hat{\boldsymbol{l}}_{AP} = \dfrac{\boldsymbol{r}_P - \boldsymbol{r}_A}{\|\boldsymbol{r}_P - \boldsymbol{r}_A\|} \\ \hat{\boldsymbol{l}}_{QB} = \dfrac{\boldsymbol{r}_B - \boldsymbol{r}_Q}{\|\boldsymbol{r}_B - \boldsymbol{r}_Q\|} \end{cases} \rightarrow \delta W_{m_n} = F_{m_n}\hat{\boldsymbol{l}}_{AP}^T\delta(\boldsymbol{r}_A - \boldsymbol{r}_P) + F_{m_n}\hat{\boldsymbol{l}}_{QB}^T\delta(\boldsymbol{r}_Q - \boldsymbol{r}_B) \tag{7.17}$$

Relating the points' virtual displacements to the virtual variation of the bodies' Lagrangian coordinates through the matrix $\boldsymbol{J}$ defined in (7.18), the muscle n virtual work δW_{m_n} is written as a function of the virtual Lagrangian displacements associated with the bodies i, j, and k.

$$\delta \boldsymbol{r} = \begin{bmatrix} \boldsymbol{I} & -\boldsymbol{R}(\theta)\tilde{\bar{u}}\,\bar{\boldsymbol{G}} \end{bmatrix} \begin{bmatrix} \delta \boldsymbol{t} \\ \delta \boldsymbol{\theta} \end{bmatrix} = \boldsymbol{J}(\boldsymbol{q},\bar{\boldsymbol{u}})\delta \boldsymbol{q}$$

$$\delta W_{m_n} = F_{m_n}\Big[\big(\hat{\boldsymbol{l}}_{AP}^T\boldsymbol{J}(\boldsymbol{q}_i,\bar{\boldsymbol{u}}_A)\big)\delta \boldsymbol{q}_i + \big(-\hat{\boldsymbol{l}}_{AP}^T\boldsymbol{J}(\boldsymbol{q}_k,\bar{\boldsymbol{u}}_P) + \hat{\boldsymbol{l}}_{QB}^T\boldsymbol{J}(\boldsymbol{q}_k,\bar{\boldsymbol{u}}_Q)\big)\delta \boldsymbol{q}_k + \big(-\hat{\boldsymbol{l}}_{QB}^T\boldsymbol{J}(\boldsymbol{q}_j,\bar{\boldsymbol{u}}_B)\big)\delta \boldsymbol{q}_j\Big] \tag{7.18}$$

The vectors multiplying the virtual displacements in (7.18) contribute to a specific row of the muscle Jacobian $\boldsymbol{\Phi}_q$ associated with the muscle n: summing up the virtual work done by all the considered n_m muscles, the generalized muscle force vector $\boldsymbol{Q}_m$ is introduced as the muscle force vector $\boldsymbol{F}_m$ oriented through the muscle Jacobian $\boldsymbol{\Phi}_q$ in the multibody space, as reported in Eq. (7.19).

$$\begin{aligned} \delta W_{m_n} &= F_{m_n}(\boldsymbol{\Phi}_{q_{ni}}\delta \boldsymbol{q}_i + \boldsymbol{\Phi}_{q_{nk}}\delta \boldsymbol{q}_k + \boldsymbol{\Phi}_{q_{nj}}\delta \boldsymbol{q}_j) \\ &= F_{m_n}\begin{bmatrix} \cdots & \boldsymbol{\Phi}_{q_{ni}} & \cdots & \boldsymbol{\Phi}_{q_{nk}} & \cdots & \boldsymbol{\Phi}_{q_{nj}} & \cdots \end{bmatrix}\delta \boldsymbol{q} \end{aligned}$$

$$\sum_{n=1}^{n_m} \delta W_{m_n} = \boldsymbol{F}_m^T\boldsymbol{\Phi}_q\delta \boldsymbol{q} = \boldsymbol{Q}_m^T\delta \boldsymbol{q} \rightarrow \boldsymbol{Q}_m = \boldsymbol{\Phi}_q^T\boldsymbol{F}_m \tag{7.19}$$

7.2.4 The Geodesic Muscle Wrapping

The kinematics of the wrapping points P and Q for each wrapping muscle is evaluated by a geodesic wrapping algorithm which detects the possible intersections between the straight line muscle AB and the analyzed wrapping object, as shown in Figure 7.4 for the cylinder case. Since the local position of the points belonging to the wrapping surface can be described by an explicit function $\boldsymbol{x}(\boldsymbol{u})$, where $\boldsymbol{u}$ is the vector of the surface parameters u and v, the algorithm varies $\boldsymbol{u}_P$ and $\boldsymbol{u}_Q$ so that the geodesics starting from the points P and Q, i.e., $\boldsymbol{c}_P$ and $\boldsymbol{c}_Q$, will join collinearly in correspondence to their closest points P^* and Q^*, by means of manipulating the curve tangent $\hat{\boldsymbol{t}}$, normal $\hat{\boldsymbol{n}}$, and binormal $\hat{\boldsymbol{b}}$ unit vectors along the curvilinear abscissa s defined in Eq. (7.20) (the derivative of a quantity x with respect to s is denoted with x').

$$\hat{\boldsymbol{t}} = \boldsymbol{x}' = \boldsymbol{x}_u u' + \boldsymbol{x}_v v' = [\boldsymbol{x}_u \quad \boldsymbol{x}_v]\begin{bmatrix} u' \\ v' \end{bmatrix} = \boldsymbol{J}\boldsymbol{u}' \qquad \hat{\boldsymbol{n}} = \frac{\boldsymbol{x}_u \times \boldsymbol{x}_v}{\|\boldsymbol{x}_u \times \boldsymbol{x}_v\|} \qquad \hat{\boldsymbol{b}} = \hat{\boldsymbol{t}} \times \hat{\boldsymbol{n}} \tag{7.20}$$

Starting from the current guesses $\boldsymbol{u}_P$ and $\boldsymbol{u}_Q$ in Eq. (7.21), the algorithm evaluates the directions $\hat{\boldsymbol{t}}_{AP}$ and $\hat{\boldsymbol{t}}_{BQ}$ which are projected on the surface

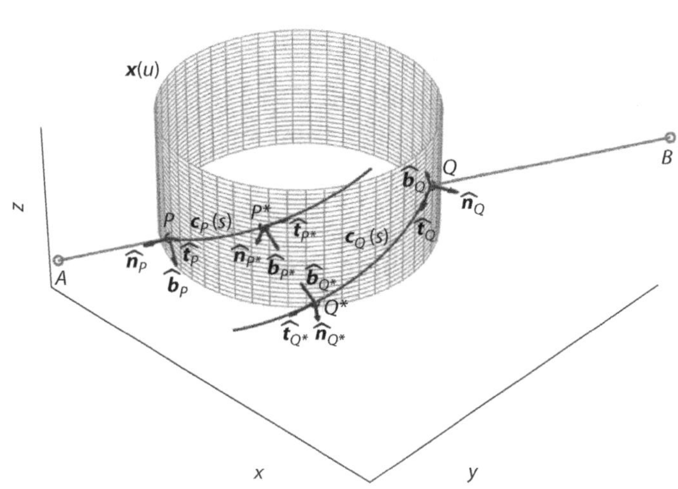

Figure 7.4 Geodesics starting from the muscle AB intersection points.

in order to obtain the tangent unit vectors $\hat{\boldsymbol{t}}_P$ and $\hat{\boldsymbol{t}}_Q$, where the latter are necessary to elaborate the geodesics.

$$\begin{cases} \boldsymbol{r}_P = x(\boldsymbol{u}_P) \\ \boldsymbol{r}_Q = x(\boldsymbol{u}_Q) \end{cases} \rightarrow \begin{cases} \hat{\boldsymbol{t}}_{AP} = \dfrac{\boldsymbol{r}_P - \boldsymbol{r}_A}{\|\boldsymbol{r}_P - \boldsymbol{r}_A\|} \\ \hat{\boldsymbol{t}}_{BQ} = \dfrac{\boldsymbol{r}_Q - \boldsymbol{r}_B}{\|\boldsymbol{r}_Q - \boldsymbol{r}_B\|} \end{cases} \rightarrow \begin{cases} \hat{\boldsymbol{t}}_P = \dfrac{\boldsymbol{J}_P\left(\boldsymbol{J}_P^T\boldsymbol{J}_P\right)^{-1}\left(\boldsymbol{J}_P^T\hat{\boldsymbol{t}}_{AP}\right)}{\left\|\boldsymbol{J}_P\left(\boldsymbol{J}_P^T\boldsymbol{J}_P\right)^{-1}\left(\boldsymbol{J}_P^T\hat{\boldsymbol{t}}_{AP}\right)\right\|} \\ \hat{\boldsymbol{t}}_Q = \dfrac{\boldsymbol{J}_Q\left(\boldsymbol{J}_Q^T\boldsymbol{J}_Q\right)^{-1}\left(\boldsymbol{J}_Q^T\hat{\boldsymbol{t}}_{BQ}\right)}{\left\|\boldsymbol{J}_Q\left(\boldsymbol{J}_Q^T\boldsymbol{J}_Q\right)^{-1}\left(\boldsymbol{J}_P^T\hat{\boldsymbol{t}}_{BQ}\right)\right\|} \end{cases} \tag{7.21}$$

The geodesic curve starting from the intersection is simulated and reported in Eq. (7.22) as the forward dynamics motion $\boldsymbol{c}(s)$ of a particle of arbitrary mass m constrained to belong to the surface $\boldsymbol{x}(\boldsymbol{u})$ in the absence of external forces with initial position and velocity, respectively, $\boldsymbol{r}$ and $\hat{\boldsymbol{t}}$: the constraint equation $\boldsymbol{C}(\boldsymbol{c}) = \boldsymbol{0}$ is derived from the implicit surface equation coming from $\boldsymbol{x}(\boldsymbol{u})$, i.e., $f(\boldsymbol{x}) = 0$, so that the associated Lagrange multiplier $\boldsymbol{\lambda}$ is the surface normal force that keeps the particle's motion on the surface. From a numerical point of view, the final non-linear ODE problem in (7.22) is solved with marching techniques.

$$\begin{cases} m\boldsymbol{I}\boldsymbol{c}'' + \boldsymbol{C}_c^T\boldsymbol{\lambda} = \boldsymbol{0} \\ \boldsymbol{c}(0) = r \\ \boldsymbol{c}'(0) = \hat{\boldsymbol{t}} \\ \boldsymbol{C}(0) = f(0) = \boldsymbol{0} \end{cases} \rightarrow \begin{cases} \boldsymbol{C}'(\boldsymbol{c},\boldsymbol{c}') = \boldsymbol{C}_c\boldsymbol{c}' = \boldsymbol{0} \\ \boldsymbol{C}''(\boldsymbol{c},\boldsymbol{c}',\boldsymbol{c}'') = \boldsymbol{C}_c'\boldsymbol{c}' + \boldsymbol{C}_c\boldsymbol{c}'' = \boldsymbol{0} \end{cases} \rightarrow \begin{cases} \begin{bmatrix} m\boldsymbol{I} & \boldsymbol{C}_c^T \\ \boldsymbol{C}_c & \boldsymbol{0} \end{bmatrix}\begin{bmatrix} \boldsymbol{c}'' \\ \lambda \end{bmatrix} = \begin{bmatrix} \boldsymbol{0} \\ -\boldsymbol{c}_c'\boldsymbol{c}' \end{bmatrix} \\ \boldsymbol{c}'(0) = \hat{\boldsymbol{t}} \\ \boldsymbol{c}'(0) = \boldsymbol{r} \end{cases} \tag{7.22}$$

Once the two geodesics $\boldsymbol{c}_P$ and $\boldsymbol{c}_Q$ are calculated through (7.22), the reference triad $\{\hat{\boldsymbol{t}},\hat{\boldsymbol{n}},\hat{\boldsymbol{b}}\}$ is evaluated in correspondence to the minimum distance points P^* and Q^*, so that the geodesics' joining function $\boldsymbol{f}(\boldsymbol{\xi})$ is assembled: the problem $\boldsymbol{f}(\boldsymbol{\xi}) = \boldsymbol{0}$ assembled in (7.23) ensures the tangency between the split segments AP and QB with respect to the surface $\boldsymbol{x}(\boldsymbol{u})$ and a smooth connection between the geodesics $\boldsymbol{c}_P$ and $\boldsymbol{c}_Q$ in correspondence to the collinearity points P^* and Q^*. The non-linear problem (7.23) is solved through the Newton iterative method in (7.24) by assembling numerically the Jacobian of $f(\xi)$ through a small constant ε: once the problem is solved, the curves from P to P^* and from Q^* to Q are joined in the resulting curved muscle path c_{PQ}, as shown in Figure 7.5.

$$\xi = \begin{bmatrix} u_P \\ v_P \\ u_Q \\ v_Q \end{bmatrix} \rightarrow f(\xi) = \begin{bmatrix} \hat{t}_{AP}^T \hat{n}_P \\ \hat{t}_{BQ}^T \hat{n}_Q \\ \hat{b}_{P^*}^T \left(c_{Q^*} - c_{P^*} \right) \\ \hat{b}_{P^*}^T \hat{t}_{Q^*} \end{bmatrix} = 0 \tag{7.23}$$

$$\xi^{(k)} = \xi^{(k-1)} - f_\xi^{-1}(\xi^{(k-1)}) f(\xi^{(k-1)}) \quad [f_\xi]_{ij} = \frac{f_i(\xi_j(1+\varepsilon)) - f_i(\xi_j)}{\varepsilon \xi_j} \tag{7.24}$$

The described algorithm allows the calculation of the local vectors $\bar{u}_P$ and $\bar{u}_Q$ of Eq. (7.17), so that the kinematics of all the muscle points, both attachment and wrapping ones, is known. The next step consists in evaluating the length l_{mt} and the deformation velocity v_{mt} of a generic muscle actuator composed of n_s straight lines from A_i to B_i and n_c curved lines from P_j to Q_j, because they are the actuator states that produce the contractive

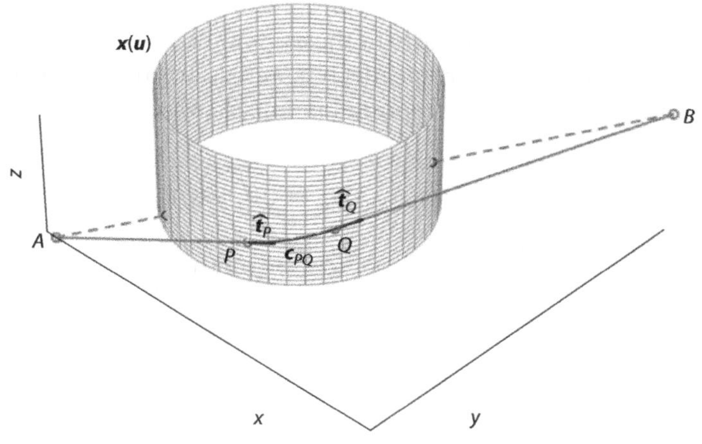

Figure 7.5 The geodesic wrapping muscle.

force depending on the used muscle–tendon model. Considering the calculation of the velocity vector $\dot{\boldsymbol{r}}_P$ of a point P fixed to the body i in (7.25), the muscle–tendon states l_{mt} and v_{mt} are given.

$$\dot{\boldsymbol{r}}_P = \boldsymbol{J}(\boldsymbol{q}_i, \overline{\boldsymbol{u}}_P)\dot{\boldsymbol{q}}_i \rightarrow \begin{cases} l_{mt} = \sum_{i=1}^{n_s} \left\| \boldsymbol{r}_{B_i} - \boldsymbol{r}_{A_i} \right\| + \sum_{j=1}^{n_c} \int_{s_{P_j}}^{s_{Q_j}} \left\| \boldsymbol{c}'_{P_jQ_j}(s) \right\| ds \\ v_{mt} = \dot{l}_{mt} = \sum_{i=1}^{n_s} \left(\dot{\boldsymbol{r}}_{B_i} - \dot{\boldsymbol{r}}_{A_i} \right)^T \hat{\boldsymbol{l}}_{A_iB_i} + \sum_{j=1}^{n_c} \dot{\boldsymbol{r}}_{Q_j}^T \hat{\boldsymbol{t}}_{Q_j} - \dot{\boldsymbol{r}}_{P_j}^T \hat{\boldsymbol{t}}_{P_j} \end{cases} \tag{7.25}$$

7.2.5 The Hill Muscle–Tendon Model

The muscle–tendon model used in this work is the Hill one (Figure 7.6), which considers the actuator unit as composed of the tendon (series element, SE) with length l_t in series with the muscle with length l_m inclined through the pennation angle α_p with respect to the muscle–tendon action line, in order to simulate the action of the pennated muscles; the muscle is represented by a parallel between the contractile element, CE, i.e., the real force generator through the sarcomeres' contraction, and the passive element, PE, which plays the role of the rounding connective tissues' resistance to the muscle lengthening.

Knowing l_{mt} and v_{mt} from Eq. (7.25), the assumptions of the Hill model are to consider a rigid tendon, i.e., l_t constant, and a constant muscle width w_m equal to the width particularly in the case of the optimal muscle fibers' length and pennation angle, respectively, l_0 and α_0, in order to calculate the muscle states l_m, v_m, and α_p, as reported in Eq. (7.26).

$$\begin{cases} l_{mt} = l_t + l_m \cos\alpha_p \\ w_m = l_m \sin\alpha_p = l_0 \sin\alpha_0 \end{cases} \rightarrow \begin{cases} l_m = \sqrt{(l_0 \sin\alpha_0)^2 + (l_{mt} - l_t)^2} \\ \alpha_p = \arctan\left(\dfrac{l_0 \sin\alpha_0}{l_{mt} - l_t} \right) \\ v_m = \dot{l}_m = v_{mt} \cos\alpha_p \end{cases} \tag{7.26}$$

Because of the series between the tendon and the muscle, the force F_m exerted by the whole muscle–tendon unit is the sum of the one generated

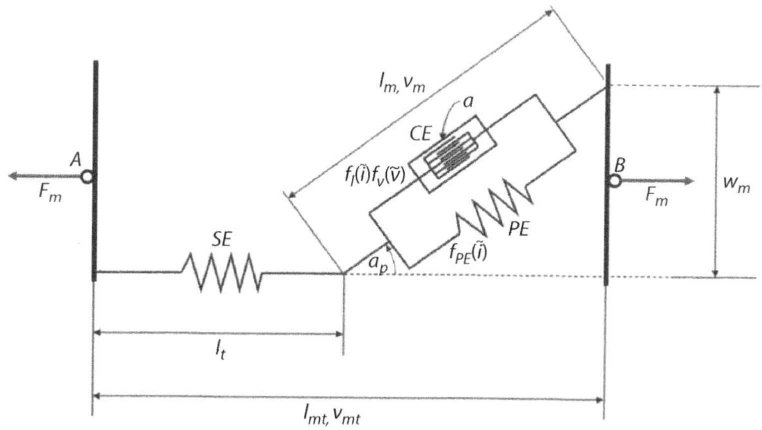

Figure 7.6 The Hill muscle–tendon model.

by the contractile element F_{CE} and the one generated by the passive element F_{PE} projected on the action line. The two forces are a fraction of the maximum isometric muscle force F_0 described in Eq. (7.26) and in Figure 7.7:

- The contractile element force is given by the muscle activation level a and by two dimensionless functions which describe the response of the muscle to its length, f_l, and to its deformation velocity, f_v; in particular, f_l provides the maximum force F_0 when the muscle fibers are long l_0, around which it decreases, while f_v provides zero force for contraction velocity greater than the maximum one v_{max} and increasing force with the lengthening of the muscle fibers;
- The passive element produces an increasing force only when the muscle fibers are lengthening above the optimal fibers length, through the dimensionless function f_{PE}.

$$\begin{matrix} \tilde{l} = \dfrac{l_m}{l_0} \\ \tilde{v} = \dfrac{v_m}{v_{max}} \end{matrix} \rightarrow \begin{cases} F_{CE} = a f_l(\tilde{l}) f_v(\tilde{v}) F_0 \\ F_{PE} = f_{PE}(\tilde{l}) F_0 \end{cases} \rightarrow F_m = (F_{CE} + F_{PE}) \cos\alpha_p \tag{7.27}$$

Therefore, since each muscle force F_{m_n} can be written as the sum between a contractile contribution $\overline{F}_{CE_n}$ multiplied by its activation level

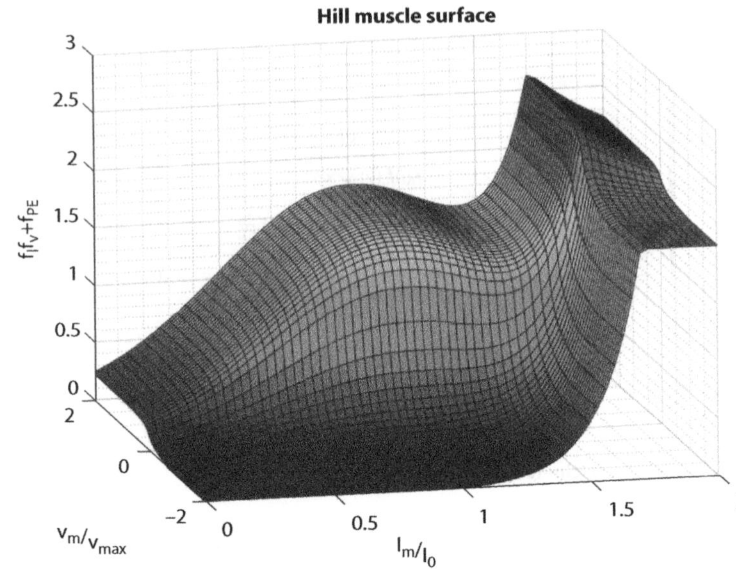

Figure 7.7 The Hill muscle–tendon surface.

a_n and a passive contribution $\bar{F}_{PE_n}$, the whole muscle force vector $\boldsymbol{F}_m$ of Eq. (7.19) can be viewed in (7.28) as a linear function of the activation vector $\boldsymbol{a}$ through the diagonal matrix $\boldsymbol{D}_{CE}$, obtained with $\bar{\boldsymbol{F}}_{CE}$, summed to a passive muscle force vector $\bar{\boldsymbol{F}}_{PE}$, so that the generalized muscle force vector $\boldsymbol{Q}_m$ can be defined through the contractile Jacobian $\boldsymbol{\Phi}_{qCE}$ and the passive muscle force generalized vector $\boldsymbol{Q}_{PE}$, introduced in Eq. (7.29).

$$\begin{cases} \bar{F}_{CE_n} = f_l(\tilde{l}_n) f_v(\tilde{v}_n) F_{0_n} \cos\alpha_{p_n} \\ \bar{F}_{PE_n} = f_{PE}(\tilde{l}_n) F_{0_n} \cos\alpha_{p_n} \end{cases} \rightarrow F_{m_n} = a_n \bar{F}_{CE_n} + \bar{F}_{PE_n} \rightarrow \boldsymbol{F}_m = \boldsymbol{D}_{CE}\boldsymbol{a} + \bar{\boldsymbol{F}}_{PE} \tag{7.28}$$

$$\boldsymbol{Q}_m = \boldsymbol{\Phi}_q^T \boldsymbol{F}_m = \boldsymbol{\Phi}_q^T \boldsymbol{D}_{CE}\boldsymbol{a} + \boldsymbol{\Phi}_q^T \bar{\boldsymbol{F}}_{PE} = \boldsymbol{\Phi}_{qCE}^T \boldsymbol{a} + \boldsymbol{Q}_{PE} \tag{7.29}$$

7.2.6 The Static Optimization

At this level of detail, the generalized force related to the rheonomic constraints in the equation of motion (7.16) can be substituted by the real driving forces $\boldsymbol{Q}_m$, introducing the linear system in (7.30).

$$\begin{cases} Q_m - C_{q_s}^T \lambda_s - C_{q_b}^T \lambda_b = M\ddot{q} - Q_v - Q_e \\ Q_m = \Phi_{q_{CE}}^T a + Q_{PE} \end{cases} \rightarrow \begin{bmatrix} \Phi_{q_{CE}}^T & -C_{q_s}^T \end{bmatrix} \begin{bmatrix} a \\ \lambda_s \end{bmatrix} = M\ddot{q} - Q_v - Q_e - Q_{PE} + C_{q_b}^T \lambda_b \tag{7.30}$$

In (7.30), the driving forces are related to the real contractile element actuators; furthermore, the passive muscle forces are considered out of the driving action; hence, they are going to load the joints, so that the scleronomic Lagrange multipliers $\boldsymbol{\lambda}_s$ are actually the joint reactions. Due to the muscle actuation system redundancy, the linear problem (7.30) is characterized by more unknowns $\boldsymbol{x}$ than equations; therefore, it has to be solved through an optimization algorithm that minimizes a chosen objective function $f(\boldsymbol{x})$ coherent to the muscular recruitment, in this case represented by the squared norm of the muscles' activation vector $\boldsymbol{a}$, addressed with equality constraints $\boldsymbol{c}(\boldsymbol{x})$ in order to satisfy the equation of motion and inequality constraints $\boldsymbol{g}(\boldsymbol{x})$ in order to bind the muscle activation level between 0 and 1. The assembled minimization problem in (7.31) is called static optimization.

$$\begin{cases} x = \begin{bmatrix} a \\ \lambda_s \end{bmatrix} \\ A = \begin{bmatrix} \Phi_{q_{CE}}^T & -C_{q_s}^T \end{bmatrix} \\ b = M\ddot{q} - Q_v - Q_e - Q_{PE} + C_{q_b}^T \lambda_b \end{cases} \rightarrow \begin{cases} f(x) = a^T a \\ c(x) = Ax - b \\ g(x) = \begin{bmatrix} -a \\ a - 1 \end{bmatrix} \end{cases} \rightarrow \begin{cases} \min\limits_{x} f(x) \\ c(x) = 0 \\ g(x) \leq 0 \end{cases} \tag{7.31}$$

7.3 The Lubrication/Contact Model

The resolution of the inverse dynamics associated with the musculoskeletal multibody system leads to the evaluation of the Lagrangian coordinates, velocities, and accelerations; the muscles' activation level; and the joint reactions due to a certain input kinematics governed by the known time evolution of the degrees of freedom: therefore, given a musculoskeletal system moving during a particular physical activity, the multibody output associated with the joint reactions and the relative velocities related to a chosen joint represent the input for the tribological analysis of that joint. In particular, in this work, the hip joint is analyzed.

7.3.1 The Hip Joint

From a kinematical point of view, the hip spherical joint allows three successive rotations (flexion/extension θ_{FE} around the medial/lateral axis z_{ML}, adduction/abduction θ_{AA} around the anterior/posterior axis x_{AP}, and internal/external rotation θ_{IER} around the proximal/distal axis y_{PD}) between the linked pelvis and femur bones; the associated artificial hip implant is composed of the femoral head fixed to the femur and the acetabular cup fixed to the pelvis, where the latter, shaped as a hollow hemisphere, is characterized by the cup axis inclined with respect to the sagittal plane $x_{AP}y_{PD}$ through the inclination angle α_{in}, then its projection on the sagittal plane is inclined with respect to the longitudinal axis x_{AP} through the anteversion angle β_{av}. Hence, once the hip load and the angular velocity vectors, i.e., $\boldsymbol{N}_J$ and $\boldsymbol{\omega}_p$, are calculated in the multibody space, they have to be rotated into the acetabular cup reference frame *Oxyz* shown in Figure 7.8, through the rotation matrices in Eq. (7.32) associated with a rotation around the x- and the z-axes, i.e., $\boldsymbol{R}_x$ and $\boldsymbol{R}_z$.

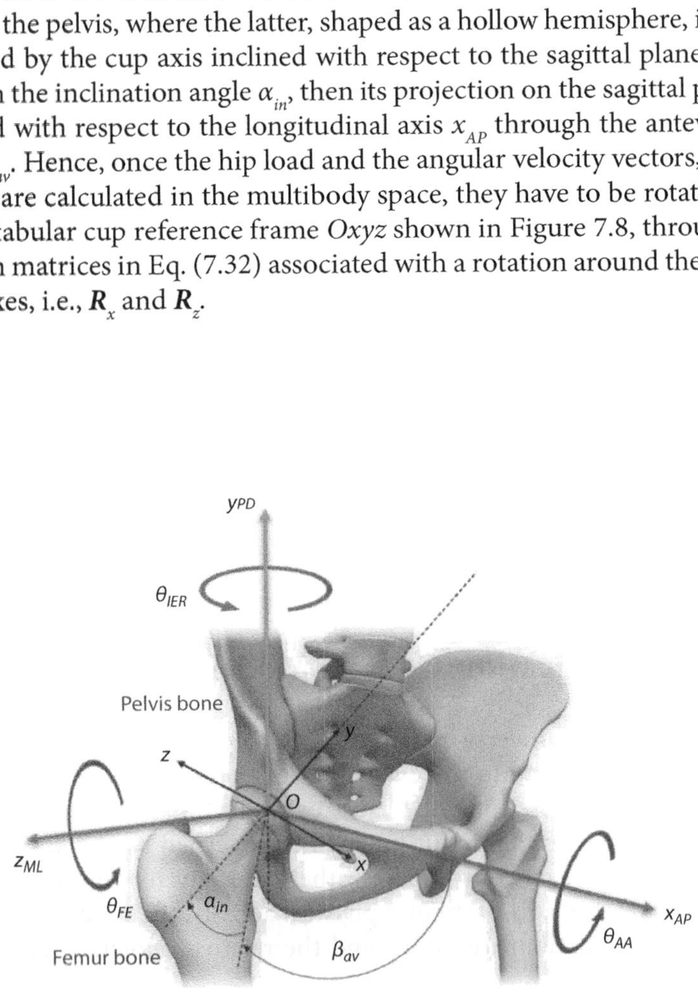

Figure 7.8 Hip reference frames.

$$\boldsymbol{R}_c = \boldsymbol{R}_z\left(\frac{\pi}{2} - \beta_{av}\right)\boldsymbol{R}_x(-\alpha_{in}) \rightarrow \begin{array}{l} \boldsymbol{N} = \boldsymbol{R}_c^T \boldsymbol{N}_J \\ \boldsymbol{\omega} = \boldsymbol{R}_c^T \boldsymbol{\omega}_J \end{array} \tag{7.32}$$

7.3.2 The Reynolds Equation

The lubrication of the hip joint is simulated through the Reynolds equation in (7.33), which governs the lubricant pressure field p over the 2D domain Ω along time t due to the surfaces' separation h and entraining velocity vector $\boldsymbol{v}$ fields; the lubricant, i.e., the synovial fluid, is characterized by the rheological properties of density ρ and dynamical viscosity μ; a constant pressure p_0 is imposed at the boundaries of the domain $\partial\Omega$.

$$\begin{cases} \nabla \cdot \left(\dfrac{\rho h^3}{12\mu} \nabla p \right) = \nabla \cdot (\rho h \boldsymbol{v}) + \dfrac{\partial(\rho h)}{\partial t} \\ p_{\partial\Omega} = p_0 \end{cases} \tag{7.33}$$

With reference to Figure 7.9, the acetabular cup lubrication domain in the reference frame *Oxyz* is defined as a function of the spherical angles θ and φ: a local reference triad is defined through the two tangent and the radial unit vectors, respectively, $\hat{\boldsymbol{t}}_\theta, \hat{\boldsymbol{t}}_\varphi$ and $\hat{\boldsymbol{r}}$ in (7.34), which are necessary to evaluate the surfaces' gap h, the entraining motion $\boldsymbol{v}$, and to write the spherical coordinates Reynolds equation.

$$\begin{bmatrix} \hat{\boldsymbol{t}}_\theta & \hat{\boldsymbol{t}}_\varphi & \hat{\boldsymbol{r}} \end{bmatrix} = \begin{bmatrix} \cos\theta\cos\varphi & -\sin\varphi & \sin\theta\cos\varphi \\ \cos\theta\sin\varphi & \cos\varphi & \sin\theta\sin\varphi \\ -\sin\theta & 0 & \cos\theta \end{bmatrix} \quad \begin{array}{l} 0 \le \theta \le \pi \\ 0 \le \varphi \le \pi \end{array} \tag{7.34}$$

The surfaces' separation h is composed of the sum of:

- A geometrical approach contribution h_g in (7.35) due to the eccentricity vector $\boldsymbol{e}$, locating the femoral head center with respect to the cup frame, depending on the cup and head radii, respectively, R and r, and in particular, on their radial clearance $c = R - r$;

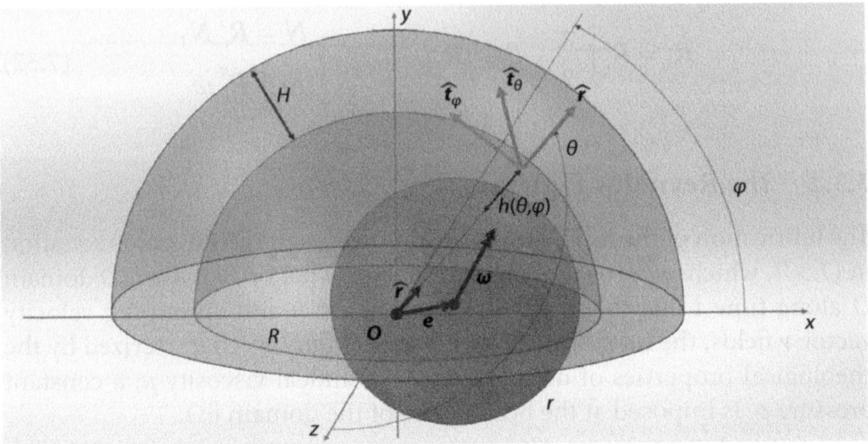

Figure 7.9 Hip tribosystem reference frame.

$$h_g = c - \boldsymbol{e}^T \hat{\boldsymbol{r}} \tag{7.35}$$

- The elastic deformation δ in (7.36) due to the fluid pressure p, calculated through a constrained column model assuming a proportionality between δ and p through a constant k_d dependent on the cup geometrical and mechanical properties (radius R, thickness H, Young modulus E, and Poisson ratio v), assumed to be much softer than the femoral head (hard-on-soft prosthesis);

$$k_d = \frac{R\left[\left(1+\frac{H}{R}\right)^3 - 1\right]}{E\left[\frac{1}{1-2v} + \frac{2}{1+v}\left(1+\frac{H}{R}\right)^3\right]} \rightarrow \delta = k_d p \tag{7.36}$$

- The contact deformation δ_c in (7.37) due to the contact pressure p_c rising to avoid the surfaces' interpenetration above a small boundary layer Δ_b;

$$\delta_c = \begin{cases} \Delta_b - (h_g + \delta + u_w) & if\ h < \Delta_b \\ 0 & if\ h \geq \Delta_b \end{cases} \tag{7.37}$$

- The wear penetration depth u_w in (7.38) cumulated over time, evaluated through the Archard law which assumes that the linear wear rate is proportional to the product between the total pressure and the surfaces' sliding velocity v_{sl} through a wear factor k_w (assumed in this work to increase with respect to its nominal value k_{w_0} with a power (α_w) law growth when the ratio between h and the surfaces' arithmetic average roughness R_a decreases).

$$k_w = k_{w_0}\left(\frac{h}{R_a}\right)^{-\alpha_w} \rightarrow u_w = \int_0^t k_w(p + p_c)v_{sl}dt \tag{7.38}$$

The surfaces' entraining motion field is given by the arithmetic average between the velocities of two points belonging respectively to the cup and to the head located by the same spherical angles θ and φ. The femoral head point velocity vector $\boldsymbol{v}$ field is composed in (7.39) by a translational contribution due to the variation over time of the eccentricity vector $\boldsymbol{e}$ and by a rotational one due to the angular velocity vector $\boldsymbol{\omega}$ of the femoral head with respect to the acetabular cup.

$$\boldsymbol{v} = \dot{\boldsymbol{e}} + \boldsymbol{\omega} \times [(R - h)\hat{\boldsymbol{r}} - \boldsymbol{e}] \tag{7.39}$$

Considering that the acetabular cup is fixed to the *Oxyz* reference frame, the two possible entraining velocities U_θ and U_φ are obtained in (7.40) by projecting $\boldsymbol{v}$ (after the arithmetic average) along the two directions given by the unit vectors $\hat{\boldsymbol{t}}_\theta$ and $\hat{\boldsymbol{t}}_\varphi$; moreover, the sliding velocity v_{sl} for this configuration is reported.

$$U_\theta = \hat{\boldsymbol{t}}_\theta^T \frac{\boldsymbol{v}}{2} \qquad U_\varphi = \hat{\boldsymbol{t}}_\varphi^T \frac{\boldsymbol{v}}{2} \qquad v_{sl} = 2\sqrt{U_\theta^2 + U_\varphi^2} \tag{7.40}$$

Since the mixed elastohydrodynamic lubrication regime is considered, the synovial fluid rheological properties of viscosity μ and density ρ are assumed to vary with the pressure p through, respectively, the Barus and the Dowson–Higginson relationships [51]. Moreover, due to the synovial fluid shear-thinning behavior caused by the aggregation of hyaluronic acid macromolecules included in its composition, the viscosity is further modified through the non-Newtonian Cross model [42, 51]. The used rheological models are reported in Eq. (7.41) and they are governed by the parameters μ_0, μ_∞, k_μ, n_μ, α_μ, ρ_0, a_ρ, and b_ρ.

$$\mu = \left[\mu_\infty + \frac{\mu_0 - \mu_\infty}{1 + k_\mu \left(\frac{v_{sl}}{h} \right)^{n_\mu}} \right] e^{\alpha_\mu p} \qquad \rho = \rho_0 \frac{a_\rho + b_\rho p}{a_\rho + p} \tag{7.41}$$

Finally, the mixed Reynolds equation in spherical coordinates is completely addressed and reported in Eq. (7.42), in which the input of eccentricity $\boldsymbol{e}$ and the angular velocity $\boldsymbol{\omega}$ vectors are highlighted, the cavitation constraint imposing a positive pressure is included, and the contact pressure p_c calculation is considered.

$$\boldsymbol{e} \rightarrow h = h_g + \delta + \delta_c + u_w \qquad \boldsymbol{\omega} \rightarrow \begin{bmatrix} U_\theta \\ U_\varphi \end{bmatrix} = \frac{1}{2} \begin{bmatrix} \hat{\boldsymbol{t}}_\theta^T \\ \hat{\boldsymbol{t}}_\varphi^T \end{bmatrix} \boldsymbol{v}$$

$$\begin{cases} \sin\theta \dfrac{\partial}{\partial\theta}\left(\sin\theta \dfrac{\rho h^3}{12\mu} \dfrac{\partial p}{\partial\theta} \right) + \dfrac{\partial}{\partial\varphi}\left(\dfrac{\rho h^3}{12\mu} \dfrac{\partial p}{\partial\varphi} \right) = R\sin\theta \left[\dfrac{\partial}{\partial\theta}(\sin\theta \rho h U_\theta) + \dfrac{\partial}{\partial\varphi}(\rho h U_\varphi) + R\sin\theta \dfrac{\partial(\rho h)}{\partial t} \right] \\ p(0,\varphi,t) = p(\pi,\varphi,t) = p(\theta,0,t) = p(\theta,\pi,t) = 0 \\ p(\theta,\varphi,t) \geq 0 \end{cases}$$

$$p_c = \frac{\delta_c}{k_d} \tag{7.42}$$

Once the problem (7.42) is solved, the load vector $\boldsymbol{N}$ and the wear volume V_w can be calculated by integrating over the surface domain, respectively, the total pressure oriented by $\hat{\boldsymbol{r}}$ and the wear penetration depth fields, as shown in (7.43).

$$\boldsymbol{N}(t) = \int_0^\pi \int_0^\pi (p(\theta,\varphi,t) + p_c(\theta,\varphi,t)) \hat{\boldsymbol{r}}(\theta,\varphi) R^2 \sin\theta \; d\theta d\varphi$$

$$V_w(t) = \int_0^\pi \int_0^\pi u_w(\theta,\varphi,t) R^2 \sin\theta d\theta d\varphi \tag{7.43}$$

7.3.3 Numerical Resolution

Problem (7.42) is a highly non-linear elliptic partial differential equation (PDE) for each time step, and it has to be solved through numerical iterative techniques. With reference to the authors' work [51] and to Figure 7.10, the Reynolds equation is written through the finite differences on the discretized domain $\{\theta_i\ \varphi_j\}$ by using a cross-stencil involving central spatial derivatives, due to the boundary conditions. The cross-stencil vector $\boldsymbol{x}_s$ of a quantity $x(\theta_i\ \varphi_j)$ is given in (7.44).

$$\boldsymbol{x}_s = [x_{i-1,j} \quad x_{i,j-1} \quad x_{ij} \quad x_{i,j+1} \quad x_{i+1,j}]^T \tag{7.44}$$

With the above numerical assumptions and with the chosen deformation model, the discrete Reynolds equation in each grid point depends on the values assumed by the pressure on the local cross stencil: then, knowing the analytical derivatives of all the involved quantities with respect to pressure p, the discretized Reynolds equation Jacobian will be a five-diagonal band matrix. The Reynolds equation on a grid point I, i.e., R_I (lexicographical order involving i and j), and its derivative with respect to the local cross-stencil pressure vector are given in (7.45), through the definitions of some groups of vectors and matrices associated with both the finite

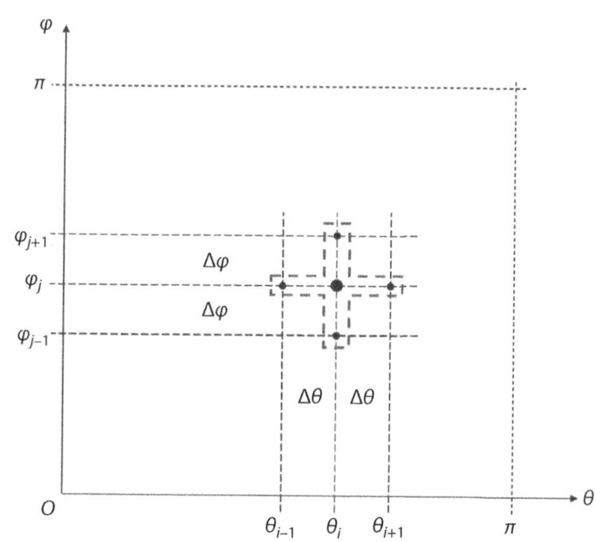

Figure 7.10 The Reynolds equation spherical discrete domain and cross stencil.

differences and the involved quantities: scanning for each I, the equation system $\boldsymbol{R}$ and its Jacobian $\boldsymbol{J}_R$ are assembled.

$$R_I(\boldsymbol{p}_s)=\boldsymbol{f}_s^T\boldsymbol{D}_{2s}\boldsymbol{p}_s-\boldsymbol{D}_{\theta_s}^T\boldsymbol{u}_{\theta_s}-\boldsymbol{D}_{\varphi_s}^T\boldsymbol{u}_{\varphi_s}-\boldsymbol{D}_{t_I}^T\boldsymbol{u}_{t_I}=0\rightarrow\boldsymbol{R}(\boldsymbol{p})=0$$
$$\frac{\partial R_I}{\partial\boldsymbol{p}_s}(\boldsymbol{p}_s)=\boldsymbol{f}_s^T\boldsymbol{D}_{2s}+\boldsymbol{p}_s^T\boldsymbol{D}_{2s}^T\frac{\partial\boldsymbol{f}_s}{\partial\boldsymbol{p}_s}-\boldsymbol{D}_{\theta_s}^T\frac{\partial\boldsymbol{u}_{\theta_s}}{\partial\boldsymbol{p}_s}-\boldsymbol{D}_{\varphi_s}^T\frac{\partial\boldsymbol{u}_{\varphi_s}}{\partial\boldsymbol{p}_s}-\boldsymbol{D}_{t_I}^T\frac{\partial\boldsymbol{u}_{t_I}}{\partial\boldsymbol{p}_s}=\boldsymbol{J}_{R_I}(\boldsymbol{p}_s)\rightarrow\frac{\partial\boldsymbol{R}}{\partial\boldsymbol{p}}=\boldsymbol{J}_R(\boldsymbol{p})$$
$$(7.45)$$

Therefore, the problem $\boldsymbol{R}(\boldsymbol{p}) = \boldsymbol{0}$ in (7.45) is solved through the Newton method by introducing a relaxation factor λ in order to ensure the convergence, as shown in (7.46).

$$\boldsymbol{p}^{(k)}=\boldsymbol{p}^{(k-1)}-\lambda\boldsymbol{J}_R^{-1}(\boldsymbol{p}^{(k-1)})\boldsymbol{R}(\boldsymbol{p}^{(k-1)}) \tag{7.46}$$

As discussed above, the multibody model provides the hip loads and angular velocities to the developed lubrication/contact model, then, since the femoral head eccentricity vector $\boldsymbol{e}$ is unknown, an outer Newton iterative cycle has to be implemented, which imposes the equality between the reference loads $\boldsymbol{N}_{ref}$ obtained by the musculoskeletal simulation and the ones, $\boldsymbol{N}$, obtained by integrating the pressure elaborated through the inner iterative cycle (7.46) for a given eccentricity vector guess. Hence, the function $\boldsymbol{F}$ is written in (7.47) together with the numerical definition of its Jacobian $\boldsymbol{J}_F$.

$$\boldsymbol{F}(\boldsymbol{e})=\boldsymbol{N}(\boldsymbol{e})-\boldsymbol{N}_{ref}=\boldsymbol{0}$$
$$\boldsymbol{J}_F(\boldsymbol{e})=\frac{\partial\boldsymbol{F}}{\partial\boldsymbol{e}}(\boldsymbol{e})\rightarrow[\boldsymbol{J}_F]_{ij}=\frac{F_i(e_j(1+\varepsilon))-F_i(e_j)}{\varepsilon e_j} \tag{7.47}$$

The eccentricity Newton update is reported in (7.48), and it is denoted with the outer iteration index (h).

$$\boldsymbol{e}^{(h)}=\boldsymbol{e}^{(h-1)}-\boldsymbol{J}_F^{-1}(\boldsymbol{e}^{(h-1)})\boldsymbol{F}(\boldsymbol{e}^{(h-1)}) \tag{7.48}$$

7.3.4 Coupling Models

Both models governing the equations are written and simulated in the Matlab computational environment. Summing up the whole pipeline that the developed models follow, with reference to Figure 7.11:

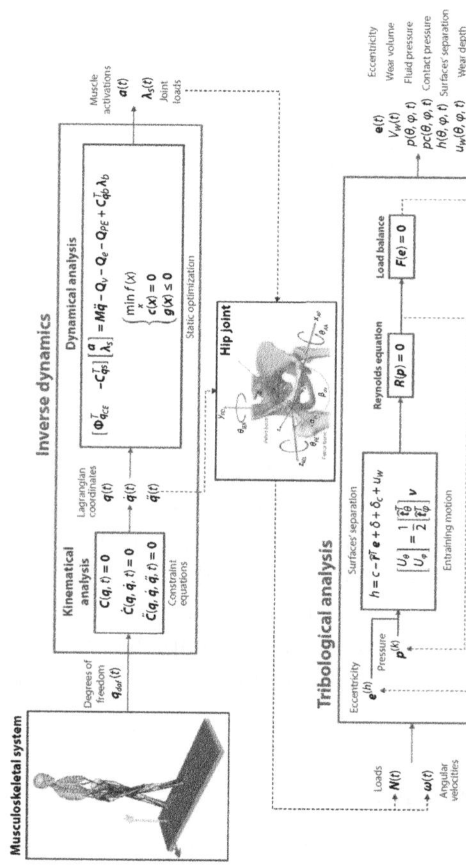

Figure 7.11 Coupling the multibody and the tribological models.

- The algorithm is supplied by a musculoskeletal model moving with a set of known degrees of freedom;
- Then, the constraint equations are assembled and solved in order to elaborate the Lagrangian coordinates' time evolution during the selected kinematics;
- Taking into account the kinematics of the muscles' attachment/wrapping points and the external forces, the equation of motion is assembled and solved through the static optimization simulating a muscles' recruitment criterion;
- The joint reactions and the relative angular velocities of the hip joint are used as input in the lubrication/contact model, in which the two nested iterative cycles associated with, respectively, the Reynolds equation and the load balance solutions, are performed, leading to the final tribological outputs.

7.4 Simulations

The following simulations are associated with a lower limb musculoskeletal system moving during one gait cycle, taken from a model available in the OpenSim software, in which 23 degrees of freedom and 92 muscles are considered, while the calcaneus bodies are subjected to the ground reaction forces. The authors added wrapping objects taken from another OpenSim model, with both cylindrical and ellipsoidal shapes. The artificial hip joint considered was a hard-on-soft implant, made of a ceramic femoral head against polymeric acetabular cup. All the parameters used to characterize the simulations are the same selected in the authors' reference works [50, 51].

7.4.1 Gait Cycle Results

The global multibody output is shown in Figure 7.12, where the whole musculoskeletal lower limb system is reported during the gait cycle, starting when the right heel strikes the ground, then the stance phase goes from 0% to about 60% of the cycle when the right foot toes leave the ground and the remaining is covered by the swing phase; the shown musculoskeletal system is composed of:

- the black rigid segments representing the bodies;
- the black circles representing the joints;

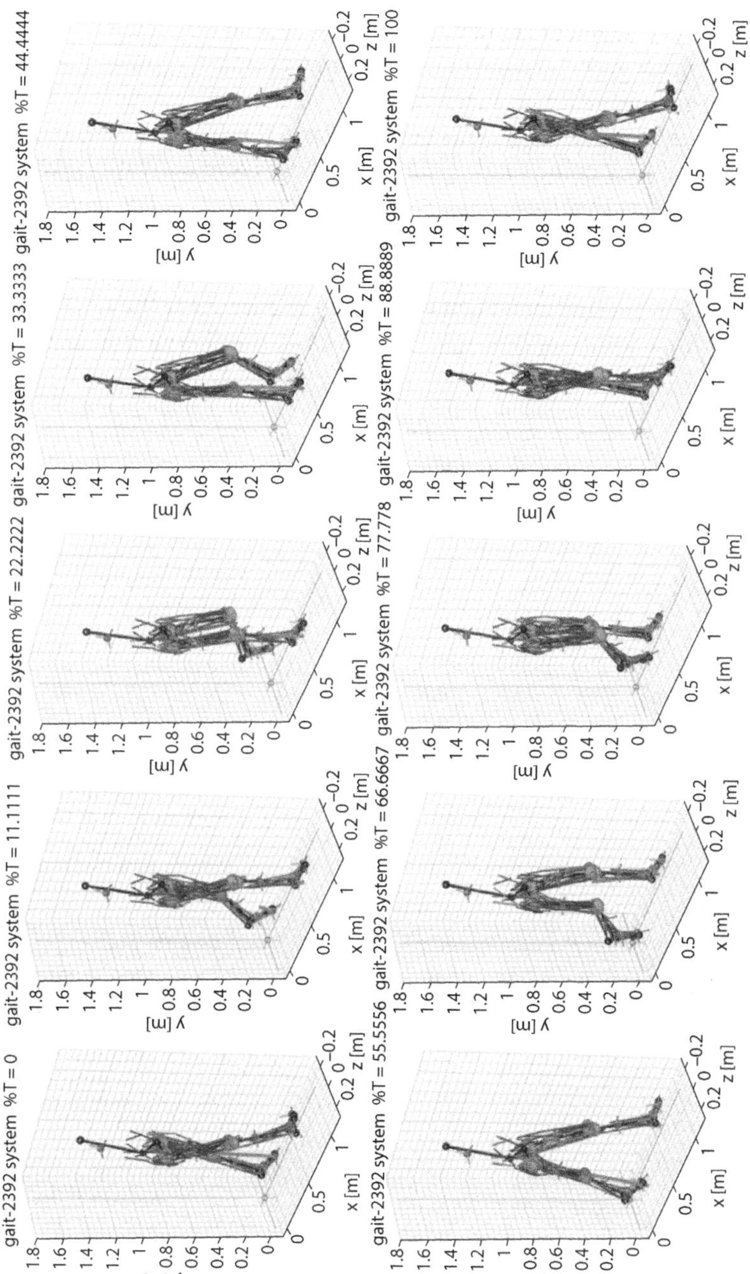

Figure 7.12 The lower limb system during the gait cycle.

- the green circles representing the bodies' center of mass together with the local reference frame;
- the cyan surfaces representing the wrapping objects;
- the thick green arrow on the calcaneus bodies, representing the ground reaction force vector;
- the deformable segments representing the muscles characterized by a color shade going from blue (zero activation) to red (full activation).

Selecting a joint, the algorithm shows the time evolution of the associated Lagrangian multipliers. For example, in Figure 7.13, the ones related to the right knee joint are reported: in this case, the rheonomic multiplier

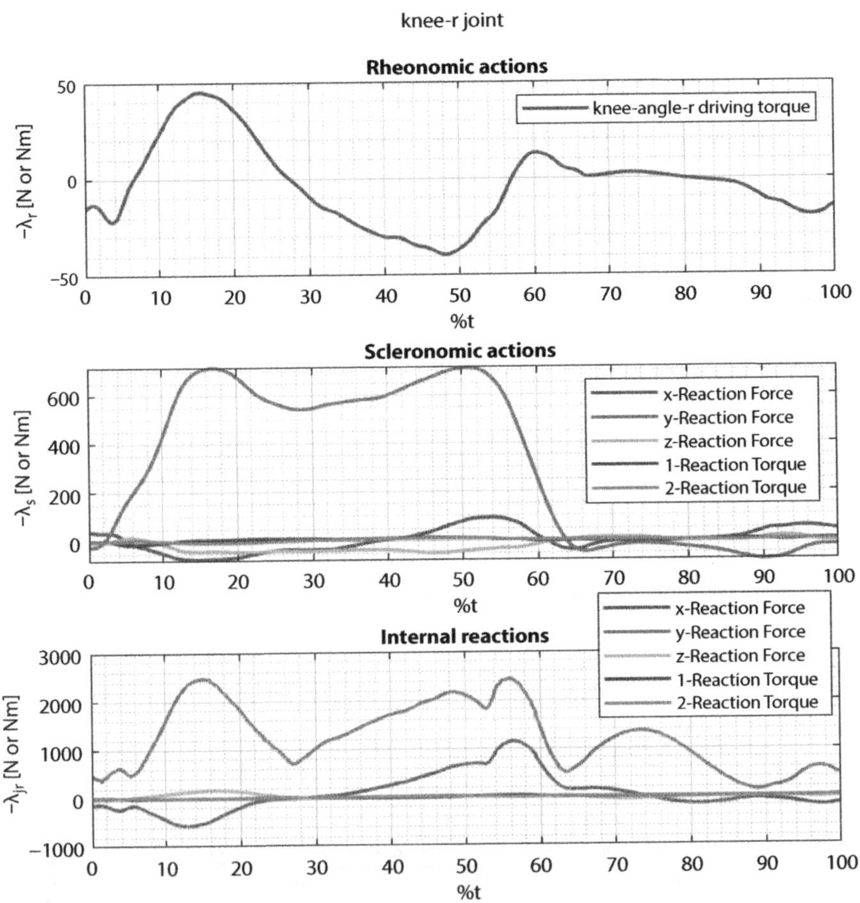

Figure 7.13 The right knee joint Lagrange multipliers.

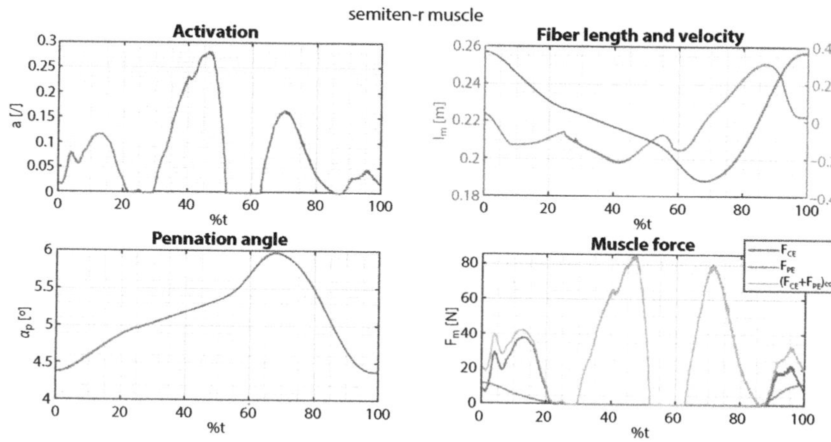

Figure 7.14 The right semitendinosus muscle states.

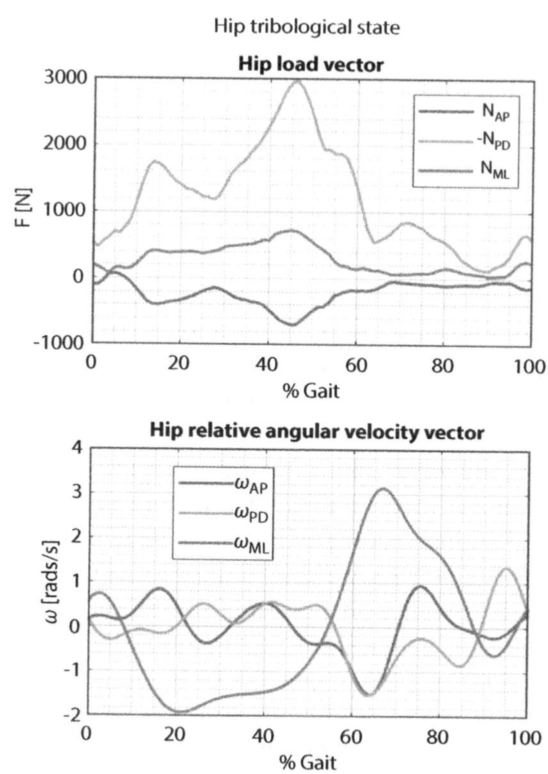

Figure 7.15 The right hip tribological input.

is referred to the net torque that moves the knee angle according to the known kinematics; the first three scleronomic multipliers are associated with the three forces that prevent the relative translation (allowing the ones related to the cam joint), while the last two scleronomic multipliers represent the torques locking the prevented relative rotations; the internal reactions are the scleronomic multipliers after the muscle actuation system is characterized (Eq. (7.30)), so they are the effective knee joint loads.

Moreover, selecting a muscle, the algorithm shows the time evolution of its states (activation, fibers' length and deformation velocity, and pennation angle) and force contributions (contractile, passive, and total muscle forces). For example, in Figure 7.14, the quantities related to the right semitendinosus muscle are depicted, which is a long muscle located at the back of the thigh involved in the knee flexion and in the hip rotations.

The analysis of the tribological configuration of the right hip is performed by selecting the hip joint reactions and angular velocities and rotating them in the joint reference frame, leading to the hip load and motion reported in Figure 7.15, referring to the axes described in Figure 7.8.

7.4.2 Tribological Results

Once the multibody model has elaborated the hip joint tribological configuration, the algorithm provides them with the developed lubrication/contact model. In Figure 7.16, the fields of fluid pressure, contact pressure, surfaces' separation, and wear depth are shown over the acetabular cup inner surface in correspondence to a time instant during the stance phase: here the transition between the full film to the mixed lubrication is clearly visualized looking both at the pressure and at the surfaces' separation fields; furthermore, the starting of the surfaces' wear can be appreciated.

The gait cycle moving the analyzed lower limb musculoskeletal system produces the time evolution of the eccentricity, maximum pressure, minimum separation, and wear volume shown in Figure 7.17: the eccentricity vector (reported in its dimensionless form with respect to the radial clearance) evolves so that the y-component pushes the femoral head against the acetabular cup producing a high contribution of contact pressure during the stance phase, following a shape similar to the vertical load trend; also the maximum pressure achieved within the couple follows the vertical load time evolution reaching a peak magnitude order of about 15 MPa, while the minimum separation decreases rapidly from the beginning of the stance phase until the mixed lubrication phase (direct contact) is established for almost the entire gait cycle; the wear volume results to be negligible in the

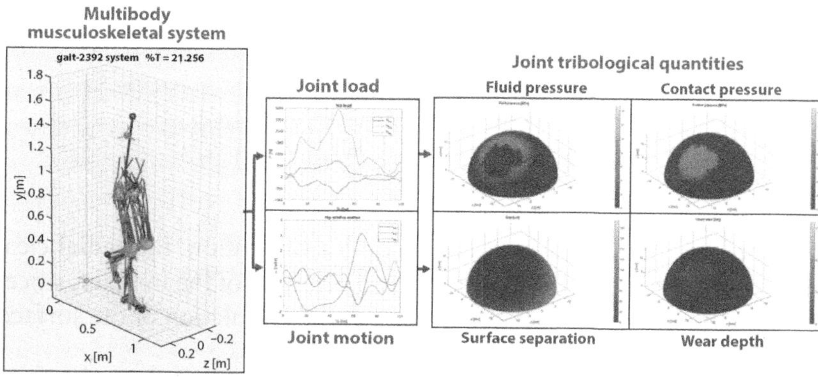

Figure 7.16 The hip joint tribological output fields.

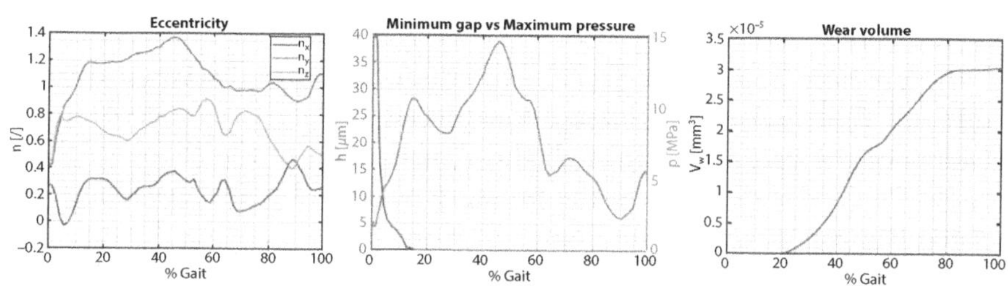

Figure 7.17 The tribological outcomes during the gait cycle.

initial full film lubrication phase and then it starts to rapidly grow with a decreasing slope toward the end of the swing phase.

While the pressure time evolution can be useful during the structural design of the implant, the wear volume can be a powerful predictive indicator of the prosthesis duration. Moreover, the whole set of the analyzed outputs can be referred to a particular subject characterized by the measured musculoskeletal proportions and moving by any kinematics. Finally, a sensitivity analysis can be conducted in order to analyze the implant response due to a variation of a chosen design parameter.

7.4.3 Radial Clearance Sensitivity Analysis

The sensitivity analysis conducted in this work is based on the variation of the radial clearance of the described artificial hip joint, according to Table 7.1.

Table 7.1 Values assumed by the hip radial clearance.

Radial clearance *c* [μm]				
50	75	100	125	150

For the same multibody inverse dynamics simulation, the radial clearance *c* should affect the overall tribological behavior of the implant, since it is included both directly and not directly in the calculation of the surfaces' separation *h*.

In Figure 7.18, the eccentricity vector response is shown: the general behavior described in Figure 7.17 is retrieved; in addition, a decreasing geometrical conformity due to the increasing radial clearance leads to greater pressure; hence, the femoral head approach to the acetabular cup is less severe.

The pressure and the surfaces' separation fields are discussed during both full film and mixed lubrication phases during two selected time instants of the gait cycle, through the visualization of the shape of their profiles in correspondence to the pressure peak grid point and along the related θ and φ directions. In Figure 7.19, the profiles during the full film lubrication phase are shown: in this configuration, the increasing radial clearance causes the growth of the pressure peak together to the shrinking of the pressure profile shape; at the same time, the minimum film thickness decreases and the area of the surfaces subjected to the greater elastic deformation becomes more localized, keeping the converging duct along the relative motion direction (clearly visible along the φ direction in this case).

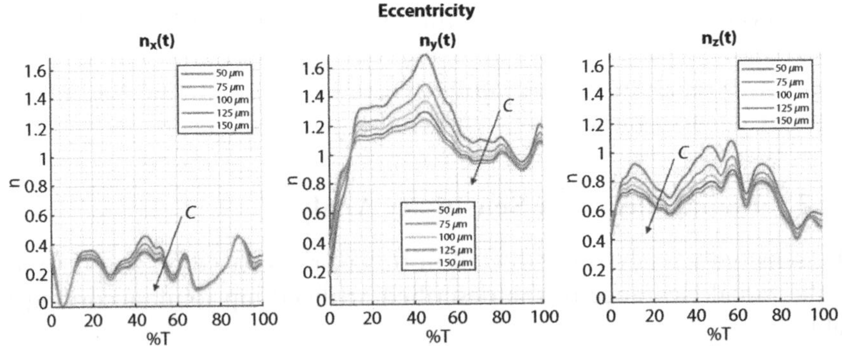

Figure 7.18 Eccentricity response to the radial clearance variation.

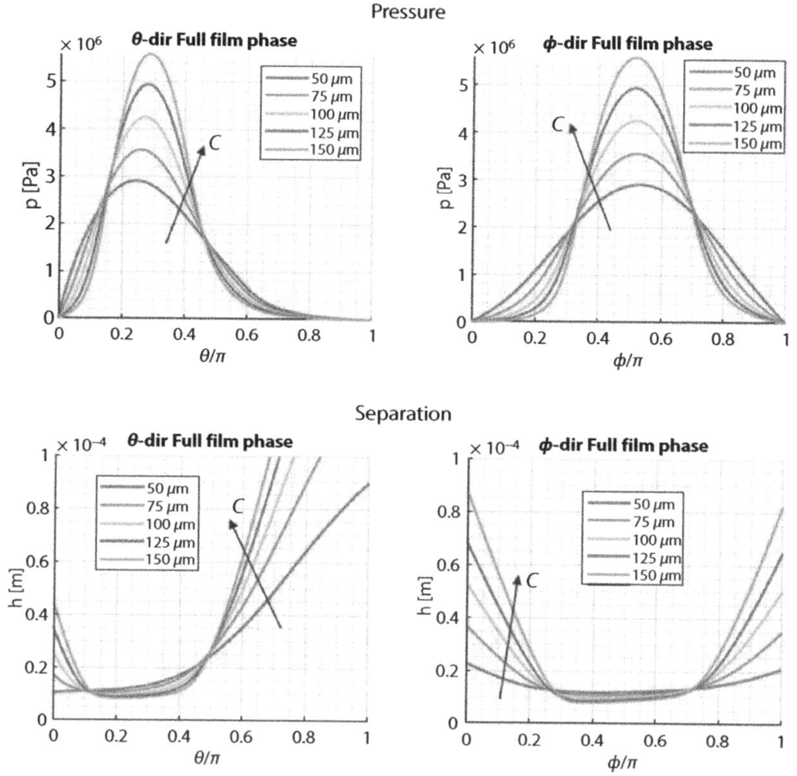

Figure 7.19 Pressure and separation response to the radial clearance variation (full film phase).

In Figure 7.20, the same profiles are reported for the mixed lubrication phase. As for the full film lubrication case, when a direct contact is established, the increase of the radial clearance produces an increasing contact pressure peak followed by a narrower pressure profile shape; as a consequence, the size of the resulting contact area decreases and the transition from the lubricated zone to the direct contact zone occurs more abruptly.

As shown in Figure 7.21, the increasing of the radial clearance produces a general increase in the total maximum pressure, while the minimum film thickness seems to reach the zero value earlier during the gait cycle. The latter behavior is better described by the contact fraction graph, i.e., the ratio between the direct contact area and the total acetabular cup inner surface area, in which the beginning of the contact area is actually anticipated, but the intensity of the ratio is not well correlated along the time with the radial clearance increase: this trend is strongly affected by the logic used by the

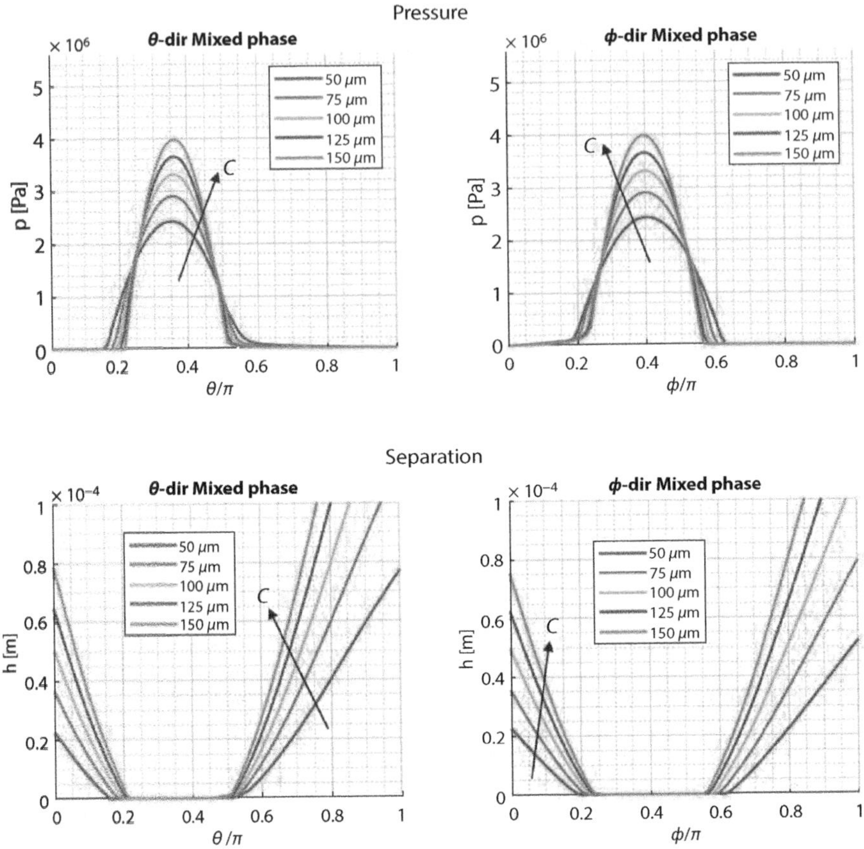

Figure 7.20 Pressure and separation response to the radial clearance variation (mixed phase).

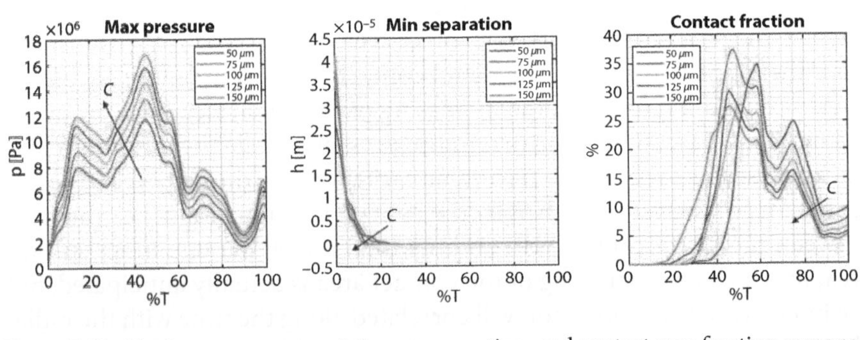

Figure 7.21 Maximum pressure, minimum separation, and contact area fraction response to the radial clearance variation.

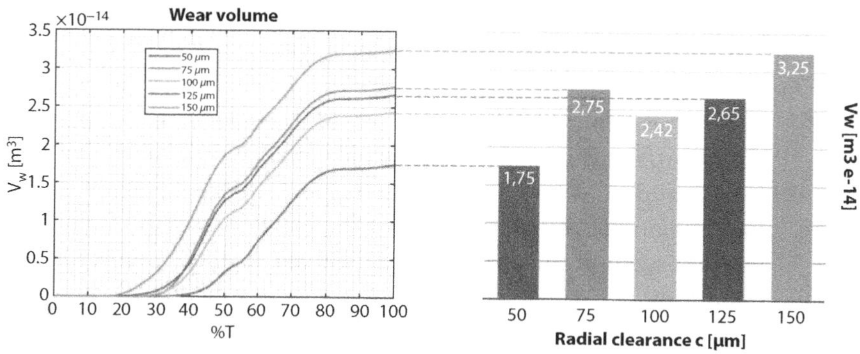

Figure 7.22 Wear volume response to the radial clearance variation.

lubrication/contact model to conceive the coexistence of both lubricated and contact areas over the domain.

The contact area response to the radial clearance is assumed to be mainly responsible for the wear volume behavior shown in Figure 7.22: although the trend reported in Figure 7.17 is retrieved, the wear volume cumulated at the end of the gait cycle is not strictly increasing with the radial clearance growth, because of a single case, i.e., in correspondence to a 75 – μm radial clearance. However, due to the specific logic used by the developed algorithm and to the coherence of all the results shown, the cumulated wear volume graph encourages to assume a general wear volume growth in correspondence to an increasing radial clearance. Moreover, this aspect solicits the authors to investigate further the approximate contact logics within the algorithm.

7.5 Conclusions

The aim of this chapter is to present an *in silico* tool developed by the authors, already proposed to the scientific community by way of their previously published works [50, 51]. The assembled algorithm is composed of two essential macromodels able to simulate respectively the inverse dynamics of multibody musculoskeletal systems and the lubrication/contact state associated with an artificial hip joint included in the original multibody apparatus.

The multibody model is supplied by the musculoskeletal system's geometrical/inertial properties, the topological relations between the bodies and the joints, the parameters associated with the muscles and the wrapping surfaces, the external forces acting on the system, and the time evolution of its degrees of freedom; it works as follows:

- The algorithm assembles the set of constraint equation based on the rheonomic and scleronomic behavior of the joints, and it solves them with respect to the unknown Lagrangian coordinates by way of the Newton method, taking into account the analytical derivatives of all the involved quantities;
- Once the non-linear position problem is solved, the algorithm solves the two linear problems associated with the time derivatives of the constraint equations, obtaining the Lagrangian velocities and accelerations;
- After the kinematical analysis is performed, the dynamical one is characterized by assembling the mass matrix, the generalized force vectors associated with the inertial and the external actions, and by including the Lagrange multipliers related to the fulfillment of the constraint equation during the known motion;
- The muscle actuation system is analyzed by evaluating the kinematics of the straight line muscles, through the knowledge of their attachment points to the bones and of the curved muscles by solving a geodesic problem generating the muscles' wrapping around the encountered surfaces;
- The knowledge of the muscles' kinematics allows to supply the Hill muscle–tendon model, in order to assemble a generalized muscle force vector dependent on their activation level and their passive resistance contribution;
- The static optimization is conducted by the minimization of an objective function describing the muscles' recruitment criteria, in order to calculate the time evolution of the muscles' activation level and the joint reactions associated with the analyzed kinematics.

The output of joint reactions and the relative angular velocities associated with the hip joint are selected and rotated into the artificial hip joint reference frame, so that the lubrication/contact model can be run.

The tribological model is based on the numerical resolution of the Reynolds lubrication equation with reference to the spherical coordinates associated with the acetabular cup of the analyzed artificial hip joint implant. The algorithm works by evaluating the surfaces' separation and the entraining motion fields due to the femoral head eccentricities and angular velocities with respect to the acetabular cup, which generates the pressure field sustaining the coupling; the Reynolds equation and the load balance are solved simultaneously with two nested Newton iterative cycles by discretizing the spherical domain and the equation with the usage of the finite differences. The mixed elastohydrodynamic lubrication regime is considered, while the sensitivity of the synovial fluid rheological properties with respect to pressure, together with its shear-thinning non-Newtonian behavior, was taken into account; furthermore, the degradation of the surfaces is implemented within the separation calculation through the Archard wear model.

The tribological output provided by the model is the fluid/contact pressure field, the surfaces' separation field, the wear penetration depth field, and in general, the wear volume which gives an idea of the longevity of the prosthesis for a certain predominant kinematics.

The developed global tool was used to evaluate the tribological performance of a hard-on-soft hip prosthesis included in a lower limb musculoskeletal system subjected to the gait cycle kinematics, obtaining several data which can be useful during the implant design, also for different musculoskeletal apparatuses and physical activities.

Finally, a sensitivity analysis of the hip implant response to a variation of its radial clearance was performed, discussing the influence of a decreasing geometrical conformity on the main tribological indicators.

The *in silico* tool presented in this chapter acts on the mechanical scale and on the tribological one, giving much information about the biotribological environment surrounding a human prosthesis. In order to achieve a more and more detailed and accurate computational instrument, the authors are deepening some aspects regarding:

- the consideration of tendon dynamics within the Hill muscle–tendon model, in order to simulate fast and intense physical activities;
- the investigation of the different objective function for the muscles' recruitment criteria during the static optimization, so that a neural dynamics could be implemented;
- the usage of the finite element method to build a more realistic deformation model able to process the tensional state of surfaces with comparable deformability and its implementation within the lubrication/contact algorithm [55, 56];

- the study of more advanced numerical techniques and approaches to solve the Reynolds/contact equation with a lower computational time or improved convergence;
- the introduction of a roughness field within the summation composing the surfaces' separation, so that the tribological couple can be characterized also with respect to some morphological parameters.

References

1. Dumbleton, J.H., *Tribology of Natural and Artificial Joints*, Elsevier, Amsterdam, Netherlands, 1981.
2. Ruggiero, A., Milestones in natural lubrication of synovial joints. *Front. Mech. Eng.*, 6, 52, 2020.
3. Popov, V.L., Poliakov, A.M., Pakhaliuk, V., II, Synovial joints. Tribology, regeneration, regenerative rehabilitation and arthroplasty. *Lubricants*, 9, 2, 15, 2021.
4. Pramanik, S., Agarwal, A.K., Rai, K., Chronology of total hip joint replacement and materials development. *Trends Biomater. Artif. Organs*, 19, 1, 15–26, 2005.
5. Carr, A.J., Robertsson, O., Graves, S., Price, A.J., Arden, N.K., Judge, A., Beard, D.J., Knee replacement. *Lancet*, 379, 9823, 1331–1340, 2012.
6. Fisher, J. and Dowson, D., Tribology of total artificial joints. *Proc. Inst. Mech. Eng. H J. Eng. Med.*, 205, 2, 73–79, 1991.
7. Harun, M., Wang, F., Jin, Z., Fisher, J., Long-term contact-coupled wear prediction for metal-on-metal total hip joint replacement. *Proc. Inst. Mech. Eng. J. J. Eng. Tribol.*, 223, 7, 993–1001, 2009.
8. Marian, M., Orgeldinger, C., Rothammer, B., Necas, D., Vrbka, M., Krupka, I., Hartl, M., Wimmer, M.A., Tremmel, S., Wartzack, S., Towards the understanding of lubrication mechanisms in total knee replacements–Part II: Numerical modeling. *Tribol. Int.*, 156, 106809, 2021.
9. Lu, X., Meng, Q., Wang, J., Jin, Z., Transient viscoelastic lubrication analyses of UHMWPE hip replacements. *Tribol. Int.*, 128, 271–278, 2018.
10. Wang, F., Wang, L., Sun, M., Tribological modelling of spherical bearings with complex spherical-based geometry and motion. *WIT Trans. Eng. Sci.*, 66, 3–15, 2010.
11. Askari, E., Flores, P., Dabirrahmani, D., Appleyard, R., Study of the friction-induced vibration and contact mechanics of artificial hip joints. *Tribol. Int.*, 70, 1–10, 2014.
12. Flores, P., Ambrósio, J., Claro, J.P., Dynamic analysis for planar multibody mechanical systems with lubricated joints. *Multibody Syst. Dyn.*, 12, 1, 47–74, 2004.

13. Tian, Q., Zhang, Y., Chen, L., Flores, P., Dynamics of spatial flexible multibody systems with clearance and lubricated spherical joints. *Comput. Struct.*, 87, 13–14, 913–929, 2009.
14. Ruggiero, A., Gomez, E., D'Amato, R., Approximate analytical model for the squeeze-film lubrication of the human ankle joint with synovial fluid filtrated by articular cartilage. *Tribol. Lett.*, 41, 2, 337–343, 2011.
15. Ruggiero, A., Gomez, E., D'Amato, R., Approximate closed-form solution of the synovial fluid film force in the human ankle joint with non-Newtonian lubricant. *Tribol. Int.*, 57, 156–161, 2013.
16. Zivic, F., Affatato, S., Trajanovic, M., Schnabelrauch, M., Grujovic, N., Choy, K.L., *Biomaterials in Clinical Practice: Advances in Clinical Research and Medical Devices*, Springer, Berlin, Germany, 2017.
17. Affatato, S., Ruggiero, A., Merola, M., Advanced biomaterials in hip joint arthroplasty. A review on polymer and ceramics composites as alternative bearings. *Compos. B Eng.*, 83, 276–283, 2015.
18. Merola, M. and Affatato, S., Materials for hip prostheses: A review of wear and loading considerations. *Materials*, 12, 3, 495, 2019.
19. Aherwar, A., Singh, A.K., Patnaik, A., Current and future biocompatibility aspects of biomaterials for hip prosthesis. *AIMS Bioeng.*, 3, 1, 23–43, 2015.
20. Mattei, L., Di Puccio, F., Piccigallo, B., Ciulli, E., Lubrication and wear modelling of artificial hip joints: A review. *Tribol. Int.*, 44, 5, 532–549, 2011.
21. Weinhandl, J.T. and Bennett, H.J., Musculoskeletal model choice influences hip joint load estimations during gait. *J. Biomech.*, 91, 124–132, 2019.
22. van Veen, B., Montefiori, E., Modenese, L., Mazzà, C., Viceconti, M., Muscle recruitment strategies can reduce joint loading during level walking. *J. Biomech.*, 97, 109368, 2019.
23. Bergmann, G., Bender, A., Dymke, J., Duda, G., Damm, P., Standardized loads acting in hip implants. *PloS One*, 11, 5, e0155612, 2016.
24. Quental, C., Folgado, J., Ambrósio, J., A window moving inverse dynamics optimization for biomechanics of motion. *Multibody Syst. Dyn.*, 38, 2, 157–171, 2016.
25. Ojeda, J., Martínez-Reina, J., Mayo, J., The effect of kinematic constraints in the inverse dynamics problem in biomechanics. *Multibody Syst. Dyn.*, 37, 3, 291–309, 2016.
26. Quental, C., Azevedo, M., Jorge, A., Goncalves, S., Folgado, J., Influence of the musculotendon dynamics on the muscle force-sharing problem of the shoulder—A fully inverse dynamics approach. *J. Biomech. Eng.*, 140, 7, 071005, 2018.
27. Delp, S.L., Loan, J.P., Hoy, M.G., Zajac, F.E., Topp, E.L., Rosen, J.M., "An interactive graphics-based model of the lower extremity to study orthopaedic surgical procedures. *IEEE Trans. Biomed. Eng.*, 37, 8, 757–767, 1990.
28. Yamaguchi, G.T. and Zajac, F.E., A planar model of the knee joint to characterize the knee extensor mechanism. *J. Biomech.*, 22, 1, 1–10, 1989.

29. Anderson, F.C. and Pandy, M.G., A dynamic optimization solution for vertical jumping in three dimensions. *Comput. Methods Biomech. Biomed. Eng.*, 2, 3, 201–231, 1999.
30. Anderson, F.C. and Pandy, M.G., Dynamic optimization of human walking. *J. Biomech. Eng.*, 123, 5, 381–390, 2001.
31. Arnold, E.M., Ward, S.R., Lieber, R.L., Delp, S.L., A model of the lower limb for analysis of human movement. *Ann. Biomed. Eng.*, 38, 2, 269–279, 2010.
32. Thelen, D.G., Adjustment of muscle mechanics model parameters to simulate dynamic contractions in older adults. *J. Biomech. Eng.*, 125, 1, 70–77, 2003.
33. Thelen, D.G., Anderson, F.C., Delp, S.L., Generating dynamic simulations of movement using computed muscle control. *J. Biomech.*, 36, 3, 321–328, 2003.
34. Seth, A., Hicks, J., Uchida, T., Habib, A., Dembia, C., Dunne, J., Ong, C., DeMers, M., Rajagopal, A., Millard, M., Hamner, S., OpenSim: Simulating musculoskeletal dynamics and neuromuscular control to study human and animal movement. *PloS Comput. Biol.*, 14, 7, e1006223, 2018.
35. Wang, F., Brockett, C., Williams, S., Udofia, I., Fisher, J., Jin, Z., Lubrication and friction prediction in metal-on-metal hip implants. *Phys. Med. Biol.*, 53, 5, 1277, 2008.
36. Jin, Z. and Dowson, D., A full numerical analysis of hydrodynamic lubrication in artificial hip joint replacements constructed from hard materials. *Proc. Inst. Mech. Eng. C J. Mech. Eng. Sci.*, 213, 4, 355–370, 1999.
37. Askari, E. and Andersen, M.S., A modification on velocity terms of Reynolds equation in a spherical coordinate system. *Tribol. Int.*, 131, 15–23, 2019.
38. Jalali-Vahid, D., Jin, Z., Dowson, D., Elastohydrodynamic lubrication analysis of hip implants with ultra high molecular weight polyethylene cups under transient conditions. *Proc. Inst. Mech. Eng. C J. Mech. Eng. Sci.*, 217, 7, 767–777, 2003.
39. Gao, L., Wang, F., Yang, P., Jin, Z., Effect of 3D physiological loading and motion on elastohydrodynamic lubrication of metal-on-metal total hip replacements. *Med. Eng. Phys.*, 31, 6, 720–729, 2009.
40. Udofia, I.J. and Jin, Z.M., Elastohydrodynamic lubrication analysis of metal-on-metal hip-resurfacing prostheses. *J. Biomech.*, 36, 4, 537–544, 2003.
41. Wang, F. and Jin, Z., Transient elastohydrodynamic lubrication of hip joint implants. *J. Tribol.*, 130, 1, 011007, 2008.
42. Gao, L., Dowson, D., Hewson, R.W., A numerical study of non-Newtonian transient elastohydrodynamic lubrication of metal-on-metal hip prostheses. *Tribol. Int.*, 93, 486–494, 2016.
43. Jalali-Vahid, D. and Jin, Z.M., Transient elastohydrodynamic lubrication analysis of ultra-high molecular weight polyethylene hip joint replacements. *Proc. Inst. Mech. Eng. C J. Mech. Eng. Sci.*, 216, 4, 409–420, 2001.
44. Jalali-Vahid, D., Jagatia, M., Jin, Z., Dowson, D., Prediction of lubricating film thickness in a ball-in-socket model with a soft lining representing human

natural and artificial hip joints. *Proc. Inst. Mech. Eng. J. J. Eng. Tribol.*, 215, 4, 363–372, 2001.
45. Wang, F. and Jin, Z., Prediction of elastic deformation of acetabular cups and femoral heads for lubrication analysis of artificial hip joints. *Proc. Inst. Mech. Eng. J. J. Eng. Tribol.*, 218, 3, 201–209, 2004.
46. Jalali-Vahid, D., Jagatia, M., Jin, Z.M., Dowson, D., Prediction of lubricating film thickness in UHMWPE hip joint replacements. *J. Biomech.*, 34, 2, 261–266, 2001.
47. Wang, F. and Jin, Z., Lubrication modelling of artificial hip joints: From fluid film to boundary lubrication regimes, in: *ASME*, vol. pp, pp. 605–611, 2004.
48. Zhu, D., On some aspects of numerical solutions of thin-film and mixed elastohydrodynamic lubrication. *Proc. Inst. Mech. Eng. J. J. Eng. Tribol.*, 221, 5, 561–579, 2007.
49. Zhu, D. and Hu, Y.Z., The study of transition from elastohydrodynamic to mixed and boundary lubrication, in: *The Advancing Frontier of Engineering Tribology, Proceedings of the 1999 STLE/ASME HS Cheng Tribology Surveillance*, pp. 150–156, 1999.
50. Ruggiero, A. and Sicilia, A., Musculoskeletal multibody algorithm based on a novel rheonomic constraints definition applied to the lower limb. *J. Biomech. Eng.*, 144, 8, 081010, 2022.
51. Ruggiero, A. and Sicilia, A., A mixed elasto-hydrodynamic lubrication model for wear calculation in artificial hip joints. *Lubricants*, 8, 7, 72, 2020.
52. Ruggiero, A. and Sicilia, A., A novel explicit analytical multibody approach for the analysis of upper limb dynamics and joint reactions calculation considering muscle wrapping. *Appl. Sci.*, 10, 21, 7760, 2020.
53. Ruggiero, A. and Sicilia, A., Lubrication modeling and wear calculation in artificial hip joint during the gait. *Tribol. Int.*, 142, 105993, 2020.
54. Ruggiero, A., Sicilia, A., Affatato, S., *In silico* total hip replacement wear testing in the framework of ISO 14242-3 accounting for mixed elasto-hydrodynamic lubrication effects. *Wear*, 460, 203420, 2020.
55. Ruggiero, A. and Sicilia, A., Mathematical development of a novel discrete hip deformation algorithm for the *in silico* elasto-hydrodynamic lubrication modelling of total hip replacements. *Lubricants*, 9, 4, 41, 2021.
56. Ruggiero, A. and Sicilia, A., Implementation of a finite element deformation model within an elasto-hydrodynamic lubrication numerical solver for a ball in socket tribopair. *Front. Mech. Eng.*, 8, 909156, 2022.

8

Techniques of Biopolymer and Bioceramic Coatings on Prosthetic Implants

Sikta Panda[1], Chandan Kumar Biswas[1] and Subhankar Paul[2]*

[1]Department of Mechanical Engineering, National Institute of Technology, Rourkela, Odisha, India
[2]Department of Biotechnology and Medical Engineering, National Institute of Technology, Rourkela, Odisha, India

Abstract
Bone injuries are common consequences after sudden shocks like accidents and diseases in the human skeletal system. In general, bone injuries are curable if they occur at a younger age, but with aging, the injuries often become unrecoverable and lead to bone replacements or synthetic bone implants. Three types of materials are used as prosthetic implant materials, i.e., metals, polymers, and ceramics. Titanium (Ti) is known to possess mechanical properties (bending strength and fatigue resistance) similar to the human bone and bears a very high strength-to-weight ratio. Moreover, being bioinert and non-toxic, it is used as a biomaterial in orthopedic as well as dental implants. Generally, Ti alloys such as Ti6Al4V ELI (Ti grade 23) commercially pure titanium (CP Ti) and Ti6Al4V (Ti grade 5) are considered for this purpose. However, titanium implants like other metallic implants take much time for cell adhesion and proliferation (lower osseointegration) and sometimes become a source of toxic ions if exposed to corrosive body fluids. Dealing with these issues led to the application of protective bioactive coating on the titanium substrates. All the available coating techniques and bioactive materials have their own sets of merits and demerits. These challenges include delamination, toxicity, drug loading capacity, mechanical abrasion, difficult rigid sanction pathways, and high production overheads. In this chapter, the performances of various types of coating techniques and the suitability of materials to be deposited are reviewed based on their biocompatibility, corrosion resistance, stability, and cost-effectiveness based on the literature available in recent years.

**Corresponding author*: spaul@nitrkl.ac.in

Jay Prakash Srivastava, Dražan Kozak, Vinayak Ranjan, Pankaj Kumar, Ranjan Kumar and Shubham Tayal (eds.) Mechanical Engineering in Biomedical Applications: Bio-3D Printing, Biofluid Mechanics, Implant Design, Biomaterials, Computational Biomechanics, Tissue Mechanics, (231–260) © 2024 Scrivener Publishing LLC

Keywords: Orthopedic implant, coating, biocompatibility, corrosion resistance

8.1 Introduction

The American Academy of Orthopaedic Surgeons has proposed that nearly 6.3 million bone injuries occur each year in the U.S. and men are more susceptible to bone injuries than women [1]. India has also witnessed amplified figures of skeletal injuries in recent years and ranks second in Asia in terms of the number of bone injuries [2]. These injuries may occur due to accidents or ailments in human skeletal systems. Bone injuries are generally irrecoverable, particularly in older patients. In such circumstances, the patients are required to undergo replacement surgery with synthetic bone implants. The requirement for a material to be used as an implant is as follows. It should be fracture-resistant, strong, lightweight, corrosion-resistant, biocompatible, long-lasting, non-toxic, osseointegrated (bone's bonding with the implant), flexible, and elastic as the human bone. The modulus of elasticity of metal necessarily is nearer to actual human bone tissue.

Owing to these properties, the metals used in medical practices for the commercial production of artificial bone implants are titanium and its alloys, stainless steels (SS), cobalt–chromium (Co–Cr) alloys, etc. Along with these metals, refractory metals, viz., hafnium, tantalum, niobium, and rhenium, can be used for the same purpose in the future according to Matsuno *et al.* [3]. However, the integration of metal implants with the surrounding tissues is slower (lesser osseointegration property) and may cause a chemical reaction when the metallic particles leached into the body environment. Other problems associated with implants are infection and inflammation. These can cause adverse effects on the implanted body as well as lessen the lifespan of the implant. These problems can be overcome by diverse approaches such as changing the surface properties by various surface modification techniques. Coating stands among the most used methods for surface modification that has been discussed here.

The coating of implants may employ numerous methods like 3D printing, spin coating, dip coating, biomimetic deposition, drop casting, plasma spraying, sol–gel deposition, chemical vapor deposition (CVD), physical vapor deposition (PVD), electrospinning, electrophoretic deposition, and anodization [4–9]. All these processes are used with respect to their ease of application and depending on the substrate and coating material. A good process of coating should provide uniform coating thickness and a long-lasting stable coating onto the substrate. It should not leave any stress on the implant surface and should be cost-effective and easy to be used commercially. Coatings are often used as carriers for various drugs to deal with infection and inflammation. Different biopolymers and bioceramics which can act as bioactive coatings for implants will be discussed here. Such coatings include polyethylene glycol (PEG), polycarbolactone (PLA) and polyglycolic acid (PGA) chitosan, poly(lactic-co-glycolic acid) (PLGA), and hydroxyapatite (HAP) [10]. Coatings may also use composites (formed with polymers and ceramics) based on the requirement of the implant. In the following section, the need for bioactive coating on a metal implant surface will be discussed in detail and the various processes of coating these implants will also be discussed. The advantages and disadvantages of these procedures will also be presented taking into account recent articles published by the concerned researchers.

8.2 Driving Factors for the Application of Coatings

In the previous section, the need for implants was discussed. However, it is a well-known fact that uncoated metal implants pose risks of bacterial colonization and implant rejection. When an implant is coated with a bioactive substance, the rate at which the coating dissolves could extend its lifespan. Osteoclasts reabsorb apatite (mainly HAP) coatings that dissolve quickly. The osteoblasts replace the resorbed coverings with new bone as part of the regular bone remodeling process. The rate at which coatings dissolve is important because a coating that dissolves can promote bone development. Rapid coating deterioration, however, could potentially result in implant loosening and loss of fixation. The second method of implant catastrophes is osteolysis, as a result of corrosion without the presence of surface coatings. Furthermore, certain coatings may also equip the implants with other functional abilities like controlled drug release and anticancer properties. The requirement of application of coating is thus very essential, especially for metallic implants, and it will be discussed here.

8.2.1 Corrosion of Metal Implants and Its Categories

Due to their suitable physical attributes such as high strength, ductility, and toughness, metals are the most sought for implants. Implant corrosion triggered by the leaching of constituent ions into the adjacent tissues is a noteworthy shortcoming associated with metallic implants [11]. This can give rise to health complications like metalloma. The existence of an inert oxide layer on the exterior of metal implants necessarily limits the corrosion progression by restraining the ion release. Nevertheless, the complicated physiology of body fluids along with temperature creates a saline solution with oxygen, making the surroundings unreceptive and causing implant corrosion. These metal ions seeping out to cells have been discovered to be extremely hazardous [12]. The corrosion can be reduced significantly by the use of coatings [6–9]. Various types of implant corrosion have been discussed as follows.

8.2.1.1 Uniform Attack

The term "uniform attack" describes the unavoidable corrosion that any metal experiences when exposed to electrolytic solutions. Serum proteins can cause a two- to ten-fold increase in uniform attack [13]. Titanium is the least susceptible metal to uniform attack.

8.2.1.2 Fretting Corrosion

The slightest movement of the surfaces of two metals in contact causes the passive film to break down, allowing the area to corrode quickly. This is known as fretting corrosion. Small pieces of metal are ripped off from the surfaces by this type of corrosion that occurs at the points where materials undergo stress due to vibration and sliding contact. Weinstein *et al.* [14], from their investigation of numerous implant components, found that SS implants demonstrated significant levels of fretting corrosion.

8.2.1.3 Galvanic Corrosion

Galvanic corrosion is often referred to describe the dissolution of alloys caused by macroscopic changes in electrochemical voltages, typically due to different metals being close to one another. The inappropriate use of metals, such as a screw made up of titanium or its alloy in contact with an SS joint, may cause this type of corrosion [15].

8.2.1.4 Pitting Corrosion

Pitting is another typical corrosion problem with metallic implants. Pits develop on the metal surface in this type of confined, symmetric corrosion. Metals are especially prone to pitting in surroundings with chloride. When the passive film is damaged, the little oxygen and the chloride ions significantly impair the passivation film's ability to cure itself. The pit's high current density causes pinholes to form on the metal surface, while the majority of the surface is left undamaged. SS is especially susceptible to pitting as a result of the amalgamation of different metals and the creation of grain boundaries [16]. Scratches or handling damage may also be the cause of it.

8.2.1.5 *Crevice Corrosion*

The most noticeable type of corrosion is without a doubt crevice corrosion, a sort of local corrosion brought on by variations in O_2 concentration and electrolytes or variations in pH in a small area, like the gap between a screw and a plate [17]. SS multicomponent devices frequently experience crevice corrosion, which also causes severe tissue reactions and subsequently necessitates the retraction of the device. Differential aeration of the stagnant solution is the main contributor to crevice corrosion. Corrosion moves more quickly because the oxygen-deficient areas inside the fissure turn anodic in comparison to the material as a whole.

8.2.1.6 *Leaching*

Chemical variations within the grains themselves (not at the grain boundaries) cause this type of corrosion. Leaching occurs in alloys that are multiphasic, such as the F582 cobalt-base alloy with 35% Ni content [13].

8.2.1.7 *Stress Corrosion Cracking (SCC)*

SCC is uncommon; however, cases have been documented earlier [18]. It is brought on by residual tensile strains following an implant's bending. It is a phenomenon when a metal gets stressed and fails from corrosion with a considerably lesser level of stress than the normal level in a certain environment (particularly one that is chloride-rich). Corrosion of the metal surface can be checked by applying the appropriate coatings which are discussed in Table 8.1.

8.2.2 Bioactivity of the Surface

The unhindered attachment and proliferation of the host cell on an implant avoiding the complications related to bacterial infection, angiogenesis, and immune rejection is the most important aspect of any implant-related surgery. The implant surface's bioactivity plays a vital role in its success. Metallic implants are generally bioinert, and their surfaces may also host bacteria colonies. These issues can be resolved by applying bioactive coatings which are discussed in Table 8.1.

8.2.2.1 Immune Rejection, Osteoinduction, Osteoconduction, and Osseointegration

Osteoinduction refers to the process through which undifferentiated and primitive cells are by some means encouraged to transform into bone-forming cell lineage. Osteoconduction describes how the bone develops on a surface. A surface that encourages bone development on the surface or into the pores or pipelines is said to be osteoconductive. The anchoring of an implant with the growth of bone tissue surrounding it, without the development of tumor tissue, is known as osseointegration [19]. The implants may also be rejected by the immune cells treating it as a foreign body. This type of implant failure is triggered by cytokines and immune cells especially macrophages and dendritic cells [20]. The application of coatings has been reported to increase osteoinduction, osteoconduction, osseointegration, and hemocompatibility, thereby decreasing immune rejection [21–24].

8.2.2.2 Toxicity and Bacterial Biofilm Formation

Toxic elements in an implant may lead to cell death and complications related to hemotoxicity and cytotoxicity [25, 26]. However, scientists have established that bioactive coatings help in reducing the toxicity of the implants [27–31]. Similarly, the formation of a bacterial biofilm can cause implant catastrophe and revised surgery due to infection of the surgery wound. This is avoided by functionalized and drug-loaded coatings on the implant surface [28, 32].

Table 8.1 shows some of the recent articles related to the application of coatings to minimize the corrosion of implants and enhance their bioactivity.

Table 8.1 enumerates some important driving factors for the application of bioactive coating on an implant surface which has also been schematically explained in Figure 8.1. A suitable coating results in improved topography and roughness of the surface conceals toxic materials, and strengthens the bonding strength of the material by enhancing adhesion. To facilitate the bone-creating capacity of the cells and form biological interlinking of the implant with the neighboring tissues, composite or polymeric artificial bone-like materials are introduced into the surface of the bone during rapid bone healing. Another technique for speeding up

Table 8.1 Types of coatings and their advantages.

Author	Substrate	Coating	Inference
Panda *et al.* [28]	Ti6Al4V	Sr–Nb-doped hydroxyapatite (HAP)	Increase in corrosion resistance and cell proliferation
Acheson *et al.* [33]	Mg alloy	Calcium phosphate (CaP)	The coating thickness can regulate the rate of corrosion for improved bioactivity
Panda *et al.* [34]	Ti6Al4V	Nb-doped hydroxyapatite (HAP)	Increase in corrosion resistance, biofilm inhibition capacity, and lower hemotoxicity and cytotoxicity
Panda *et al.* [35]	Ti6Al4V	TiO_2/hydroxyapatite (HAP)	Increased biofilm inhibition capacity, corrosion resistance, and higher biocompatibility
Panda *et al.* [36]	Ti6Al4V	Chitosan/bovine serum albumin and gold nanoparticle	Improved bioactivity as compared with the uncoated implant
Pobloth *et al.* [37]	Zirconia-toughened alumina	HAP/bioglass (BG)	Found to improve cancellous osseointegration
Moses *et al.* [38]	Stainless steel	Silk-bioactive glass	Improved osseointegration and bactericidal properties
Kang *et al.* [39]	Pure titanium	Zinc-TiO_2	Zn-doped TiO_2 coatings showed decent biocompatibility and antimicrobial properties

(*Continued*)

Table 8.1 Types of coatings and their advantages. (*Continued*)

Author	Substrate	Coating	Inference
Nemcakova *et al.* [40]	Ti6Al4V	Nanocrystalline diamond/BMP-7	Accelerates osseointegration *in vivo* and extracellular matrix mineralization *in vitro*
Xue *et al.* [41]	Pure titanium	Fe_2O_3–FeOOH nanograins/ Zn-doped TiO_2	Fe_2O_3–FeOOH layer progresses biointegration in an infected mouse enhanced by NIR irradiation
Zhang *et al.* [42]	Pure titanium	Yb/Er-doped titanium dioxide (TiO2@UCN/Qr/ LA)	Coating found to prevent the development of bone tumors, get rid of biofilms, and encourage angiogenesis and osteogenic differentiation
Milan *et al.* [43]	Ti6Al4V	Copper/diamond-like carbon (DLC)	Coating demonstrated improved antibacterial capabilities, increased angiogenesis and osteogenesis activities, and corrosions resistance
Geng *et al.* [44]	Pure titanium	miR-21 (biomarker) nanocapsule	Enhanced angiogenesis, bone-implant osseointegration *in vivo*, and increased osteoblast anchorage and MSC angiogenic development *in vitro*

(*Continued*)

Table 8.1 Types of coatings and their advantages. (*Continued*)

Author	Substrate	Coating	Inference
Saha *et al.* [45]	Ti6Al4V	Silk fibroin-coated TiO_2	Endorsed cell adhesion
Bakhsheshi-Rad *et al.* [46]	Mg–Zn–Ca alloy	Chitosan/silver sulfadiazine/ multiwall carbon nanotubes	Decent biocompatibility and antibacterial effects
Jensen *et al.* [47]	Stainless steel	Silica, hyperbranched polyethyleneimine, gentamicin	Averted the onset of severe osteomyelitis by *Staphylococcus aureus*
Sari *et al.* [48]	Pure titanium	Carbonated hydroxyapatite	Increased cell viability
Wang *et al.* [49]	Pure titanium	Calcium phosphate/ collagen	Increased osteogenic activity
Sivamaran *et al.* [50]	Ti6Al4V	Al_2O_3/graphene/HAP	Increased apatite nucleation, antibacterial and tribomechanical (anti-wear and microhardness) properties

bone healing is the assimilation of biological molecules like proteins onto the implant's surface.

8.3 The Development of Implant Coatings

Stainless steel and alloys based on Co, Ti, and Mg are among the most commonly utilized materials for the manufacturing of implants. AISI 316, 316L, 420, and 440 are the varieties of stainless steel that are most frequently utilized in biomedical applications. Co-based alloys, also known as CoCrMo in biomedical applications, are often used as artificial joint support systems because of their strong anticorrosive nature and wear resistance properties in bodily fluids and superior mechanical qualities.

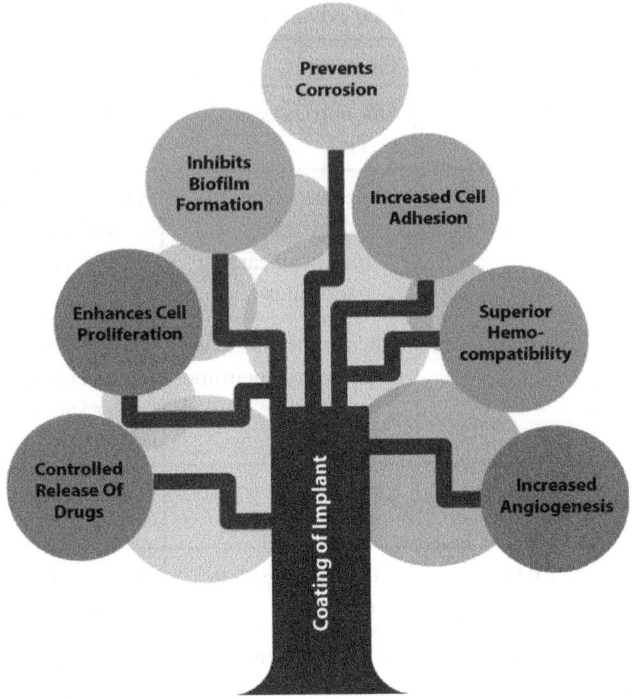

Figure 8.1 Benefits of the application of bioactive coating on implants.

Permanent implants can be made from pure Ti, Ti-based alloys (Ti6Al4V, TiNi), and Co–Cr alloys owing to their anticorrosion and mechanical load-bearing capacity. Due to their low density, Ti alloys have a lead over SS and Co–Cr alloys. The performance of metallic implant materials has been improved through the use of materials such as bioceramic, biopolymer, and other biomolecules and their composites as coatings. Generally speaking, bioactive materials possess low density and promote gene and tissue regeneration, cell division, and proliferation. Metal implants can be coated with a variety of organic and synthetic polymers, like polyether ether ketone (PEEK), polycaprolactone (PCL), PLGA, polyurethane, and polyvinyl alcohol (PVA), poly-L-lactic acid (PLLA), alginate, collagen, gelatin, cellulose, and chitosan [51, 52]. They can also support the transportation of therapeutic drugs [53]. Bioactive ceramics can bind unswervingly to the tissue and propel bone development and maintain the mechanical strength of the implant [32]. Bioactive ceramics include some calcium phosphates (CAP), well suited for orthopedic applications. β-Tricalcium phosphate (β-TCP) and synthetic hydroxyapatite (HAP), which have a Ca/P ratio of

1.67 and the chemical formula $Ca_{10}(PO_4)_6(OH)_2$ [32], are the two types of CAP that are most frequently utilized.

Various strategies to coat the implants with these bioactive materials will be discussed thoroughly in the upcoming sections.

8.3.1 Strategies for Coating the Implants

I. 3D Printing

A computer-aided design (CAD) may be read and used to create three-dimensional items using 3D printing, which was first developed in the 1980s. Three-dimensional printing (3D printing) showed significant possibilities in the biomedical and pharmaceutical industries owing to extraordinary versatility such as using a variety of materials and producing intricate geometric patterns. It exhibits a potential candidature in the growth of customized bioactive coatings for orthopedic implants due to its prospective for on-site manufacturing. There are various technologies that can be used for implant coating like drop-on-demand, selective laser sintering, digital light processing, and inkjet. Nevertheless, extrusion-based techniques like fused deposition modeling (FDM) and semisolid extrusion (SSE) are the most researched technologies for the objective of creating precise drug-eluding coatings. Layer-by-layer extrusion of fluid with the help of a nozzle with the aid of controlled air pressure or mechanical (screw and gear) movement is how SSE creates 3D structures. In contrast, a heated nozzle is used in FDM to soften and melt the filaments of the beginning material. The melted filaments are then cooled and solidified, layer-by-layer, to create the 3D object. An intriguing study used 3D printing to apply a composite containing clindamycin on metal implants (316L and Ti6Al4V) and showed the drug's controlled release for more than 2 months, making it possible to employ it to stop potential infections [54]. Adarkwa *et al.* [55] also employed the 3D printing technique to coat titanium alloy implants with PLGA and PCL entrenched with vancomycin (VA) and amorphous calcium phosphate (ACP) and demonstrated long-standing inhibitory effects on *S. aureus*. The use of this technique is still in its pioneering days when talking about implant coating, but it has witnessed great progress in the manufacturing of porous implants and scaffolds.

II. Electrospinning

The basis of electrospinning is the high-voltage discharge of a liquid droplet of the material solution from a nozzle. When the voltage reaches a critical level, the droplet body charges up and extends as a result of electrostatic repulsion and surface tension working against each other to create a Taylor's cone. The collector serves as the counter electrode, while the syringe's tip serves as an electrode. Polymer content, solution viscosity, flow velocity, voltage, standoff distance, and environmental factors like temperature, air pressure, and humidity are the factors that affect electrospinning. A high-voltage supplier, a syringe coupled with a nozzle, a pump, and a collection scheme make up the electrospinning apparatus. Bioactive coatings have been created on Mg alloys, Ti alloys, and SS, using the electrospinning method. Normally, the solution is the intended coating material like PCL, PLGA, and PLA which, when combined with other different agents, produces efficient exteriors that can also facilitate the localized release of drugs. Typically, this regulated release can continue for more than 30 days [56–60]. While polymers are frequently utilized as the raw material for making fibers, composites, metals, or ceramics may also be employed.

III. Electrophoretic Deposition

Coatings on implants can be formed using a simple process termed electrophoretic deposition (EPD). Its foundation is the formation of a potential difference between two electrodes while they are immersed in a colloidal dispersion. The suspended charged ions from the solution migrate sequentially to the oppositely charged electrode (substrate), where they get deposited and form a coating up to 2.00 mm in thickness. Negatively charged electrodes (cathode) attract positively charged particles, while positively charged electrodes attract negatively charged particles (anode). Panda *et al.* [36] used EPD to coat chitosan/bovine serum albumin and gold nanoparticle on Ti6Al4V and showed improved bioactivity as compared with the uncoated implant. This technique was also used by Sasireka *et al.* [61] to coat Zn^{2+}-substituted calcium silicate infused with graphene oxide on TNT alloy. They found that the composite coating exhibited biocompatibility and active bactericidal nature. Similar results were also reported by other researchers [62–65]. Hadzhieva and Boccaccini [66] have emphasized the bactericidal properties of EPD-coated implants in their recent article. Furthermore, sodium alginate reinforced with hydroxyapatite was coated with EPD to enhance the corrosion resistance of titanium alloy by Moskalewicz *et al.* [67]. The only disadvantage of this method is that it cannot be used for non-conductors.

IV. Dip Coating

Probably one of the simplest methods for the deposition of bioactive coating is dip coating. Its popularity comes from the fact that it is suitable for most of the coating materials and substrates, and it admits complex contours. The method involves the process of dipping the substrate in the solution of coating (sol) at a specified speed, keeping the substrate submerged for a specific time and drying it by evaporation. This method is repeated often to get a dense coating. The dipping may either be achieved manually or with a mechanized device. The number of immersion cycles, immersion time, withdrawal speed, temperature, pH, and viscosity of the solution are the deciding factors for the uniformity and thickness of the coating. The lack of uniformity still remains a cause of concern in the dip-coating process, which can be overcome by subsequent sintering at elevated temperatures (>1,000°C). In a recent study, ultra-high molecular weight polyethylene (UHMWPE) and octadecyl phosphonic acid (ODPA) composite film were deposited on Ti6Al4V alloys with the aid of a mechanized dip coater equipped with various extraction speeds to improve corrosion protection [68]. Similarly, CP Ti was coated with super amphiphobic TiO_2 film by dip coating which could improve the corrosion resistance as well as cytocompatibility of the Ti substrate [69]. Moreover, Zr-substituted HAP was coated on Ti6Al4V ELI for enhancing the implants' osseointegration process [70].

V. Dropcasting

This method also involves a sol, but unlike the dip-coating method, the sol droplets comprised of a suspension of the coating of interest are dropped onto the surface of the substrate and dried by evaporation [71, 72]. Dropcasting can also accommodate complex shapes and is a simpler method with respect to the technicalities involved. This process cycle can also be repeated to achieve a denser coating. Erturk *et al.* [73] fabricated collagen:gelatin nanopillars on a dental implant (nanoporous anodic aluminum oxide membranes) by dropcasting which supported the rapid growth of Saos-2 cells on the surface (>90% viability).

VI. Sol–Gel Deposition

The synthesis of a "sol," by dispersing the precursor in a solvent followed by the formation of a "gel," constitutes the two sequential steps of the sol–gel process. Catalysts may be added to this sol for accelerating the polymerization. The gel depicts a biphasic system in a solid 3D network with a liquid sol enclosed in its pores. The process is normally followed by aging to strengthen the weak bonds formed during gelation. To eventually produce the final dense thin coats, the gel must be dried and heated. The sol–gel process aids in creating fibers, powders, monoliths, coatings, etc. The main benefits of the sol–gel approach include low precursor intake and economic viability, the capability to utilize a variety of precursors, the capacity to produce various types of finished products, and the ability to achieve good substrate adhesion. Furthermore, it is feasible to coat substrates with intricate geometrical shapes *via* sol–gel deposition. However, the drawbacks are the surface fracturing of the final coating since the gel shrinks while dehydrating or following thermal treatment. Maintaining the coating thickness on surfaces with uneven shapes is sometimes challenging. This process is widely used for the application of bioceramic coating (CAP, HAP) to achieve higher corrosion protection property [74, 75]. It was also observed that the tribomechanical properties are also positively affected with the use of this process for nanocomposite (Al_2O_3/Gr/HAP) coating on Ti6Al4V [76]. Sirajunisha *et al.* [77] coated an SS 316L implant with a similar nanocomposite (Ti/rGO/HAP) using sol–gel deposition to achieve antibacterial and other enhanced biological properties along with corrosion protection. Sol–gel spin coating and dip coating are also employed to coat various substrates with functionalized bioactive coatings for bone and dental implants [78–81]. Hydrogels and aerogels are the yields of this process. Hydrogels are known for their hydrophilicity, elasticity, and expanding qualities. They are created from hydrophilic polymer precursors. Wet gels are supercritically dried to produce aerogels, which replace the fluid phase trapped in the gel matrix with a gaseous form. Aside from implant coating, this method can be used for various applications like wastewater treatment [82], manufacturing of Li-ion batteries [83], synthesis of flame-retarding fabric [84], nanomaterial fabrication [85], and production of corrosion-resistant surfaces [86].

VII. Plasma Spraying

Due to its simplicity and time efficiency, plasma spraying has become the most widely utilized marketable coating process for manufacturing CAP/HAP coatings. The creation of a DC arc embedded in a plasma torch, which consists of a cylindrical anode and a cone-shaped cathode, is the fundamental principle of plasma spraying. The method's fundamentals are based on the idea that when the coating materials (in the form of rods, wires, or powders) are exposed to the plasma jet inside or outside the nozzle, the materials melt and are subsequently deposited on the metallic substrate's surface. The particles of the coated material are impinged on the substrate at impact velocities (up to 400 m/s) and cluster into splats or lamellae with surface area and thickness in the micrometer range. Atmospheric plasma spraying (APS, atmospheric pressure), vacuum plasma spraying or low-pressure plasma spraying (LPPS, 3–7 kPa), and very low-pressure plasma spraying (VLPPS, 50–200 Pa) are the categories of plasma spraying according to the operating pressure. Antibacterial Ag-doped HAP coatings

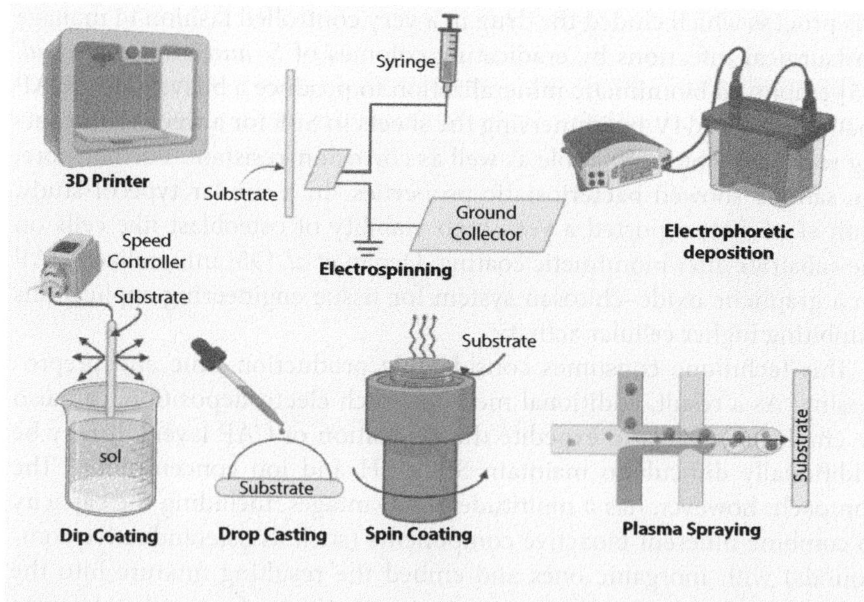

Figure 8.2 Schematic diagram of the above-discussed coating processes.

were developed by various researchers using the plasma spraying method [87–89]. Furthermore, Unabia *et al.* [90] used this method to prepare a biocompatible Cu-doped HAP coating. Rare materials like Nb having osteogenic properties were also substituted in the HAP coating for better cell adhesion using this method [91, 92]. Figure 8.2 presents a schematic diagram of the coating processes discussed so far.

VIII. Biomimetic Deposition

In this method, the biological processes are mimicked artificially to form a natural apatite layer on an implant material. It is very popular for the deposition of functionalized CAP and HAP coatings.

Treatment of the surface prior to coating, which can be accomplished by utilizing either acidic (HF or HCl) or alkaline (NaOH) solutions or pre-heat treatment, is crucial for substrate surface activation [35]. The coatings are generated by submerging a metal or polymer substrate in simulated body fluid (SBF) at pH 7.4 and a temperature of 37°C. It is crucial that the SBF has an ion concentration comparable to that of blood plasma from humans. Coatings are deposited on the substrate surface as a consequence of many weeks of immersion in SBF. Stigter *et al.* [93] produced an amorphous carbon-infused HAP loaded with tobramycin on Ti6Al4V alloy by this process which eluded the drug in a very controlled fashion to manage postsurgical infections by eradicating colonies of *S. aureus*. Panda *et al.* [35] employed biomimetic mineralization to produce a bilayer TiO_2-HAP coating on Ti6Al4V by immersing the sheets in SBF for a week. This coating was highly biocompatible as well as corrosion-resistant. Furthermore, the sample showed bacteriostatic properties. In a similar type of study, Rath *et al.* [94] reported a very high viability of osteoblast-like cells on the substrate after biomimetic coating. Depan *et al.* [95] mineralized HAP on a graphene oxide–chitosan system for tissue engineering applications exhibiting higher cellular activity.

This technique consumes considerable production time and preprocessing. As a result, additional methods, such electrodeposition, can also be employed as aids to expedite the deposition of CAP layers. It may be additionally difficult to maintain SBF's pH and ion concentration. The approach, however, has a multitude of advantages, including the capacity to combine different bioactive components (such as osteoinductive compounds) with inorganic ones and embed the resulting mixture into the coating's crystal lattice. As a consequence, in just a few weeks, deferred and controlled release of the active components can be accomplished. Furthermore, biomimetic coating is performed at low temperatures which is suitable for many biological molecules like peptides and proteins. This can be used for porous or irregularly shaped surfaces as well as substrates that are sensitive to temperature (polymers).

IX. Chemical Vapor Deposition

Another prospective deposition technique is chemical vapor deposition (CVD) which involves the deposition of a solid thin film coating from gaseous precursors via chemical reactions [96, 97]. The site of deposition is adjacent to the base material which is heated in the presence of volatile matter in a vacuum chamber. It is worth noting that the state of the volatile matter is vapor or gas which gets deposited on the substrate. This technique can be employed over a wide range from clinical laboratories to industries. The coating deposited through CVD yields high quality and elevated surface performance [98]. One disadvantage of this process is its cost which increases with the requirement of high vacuum conditions and the complexity involved with the application of higher temperatures [98]. Kuo *et al.* [99] adopted this method to deposit SiOx thin films on AZ31 magnesium alloy. They obtained a low porosity surface coating with higher corrosion resistance. Enhancement in surface corrosion by depositing aluminum on magnesium was affirmed in the studies of Christoglou *et al.* [100]. Tański *et al.* [101] followed the CVD route for coating the Mg–Al–Zn alloy with

Ti/Ti(C, N)/(Ti, Al)N which resulted in improved surface microhardness and enhanced wear resistance. In another research by Pathote *et al.* [102], a corrosion-resistant coating was obtained by depositing a 30-micron-thick metallic coating (Ta) on 316L SS by using the CVD process. The coating produced from this method showed improved functionalities such as biocompatibility, osteoblast adhesion, and proliferation [103–105]. Recently, a variety of CVD technologies, i.e., ultra-high vacuum CVD, plasma-assisted CVD, and laser-assisted CVD to name a few [106, 107], have evolved since its inception.

X. Physical Vapor Deposition

Physical vapor deposition involves producing a fine, hard coating made up of microparticles that have been mechanically extracted from a target or vaporized using plasma, an arc discharge, or a similar technique. PVD often happens in a vacuum chamber to speed up the evaporation process. For biomedical coatings, thermal evaporation and sputtering are the two PVD processes that are used. The PVD technique is founded on the formation of vapor from the target material in the form of atoms, molecules, or ions. Thermal evaporation and sputtering are two potential physical processes that generate vapor. Thermal evaporation involves supplying heat to the material till its vapor pressure surpasses the atmospheric pressure, and in sputtering, energized ions hit the surface of the solid and eject the ions, atoms, or molecules. The latter is especially helpful for ceramics and other materials that require strong adherence but are challenging to deposit on a surface through heat evaporation. Irrespective of the method's intricacy, higher cost, and overall lower capability in drug release, there are a number of benefits that come with it. Since the operating temperature is below the melting point of the substrates, the coatings produced can be applied to surfaces that are susceptible to fluctuations in temperature. The method is also environmentally sustainable and permits the manufacturing of coatings with excellent properties compared with bare substrates.

Uddin *et al.* [108] employed PVD for coating titanium substrate with TiN for better chemical and tribomechanical properties. Coating with various functionalities like antibacterial properties may also be deposited with the help of PVD [109]. Graphite coatings may also be applied for the improvement of antithrombosis and corrosion resistance [110]. Hypersensitivity to metallic ions from the metal implants can also be minimized via PVD coatings of implants [111]. Mesh implants and additively manufactured implants may also be coated via PVD apart from the bulk implants [112, 113]. It has also been reported that PVD coating performance gets affected by the surface properties of the substrate [114]. Temporary implants may also be coated with the help of PVD [115].

XI. Microwave-Assisted Coating

The most prominent advantage associated with microwave-assisted coating is the ability to simultaneously synthesize and coat apatite (CAP or HAP) on any type of substrate in a matter of a few minutes. In this process, all the intended precursors are added in a predefined ratio to make the coating solution, and the sample is introduced into the solution and irradiated in a microwave oven. The energy from the radiation helps in the breaking and making of bonds and thus facilitates rapid deposition. Calcination of the coated samples is generally performed after the coating. This process is generally adopted for apatite coating. Pretreatment of the sample plays a very vital role in the case of microwave-assisted coating. It is also a great tool for producing ionic-doped HAP coatings for various intended functionalities.

Panda *et al.* [34] have coated Ti6Al4V alloy with functionalized Nb-doped HAP for better cell adhesion and corrosion protection. In another interesting study, the authors also attempted multi-atom-substituted

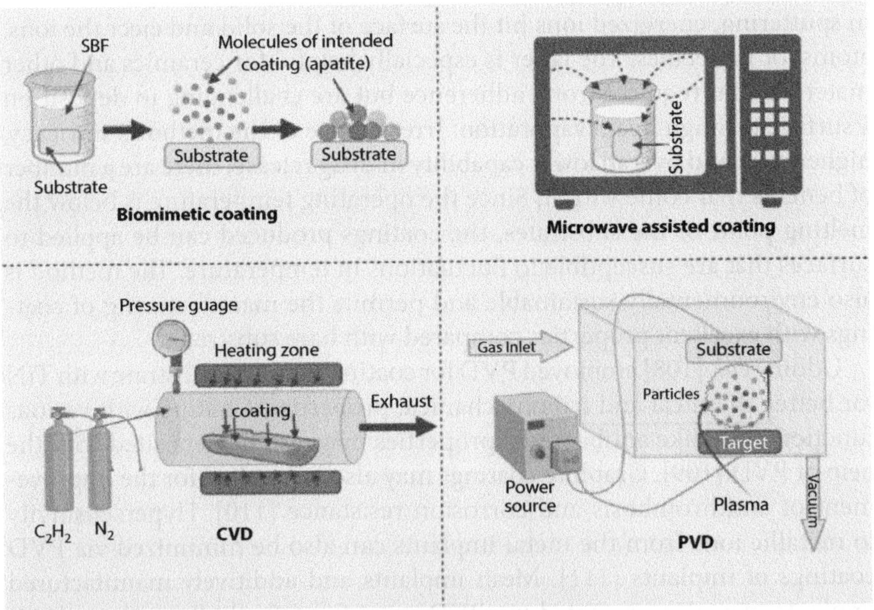

Figure 8.3 Schematic representation of the biomimetic coating, microwave-assisted coating, chemical vapor deposition, and physical vapor deposition.

(Sr and Nb) HAP coating, which is highly biocompatible, antibacterial, and anticorrosive [28]. Similar results were also reported by Zhou *et al.* [116]. Sadiq *et al.* [117] also reported that this process can produce highly anticorrosive coating with great ease. Yu *et al.* [118] also successfully coated Mg alloy using this process intended for short-term implant applications. Other researchers have also reported exciting results in agreement with these reports [119–122].

The coating processes discussed here have been schematically represented in Figure 8.3.

8.4 Conclusions

In the previous sections, we discussed the need for prosthetic implants, the necessity of their surface modification by the application of bioactive coatings, and the methods of coating. Various coating methods have their own set of merits and demerits and cannot be judged for superiority based on any single parameter. They must be chosen keeping in mind the requirements of the end product such as biocompatibility, anticorrosive properties, mechanical strength, drug loading and release requirement, coat, and time of production. Furthermore, while choosing any coating method, its environmental impact must also be analyzed. Methods like anodization must be adopted to grow native oxide layers on the implants for corrosion and wear resistance as well as better visual appeal [123]. Newer methods like plasma-assisted physical vapor deposition (PAPVD) and IonTite™ coatings must be developed with better functionalities [124, 125]. The global implant market is expanding with each passing day. So, vigorous research in the field of surface modification of these implants is the need of the hour.

References

1. Frisch, R.N., Curtis, K.M., Aenlle, K.K., Howard, G.,A., Hepatocyte growth factor and alternative splice variants-expression, regulation and implications in osteogenesis and bone health and repair. *Expert Opinion on Therapeutic Targets (EOTT)*, 20, 9, 1087–98, 2016.
2. Mithal, A., Bansal, B., Kyer, C.S., Ebeling, P., The Asia-pacific regional audit-epidemiology, costs, and burden of osteoporosis in India 2013: A report of international osteoporosis foundation. *Indian Journal of Endocrinology and Metabolism (IJEM)*, 18, 44, 449, 2014.

3. Matsuno, H., Yokoyama, A., Watari, F., Uo, M., Kawasaki, T., Biocompatibility and osteogenesis of refractory metal implants, titanium, hafnium, niobium, tantalum and rhenium. *Biomaterials*, 22, 1253–1262, 2001.
4. Mosas, K.K.A., Chandrasekar, A.R., Dasan, A., Pakseresht, A., Galusek, D., Recent advancements in materials and coatings for biomedical implants. *Gels*, 8, 5, 323, May 2022.
5. Dehghanghadikolaei, A. and Fotovvati, B., Coating techniques for functional enhancement of metal implants for bone replacement: A review. *Materials*, 12, 11, 1795, Jun. 2019.
6. Priyadarshini, B., Rama, M., Chetan, S., Vijayalakshmi, U., Bioactive coating as a surface modification technique for biocompatible metallic implants: A review. *J. Asian Ceram. Soc.*, 7, 4, 397–406, Oct. 2019.
7. Kaliaraj, G.S., Siva, T., Ramadoss, A., Surface functionalized bioceramics coated on metallic implants for biomedical and anticorrosion performance–A review. *J. Mater. Chem. B*, 9, 46, 9433–60, 2021.
8. Su, Y., Cockerill, I., Zheng, Y., Tang, L., Qin, Y.X., Zhu, D., Biofunctionalization of metallic implants by calcium phosphate coatings. *Bioact. Mater.*, 4, 196–206, Dec. 2019.
9. Hussain, M., Askari Rizvi, S.H., Abbas, N., Sajjad, U., Shad, M.R., Badshah, M.A., Malik, A., II, Recent developments in coatings for orthopedic metallic implants. *Coatings*, 11, 7, 791, Jun. 2021.
10. Kravanja, K.A. and Finšgar, M., A review of techniques for the application of bioactive coatings on metal-based implants to achieve controlled release of active ingredients. *Mater. Des.*, 238, 110653, Apr. 2022.
11. Balamurugan, A., Rajeswari, S., Balossier, G., Rebelo, A.H., Ferreira, J.M., Corrosion aspects of metallic implants—An overview. *Mater. Corros.*, 59, 11, 855–869, Nov. 2008.
12. Ma, T.M., Kumar, Y., Lin, P., Singh, Y., Role of protein adsorption in the bio corrosion of metallic implants–A review. *Colloids Surf. B Biointerfaces*, 176, 494–506, Apr. 2019.
13. Black, J., Polymers, in: *Orthopaedic Biomaterials in Research and Practice*, Churchill Livingstone Inc., New York, USA, 1988.
14. Weinstein, A., Horowitz, E., Ruff, A.W., Retrieval and analysis of orthopaedic implants, in: *Proceedings of a Symposium*, National Bureau of Standards Gaithersburg, Maryland, March 5, 1976, National Institute of Standards and Technology, Gaithersburg, MD, [online], 1977, https://doi.org/10.6028/NBS.SP.472.
15. Griffin, C.D., Buchanan, R.A., Lemons, J.E., *In vitro* electrochemical corrosion study of coupled surgical implant materials, *J. Biomed. Mater. Res.*, 17, 489, 1983.
16. Sivakumar, M., KamachiMudali, U., Rajeswari, S., Investigation of failures in stainless steel orthopaedic implant devices: Pit-induced fatigue cracks. *J. Mater. Sci. Lett.*, 14, 148–51, Jan. 1995.

17. Gilbert, J.L., Buckley, C.A., Jacobs, J.J., *In vivo* corrosion of modular hip prosthesis components in mixed and similar metal combinations. The effect of crevice, stress, motion, and alloy coupling. *J. Biomed. Mater. Res.*, 27, 12, 1533–44, Dec. 1993.
18. Cahoon, J.R. and Paxton, H.W., Metallurgical analyses of failed orthopedic implants. *J. Biomed. Mater. Res.*, 2, 1, 1–22, Mar. 1968.
19. Albrektsson, T. and Johansson, C., Osteoinduction, osteoconduction and osseointegration. *Eur. Spine J.*, 10, suppl. 2, S96–101, Oct. 2001.
20. Baseri, M., Radmand, F., Hamedi, R., Yousefi, M., Kafil, H.S., Immunological aspects of dental implant rejection. *BioMed. Res. Int.*, 9, 2020, 7279509, 2020. https://doi.org/10.1155/2020/7279509.
21. Moore, B., Asadi, E., Lewis, G., Deposition methods for microstructured and nanostructured coatings on metallic bone implants: A review. *Adv. Mater. Sci. Eng.*, 2017, 1– 9, 2016.
22. Kazimierczak, P. and Przekora, A., Osteoconductive and osteoinductive surface modifications of biomaterials for bone regeneration: A concise review. *Coatings*, 10, 10, 971, Oct. 2020.
23. Hou, P.J. *et al.*, Development of a surface-functionalized titanium implant for promoting osseointegration: Surface characteristics, hemocompatibility, and *in vivo* evaluation. *Appl. Sci.*, 10, 23, 8582, Nov. 2020.
24. Oriňaková, R. *et al.*, Evaluation of mechanical properties and hemocompatibility of open cell iron foams with polyethylene glycol coating. *Appl. Surf. Sci.*, 505, 144634, Mar. 2020.
25. Oladapo, B., II, Zahedi, S.A., Ismail, S.O., Omigbodun, F.T., 3D printing of PEEK and its composite to increase biointerfaces as a biomedical material-a review. *Colloids Surf. B Biointerfaces*, 203, 111726, Jul. 2021.
26. Zhang, M., Biocompatibility of materials, in: *Biomaterials and Tissue Engineering*, pp. 83–143, 2004.
27. Bhaskar, N. *et al.*, On the origin of better hemocompatibility of the BCxNyOz coatings. *Appl. Surf. Sci.*, 576, 151760, Feb. 2022.
28. Panda, S. *et al.*, Influence of strontium and niobium on the physical and biological performance of hydroxyapatite as a bioactive coating on implant materials. *Ceram. Int.*, 48, 22, 33256–33266, Nov. 2022.
29. Zeng, R.C. *et al.*, *In vitro* corrosion and cytocompatibility of a microarc oxidation coating and poly (l-lactic acid) composite coating on Mg–1Li–1Ca alloy for orthopedic implants. *ACS Appl. Mater. Interfaces*, 8, 15, 10014–10028, Apr. 2016.
30. Iskandar, M.E., Aslani, A., Liu, H., The effects of nanostructured hydroxyapatite coating on the biodegradation and cytocompatibility of magnesium implants. *J. Biomed. Mater. Res. A*, 101, 8, 2340–2354, Aug. 2013.
31. Xu, L. and Yamamoto, A., Characteristics and cytocompatibility of biodegradable polymer film on magnesium by spin coating. *Colloids Surf. B Biointerfaces*, 93, 67–74, May 2012.

32. Panda, S. *et al.*, A comprehensive review on the preparation and application of calcium hydroxyapatite: A special focus on atomic doping methods for bone tissue engineering. *Ceram. Int.*, 47, 20, 28122, 2021.
33. Acheson, J.G., McKillop, S., Lemoine, P., Boyd, A.R., Meenan, B.J., Control of magnesium alloy corrosion by bioactive calcium phosphate coating: Implications for resorbable orthopaedic implants. *Materialia*, 6, 100291, Jun. 2019.
34. Panda, S., Behera, B.P., Bhutia, S.K., Biswas, C.K., Paul, S., Rare transition metal doped hydroxyapatite coating prepared via microwave irradiation improved corrosion resistance, biocompatibility and anti-biofilm property of Titanium alloy. *J. Alloys Compd.*, 918, 165662, Oct. 2022.
35. Panda, S., Kazama, M., Kawai, T., Biswas, C.K., Paul, S., Controlled surface modification of Ti6Al4V using biomimetic mineralization via thermochemical route improves bioactivity. *Ceram. Int.*, 48, 8, 11286–11297, Apr. 2022.
36. Panda, S., Biswas, C.K., Paul, S., Coating of Ti-6Al-4V alloy with chitosan and BSA for enhanced biocompatibility. *Mater. Today Proc.*, 33, 5577–5581, Jan. 2020.
37. Pobloth, A.M., Mersiowsky, M.J., Kliemt, L., Schell, H., Dienelt, A., Pfitzner, B.M., Burgkart, R., Detsch, R., Wulsten, D., Boccaccini, A.R., Duda, G.N., Bioactive coating of zirconia toughened alumina ceramic implants improves cancellous osseointegration. *Sci. Rep.*, 9, 1, 1–6, Nov. 2019.
38. Moses, J.C. and Mandal, B.B., Mesoporous silk-bioactive glass nanocomposites as drug eluting multifunctional conformal coatings for improving osseointegration and bactericidal properties of metal implants. *ACS Appl. Mater. Interfaces*, 14, 13, 14961–14980, Mar. 2022.
39. Kang, B., Lan, D., Liu, L., Dang, R., Yao, C., Liu, P., Ma, F., Qi, S., Chen, X., Antibacterial activity and bioactivity of Zn-doped TiO2 coating for implants. *Coatings*, 12, 9, 1264, Aug. 2022.
40. Nemcakova, I., Litvinec, A., Mandys, V., Potocky, S., Plencner, M., Doubkova, M., Nanka, O., Olejnickova, V., Sankova, B., Bartos, M., Ukraintsev, E., Coating Ti6Al4V implants with nanocrystalline diamond functionalized with BMP-7 promotes extracellular matrix mineralization *in vitro* and faster osseointegration *in vivo*. *Sci. Rep.*, 12, 1, 5264, Mar. 2022.
41. Xue, Y., Chen, J., Ding, T., Mao, M., Zhu, S., Zhou, J., Zhang, L., Han, Y., Building biointegration of Fe2O3-FeOOH coated titanium implant by regulating NIR irradiation in an infected model. *Bioact. Mater.*, 8, 1–1, Feb. 2022.
42. Zhang, G., Wu, Z., Yang, Y., Shi, J., Lv, J., Fang, Y., Shen, Z., Lv, Z., Li, P., Yao, X., Chen, W., A multifunctional antibacterial coating on bone implants for osteosarcoma therapy and enhanced osteointegration. *Chem. Eng. J.*, 428, 131155, Jan. 2022.
43. Milan, P.B., Khamseh, S., Zarrintaj, P., Ramezanzadeh, B., Badawi, M., Morisset, S., Vahabi, H., Saeb, M.R., Mozafari, M., Copper-enriched

diamond-like carbon coatings promote regeneration at the bone-implant interface. *Heliyon*, 6, 4, e03798, Apr. 2020.
44. Geng, Z., Li, Z., Cui, Z., Wang, J., Yang, X., Liu, C., Novel bionic topography with MiR-21 coating for improving bone-implant integration through regulating cell adhesion and angiogenesis. *Nano Lett.*, 20, 10, 7716–21, Sep. 2020.
45. Saha, S., Pramanik, K., Biswas, A., Silk fibroin coated TiO2 nanotubes for improved osteogenic property of Ti6Al4V bone implants. *Mater. Sci. Eng. C*, 105, 109982, Dec. 2019.
46. Bakhsheshi-Rad, H.R., Chen, X., Ismail, A.F., Aziz, M., Abdolahi, E., Mahmoodiyan, F., Improved antibacterial properties of an Mg-Zn-Ca alloy coated with chitosan nanofibers incorporating silver sulfadiazine multiwall carbon nanotubes for bone implants. *Polym. Adv. Technol.*, 30, 5, 1333–9, May 2019.
47. Jensen, L.K., Jensen, H.E., Blirup-Plum, S.A., Bue, M., Hanberg, P., Kvich, L., Aalbæk, B., López, Y., Soto, S.M., Douloudi, M., Papageorgiou, M., Coating of bone implants with silica, hyperbranched polyethyleneimine, and gentamicin prevents development of osteomyelitis in a porcine model. *Materialia*, 24, 101473, Aug. 2022.
48. Sari, M., Ana, I.D., Yusuf, Y., Cell viability assay and surface morphology analysis of carbonated hydroxyapatite/honeycomb/titanium alloy coatings for bone implant applications. *Bioengineering*, 9, 7, 325, Jul. 2022.
49. Wang, L.N., Meng, Y.F., Feng, Y., Wang, H.C., Mao, L.B., Yu, S.H., Wang, Z.L., Amorphous precursor-mediated calcium phosphate coatings with tunable microstructures for customized bone implants. *Adv. Healthc. Mater.*, 11, 19, 2201248, Oct. 2022.
50. Sivamaran, V., Padma Rao, A., Senthil Kumar, P., Selvamani, S.T., Mandal, T.K., Sol–gel derived Al2O3/Gr/HAP nanocomposite coatings on Ti–6Al–4V alloy for enhancing tribo-mech properties and antibacterial activity for bone implants. *Appl. Phys. A*, 128, 8, 635, Aug. 2022.
51. Li, Z. and Khor, K.A., Preparation and properties of coatings and thin films on metal implants, in: *Encyclopedia of Biomedical Engineering*, R. Narayan (Ed.), pp. 203–212, Elsevier, Oxford, 2019.
52. Szaraniec, B., Pielichowska, K., Pac, E., Menaszek, E., Multifunctional polymer coatings for titanium implants. *Mater. Sci. Eng. C*, 93, 950–957, 2018.
53. Boccaccini, A.R., Erol, M., Stark, W.J., Mohn, D., Hong, Z., Mano, J.F., Polymer/bioactive glass nanocomposites for biomedical applications: A review. *Compos. Sci. Technol.*, 70, 13, 1764–1776, 2010.
54. Maver, T., Mastnak, T., Mihelic, M., Maver, U., Finšgar, M., Clindamycin-based 3D-printed and electrospun coatings for treatment of implant-related infections. *Materials*, 14, 6, 1464, Mar. 2021.
55. Adarkwa, E., Roy, A., Ohodnicki, J., Lee, B., Kumta, P.N., Desai, S., 3D printing of drug-eluting bioactive multifunctional coatings for orthopedic applications. *Int. J. Bioprinting*, 9, 2, 661, 2023.

56. Rezk, A., Mousa, H., Lee, J., Park, C., Kim, C., Composite PCL/HA/simvastatin electrospun nanofiber coating on biodegradable Mg alloy for orthopedic implant application. *J. Coat. Technol. Res.*, 15, 2, 259–272, 2018.
57. Wei, Y., Liu, Z., Zhu, X., Jiang, L., Shi, W., Wang, Y., Xu, N., Gang, F., Wang, X., Zhao, L., Lin, J., Dual directions to address the problem of aseptic loosening via electrospun PLGA@ aspirin nanofiber coatings on titanium. *Biomaterials*, 257, 120237, Oct. 2020.
58. Khosravi, F., Nouri Khorasani, S., Khalili, S., EsmaeelyNeisiany, R., RezvaniGhomi, E., Ejeian, F., Das, O., Nasr-Esfahani, M.H., Development of a highly proliferated bilayer coating on 316L stainless steel implants. *Polymers*, 12, 5, 1022, May 2020.
59. Kiran, A.S., Kizhakeyil, A., Ramalingam, R., Verma, N.K., Lakshminarayanan, R., Kumar, T.S., Doble, M., Ramakrishna, S., Drug loaded electrospun polymer/ceramic composite nanofibrous coatings on titanium for implant related infections. *Ceram. Int.*, 45, 15, 18710–20, Oct. 2019.
60. Jahanmard, F., Croes, M., Castilho, M., Majed, A., Steenbergen, M.J., Lietaert, K., Vogely, H.C., Van Der Wal, B.C., Stapels, D.A., Malda, J., Vermonden, T., Bactericidal coating to prevent early and delayed implant-related infections. *J. Control. Release*, 326, 38–52, Oct. 2020.
61. Sasireka, A., Suganthi, S., Vignesh, S., Raj, V., Oh, T.H., Preparation and characterization of zinc modified calcium silicate/polycaprolactone with graphene oxide composite coating for bone repair applications. *Ceram. Int.*, 49, 12, 20251–60, Jun 2023.
62. Deen, I., Selopal, G.S., Wang, Z.M., Rosei, F., Electrophoretic deposition of collagen/chitosan films with copper-doped phosphate glasses for orthopaedic implants. *J. Colloid Interface Sci.*, 607, 869–80, Feb. 2022.
63. Oktay, A., Yilmazer, H., Przekora, A., Yilmazer, Y., Wojcik, M., Dikici, B., Ustundag, C.B., Corrosion response and biocompatibility of graphene oxide (GO) serotonin (Ser) coatings on Ti6Al7Nb and Ti29Nb13Ta4. 6Zr (TNTZ) alloys fabricated by electrophoretic deposition (EPD). *Mater. Today Commun.*, 34, 105236, Mar. 2023.
64. Varlık, Ö., Göncü, Y., Nuran, A.Y., Electrophoretic deposition of composite titanium diboride-chitosan coating. *Mater. Chem. Phys.*, 282, 125927, Apr. 2022.
65. Kee, C.C., Ng, K., Ang, B.C., Metselaar, H.S., Synthesis, characterization and *in-vitro* biocompatibility of electrophoretic deposited europium-doped calcium silicate on titanium substrate. *J. Eur. Ceram. Soc.*, 43, 3, 1189–204, Mar. 2023.
66. Hadzhieva, Z. and Boccaccini, A.R., Recent developments in electrophoretic deposition (EPD) of antibacterial coatings for biomedical applications-a review. *Curr. Opin. Biomed. Eng.*, 21, 100367, Mar. 2022.
67. Moskalewicz, T., Warcaba, M., Łukaszczyk, A., Kot, M., Kopia, A., Hadzhieva, Z., Boccaccini, A.R., Electrophoretic deposition, microstructure and properties of multicomponent sodium alginate-based coatings incorporated with

graphite oxide and hydroxyapatite on titanium biomaterial substrates. *Appl. Surf. Sci.*, 575, 151688, Feb. 2022.

68. Anaya-Garza, K., Torres-Huerta, A.M., Domínguez-Crespo, M.A., Moreno-Palmerín, J., Ramírez-Meneses, E., Rodríguez-Salazar, A.E., Corrosion resistance improvement of the Ti6Al4V/UHMWPE systems by the assembly of ODPA molecules by dip coating technique. *Prog. Org. Coat.*, 171, 107013, Oct. 2022.
69. Aslan Çakır, M., Yetim, T., Yetim, A.F., Çelik, A., Superamphiphobic TiO2 film by sol–gel dip coating method on commercial pure titanium. *J. Mater. Eng. Perform.*, Mar. 2023. https://doi.org/10.1007/s11665-023-08049-3
70. Ardhy, S., Affi, J., Yetri, Y., Improvement in bioactivity properties of hydroxyapatite coated Ti-6Al-4V ELI with addition of zirconium oxide for orthopaedic implant. *J. Biomim. Biomater. Biomed. Eng.*, 59, 81–9, Mar. 2023.
71. Mishra, S.K. and Kannan, S., Development, mechanical evaluation and surface characteristics of chitosan/polyvinyl alcohol based polymer composite coatings on titanium metal. *J. Mech. Behav. Biomed. Mater.*, 40, 314–324, 2014.
72. De Santis, S., Sotgiu, G., Porcelli, F., Marsotto, M., Iucci, G., Orsini, M., A simplecerium coating strategy for titanium oxide nanotubes' bioactivity enhancement. *Nanomaterials*, 11, 2, 445, Feb. 2021.
73. Erturk, P.A., Altuntas, S., Irmak, G., Buyukserin, F., Bioinspired collagen/gelatin nanopillared films as a potential implant coating material. *ACS Appl. Bio Mater.*, 5, 10, 4913–4921, Oct. 2022.
74. Jaafar, A., Hecker, C., Árki, P., Joseph, Y., Sol-gel derived hydroxyapatite coatings for titanium implants: A review. *Bioengineering*, 7, 127, 2020.
75. Jaafar, A., Schimpf, C., Mandel, M., Hecker, C., Rafaja, D., Krüger, L., Árki, P., Joseph, Y., Sol–gel derived hydroxyapatite coating on titanium implants: Optimization of sol–gel process and engineering the interface. *J. Mater. Res.*, 37, 16, 2558–2570, Aug. 2022.
76. Sivamaran, V., Padma Rao, A., Senthil Kumar, P., Selvamani, S.T., Mandal, T.K., Sol–gel derived Al2O3/Gr/HAP nanocomposite coatings on Ti–6Al–4V alloy for enhancing tribo-mech properties and antibacterial activity for bone implants. *Appl. Phys. A*, 128, 8, 635, Aug. 2022.
77. Sirajunisha, H., Balakrishnan, T., Sakthivel, P., Krishnaveni, A., Enhanced corrosion resistance, antibacterial and biological properties of sol-gel derived Ti-rGO-HAp nanocomposites. *Chem. Phys. Impact*, 6, 100159, Jun. 2023.
78. Malakauskaite-Petruleviciene, M., Stankeviciute, Z., Beganskiene, A., Kareiva, A., Sol–gel synthesis of calcium hydroxyapatite thin films on quartz substrate using dip-coating and spin-coating techniques. *J. Sol-Gel Sci. Technol.*, 71, 437–446, Sep. 2014.
79. Carradò, A. and Viart, N., Nanocrystalline spin coated sol–gel hydroxyapatite thin films on Ti substrate: Towards potential applications for implants. *Solid State Sci.*, 12, 7, 1047–1050, Jul. 2010.

80. Predoi, D., Iconaru, S.L., Predoi, M.V., Buton, N., Motelica-Heino, M., Zinc doped hydroxyapatite thin films prepared by sol–gel spin coating procedure. *Coatings*, 9, 3, 156, Feb. 2019.
81. Dommeti, V.K., Roy, S., Pramanik, S., Merdji, A., Ouldyerou, A., Özcan, M., Design and development of tantalum and strontium ion doped hydroxyapatite composite coating on titanium substrate: Structural and human osteoblast-like cell viability studies. *Materials*, 16, 4, 1499, Feb. 2023.
82. Rafie, N., Khodadadi, M., Zamani, M., Zarepour, A., Zarrabi, A., Magnetic silica nanoparticles adorned with a metal-organic framework; A novel nanosorbent for elimination of aqueous Pb ions contaminant. *Environ. Res.*, 226, 115694, 2023.
83. Wu, Y., Li, M., Wahyudi, W., Sheng, G., Miao, X., Anthopoulos, T.D., Huang, K.W., Li, Y., Lai, Z., Performance and stability improvement of layered NCM lithium-ion batteries at high voltage by a microporous Al2O3 sol–gel coating. *ACS Omega*, 4, 9, 13972–13980, Aug. 2019.
84. Castellano, A., Colleoni, C., Iacono, G., Mezzi, A., Plutino, M.R., Malucelli, G., Rosace, G., Synthesis and characterization of a phosphorous/nitrogen-based sol-gel coating as a novel halogen-and formaldehyde-free flame-retardant finishing for cotton fabric. *Polym. Degrad. Stab.*, 162, 148–159, Apr. 2019.
85. Tan, W.K., Muto, H., Kawamura, G., Lockman, Z., Matsuda, A., Nanomaterial fabrication through the modification of sol–gel derived coatings. *Nanomaterials*, 11, 1, 181, Jan. 2021.
86. Farag, M.M., Ahmed, H.Y., Al-Rashidy, Z.M., Improving the corrosion resistance of magnesium alloy by magnesium phosphate/glass composite coatings using sol–gel method. *Silicon*, 15, 3841–3854, 2022.
87. Guimond-Lischer, S., Ren, Q., Braissant, O., Gruner, P., Wampfler, B., Maniura-Weber, K., Vacuum plasma sprayed coatings using ionic silver doped hydroxyapatite powder to prevent bacterial infection of bone implants. *Biointerphases*, 11, 1, 011012, Mar. 2016.
88. Chen, Y., Zheng, X., Xie, Y., Ji, H., Ding, C., Li, H., Dai, K., Silver release from silver-containing hydroxyapatite coatings. *Surf. Coat. Technol.*, 205, 7, 1892–1896, Dec. 2010.
89. Fielding, G.A., Roy, M., Bandyopadhyay, A., Bose, S., Antibacterial and biological characteristics of silver containing, and strontium doped plasma sprayed hydroxyapatite coatings. *Acta Biomater.*, 8, 8, 3144–3152, Aug. 2012.
90. Unabia, R.B., Candidato Jr., R.T., Pawłowski, L., Salvatori, R., Bellucci, D., Cannillo, V., *In vitro* studies of solution precursor plasma-sprayed copper-doped hydroxyapatite coatings with increasing copper content. *J. Biomed. Mater. Res. Part B Appl. Biomater.*, 108, 6, 2579–2589, Aug. 2020.
91. Singh, B., Singh, G., Sidhu, B.S., Bhatia, N., *In-vitro* assessment of HA-Nb coating on Mg alloy ZK60 for biomedical applications. *Mater. Chem. Phys.*, 231, 138–149, Jun. 2019.
92. Singh, B., Singh, G., Sidhu, B.S., Investigation of the *in vitro* corrosion behavior and biocompatibility of niobium (Nb)-reinforced hydroxyapatite (HA)

coating on CoCr alloy for medical implants. *J. Mater. Res.*, 34, 10, 1678–1691, May 2019.

93. Stigter, M., De Groot, K., Layrolle, P., Incorporation of tobramycin into biomimetic hydroxyapatite coating on titanium. *Biomaterials*, 23, 20, 4143–4153, Oct. 2002.
94. Rath, P.C., Besra, L., Singh, B.P., Bhattacharjee, S., Titania/hydroxyapatite bi-layer coating on Ti metal by electrophoretic deposition: Characterization and corrosion studies. *Ceram. Int.*, 38, 4, 3209–3216, May 2012.
95. Depan, D., Pesacreta, T.C., Misra, R.D., The synergistic effect of a hybrid graphene oxide–chitosan system and biomimetic mineralization on osteoblast functions. *Biomater. Sci.*, 2, 2, 264–274, Feb. 2014.
96. Behera, A., Mallick, P., Mohapatra, S.S., Nanocoatings for anticorrosion: An introduction, in: *Corrosion Protection at the Nanoscale*, S. Rajendran, T.A.N.H. Nguyen, S. Kakooei, M. Yeganeh, Y. Li (Eds.), pp. 227–243, Elsevier, 2020.
97. Barua, S., Geng, X., Chen, B., Graphene-based nanomaterials for healthcare applications, in: *Photonanotechnology for Therapeutics and Imaging*, S.K. Choi (Ed.), pp. 45–81, Elsevier, 2020.
98. Choy, K.L., Chemical vapour deposition of coatings. *Prog. Mater. Sci.*, 48, 57–170, 2003.
99. Kuang-Hui, C. and Kuo, Y.L., Atmospheric pressure plasma enhanced chemical vapor deposition of SiOx films for improved corrosion resistant properties of AZ31 magnesium alloys. *Surf. Coat. Technol.*, 283, 194–200, 2015.
100. Christoglou, C., Voudouris, N., Angelopoulos, G.N., Pant, M., Dahl, W., Deposition of aluminium on magnesium by a CVD process. *Surf. Coat. Technol.*, 184, 149–155, 2004.
101. Tański, T., Surface layers on the Mg-Al-Zn alloys coated using the CVD and PVD methods. *J. Achiev. Mater. Manuf. Eng.*, 53, 2, 89–96, 2012.
102. Pathote, D., Kumari, P., Singh, V., Jaiswal, D., Gautam, R.K., Behera, C.K., Biocompatibility evaluation, wettability, and scratch behavior of Ta-coated 316L stainless steel by DC magnetron sputtering for the orthopedic applications. *Surf. Coat. Technol.*, 459, 129392, 2023.
103. Li, X., Wang, L., Yu, X., Feng, Y., Wang, C., Yang, K., Su, D., Tantalum coating on porous Ti6Al4V scaffold using chemical vapor deposition and preliminary biological evaluation. *Mater. Sci. Eng. C*, 33, 5, 2987–2994, 2013.
104. Lu, M., Shao, D., Wang, P., Chen, D., Zhang, Y., Li, M., Zhao, J., Zhou, Y., Enhanced osteoblast adhesion on amino-functionalized titanium surfaces through combined plasma enhanced chemical vapor deposition (PECVD) method. *RSC Adv.*, 6, 86, 82688–82697, 2016.
105. Youn, Y.H., Lee, S.J., Choi, G.R., Lee, H.R., Lee, D., Heo, D.N., Kim, B.S., Bang, J.B., Correlo, V.M., Reis, R.L., Im, S.G., Kwon, I.K., Simple and facile preparation of recombinant human bone morphogenetic protein-2 immobilized titanium implant via initiated chemical vapor deposition technique

to promote osteogenesis for bone tissue engineering application. *Mater. Sci. Eng. C*, 100, 949–958, 2019.

106. Madhuri, K.V., Thermal protection coatings of metal oxide powders, in: *Metal Oxide Powder Technologies*, Y. Al-Douri (Ed.), pp. 209–231, Elsevier, 2020.
107. Cassell, A.M., Raymakers, J.A., Kong, J., Dai, H., Large scale CVD synthesis of single-walled carbon nanotubes. *J. Phys. Chem. B*, 103, 31, 6484–6492, 1999.
108. Uddin, G.M., Jawad, M., Ghufran, M., Saleem, M.W., Raza, M.A., Rehman, Z.U., Arafat, S.M., Irfan, M., Waseem, B., Experimental investigation of tribo-mechanical and chemical properties of TiN PVD coating on titanium substrate for biomedical implants manufacturing. *Int. J. Adv. Manuf. Technol.*, 102, 1391–1404, 2019.
109. Mironov, M.M., Bayazitova, L.T., Grebenshchikova, M.M., PVD barrier coatings with antimicrobial function for medical implants, in: *Journal of Physics: Conference Series*, vol. 1588, p. 012032, IOP Publishing, Jul. 2020.
110. Hassan, S., Nadeem, A.Y., Ali, M., Ali, M.N., Niazi, M.B., Mahmood, A., Graphite coatings for biomedical implants: A focus on anti-thrombosis and corrosion resistance properties. *Mater. Chem. Phys.*, 290, 126562, Oct. 2022.
111. Dempwolf, H., Proft, M., Baumann, A., Malz, S., Keßler, O., The impact of bias and nitrogen pressure on TiNbN coatings in arc-PVD processes—A multifactorial study. *Coatings*, 12, 7, 935, Jul. 2022.
112. Taran, A., Tymoshenko, O., Misiruk, I., Vorontsov, P., Garkusha, I., Gnidenko, Y., Kravchenko, Y., Polypropylene mesh implants modified by nanostructured PVD coatings, in: *2022 IEEE 12th International Conference Nanomaterials: Applications & Properties (NAP)*, pp. 1–4, Sep. 11, 2022.
113. Tillmann, W., Dias, N.F., Kokalj, D., Stangier, D., Hein, M., Hoyer, K.P., Schaper, M., Gödecke, D., Oltmanns, H., Meißner, J., Tribo-functional PVD thin films deposited onto additively manufactured Ti6Al7Nb for biomedical applications. *Mater. Lett.*, 321, 132384, Aug. 2022.
114. Taherian, M.H., Rezazadeh, M., Taji, A., Optimum surface roughness for titanium-coated PEEK produced by electron beam PVD for orthopedic applications. *Mater. Technol.*, 37, 8, 631–644, Jul. 2022.
115. Zarka, M., Dikici, B., Niinomi, M., Ezirmik, K.V., Nakai, M., Kaseem, M., The Ti3.6Nb1.0Ta0.2Zr0.2 coating on anodized aluminum by PVD: A potential candidate for short-time biomedical applications. *Vacuum*, 192, 110450, Oct. 2021.
116. Zhou, H., Kong, S., Pan, Y., Zhang, Z., Deng, L., Microwave-assisted fabrication of strontium doped apatite coating on Ti6Al4V. *Mater. Sci. Eng. C*, 56, 174–180, Nov. 2015.
117. Sadiq, T.O., Siti, N., Idris, J., A study of strontium-doped calcium phosphate coated On Ti6Al4V using microwave energy. *J. Bio-Tribo-Corros.*, 4, 1–8, Sep. 2018.

118. Yu, N., Cai, S., Wang, F., Zhang, F., Ling, R., Li, Y., Jiang, Y., Xu, G., Microwave assisted deposition of strontium doped hydroxyapatite coating on AZ31 magnesium alloy with enhanced mineralization ability and corrosion resistance. *Ceram. Int.*, 43, 2, 2495–2503, Feb. 2017.
119. Lamkhao, S., Phaya, M., Jansakun, C., Chandet, N., Thongkorn, K., Rujijanagul, G., Bangrak, P., Randorn, C., Synthesis of hydroxyapatite with antibacterial properties using a microwave-assisted combustion method. *Sci. Rep.*, 9, 1, 4015, Mar. 2019.
120. Shen, S., Cai, S., Zhang, M., Xu, G., Li, Y., Ling, R., Wu, X., Microwave assisted deposition of hydroxyapatite coating on a magnesium alloy with enhanced corrosion resistance. *Mater. Lett.*, 159, 146–149, Nov. 2015.
121. Shen, S., Cai, S., Li, Y., Ling, R., Zhang, F., Xu, G., Wang, F., Microwave aqueous synthesis of hydroxyapatite bilayer coating on magnesium alloy for orthopedic application. *Chem. Eng. J.*, 309, 278–287, Feb. 2017.
122. Ren, Y., Zhou, H., Nabiyouni, M., Bhaduri, S.B., Rapid coating of AZ31 magnesium alloy with calcium deficient hydroxyapatite using microwave energy. *Mater. Sci. Eng. C*, 49, 364–372, Apr. 2015.
123. Panda, S., Biswas, C.K., Paul, S., Anodization of implantable metal and alloy surfaces: Purpose and scope, in: *Surface Engineering*, pp. 169–189, CRC Press, Taylor & Francis Group, Dec. 2022.
124. Matthews, A. and Leyland, A., Hybrid techniques in surface engineering. *Surf. Coat. Technol.*, 71, 2, 88–92, Mar. 1995.
125. Sato, M., Sambito, M.A., Aslani, A., Kalkhoran, N.M., Slamovich, E.B., Webster, T.J., Increased osteoblast functions on undoped and yttrium-doped nanocrystalline hydroxyapatite coatings on titanium. *Biomaterials*, 27, 11, 2358–2369, 2006.

9

Mechanical Behavior of Bioglass Materials for Bone Implantation

Md Ershad* **and Ranjan Kumar**

Swami Vivekananda University, Kolkata, India

Abstract

This study aims to investigate the *in vitro* biomechanical behavior of bioactive glasses containing Rare Earth Elements (REEs) like CeO_2, La_2O_3, and Sm_2O_3 oxides. Previous studies have shown that these REEs possess good biological properties; are biocompatible, easily processed, and high-temperature stable; and have tunable release kinetics. The current study focuses on the preparation of bioactive glasses and their ceramic derivatives containing CeO_2, La_2O_3, Sm_2O_3, and alumina substitutions for Si^{4+} ions to create a non-charge-balanced (NCB) bioactive glass. The bioactivity of the glass and glass ceramic samples was assessed through their hydroxyapatite layer formation ability in simulated body fluid (SBF), which was confirmed through FTIR spectrometry, pH measurement, and SEM. Cell culture studies were conducted to assess the cytotoxicity of the samples against mouse fibroblast, L929 cell lines. Bioactive glass ceramics were handled through controlled crystallization of glasses, and the resulting samples were characterized for their density, hardness, compressive and bending strengths, and elastic properties. The results indicate that the incorporation of CeO_2, La_2O_3, Sm_2O_3, and La_2O_3 in 45S5 and 1393 bioactive glass and glass ceramic enhances their mechanical properties.

***Keywords*:** Bone, implantation, bioglass materials, composites, biomechanical

**Corresponding author*: mdershad.rs.cer13@iitbhu.ac.in

Jay Prakash Srivastava, Dražan Kozak, Vinayak Ranjan, Pankaj Kumar, Ranjan Kumar and Shubham Tayal (eds.) Mechanical Engineering in Biomedical Applications: Bio-3D Printing, Biofluid Mechanics, Implant Design, Biomaterials, Computational Biomechanics, Tissue Mechanics, (261–276) © 2024 Scrivener Publishing LLC

9.1 Introduction on Bio Materials

The first time frame involves the replacement of tissues or organs, such as artificial heart valves, joint replacements, or pacemakers, where the biomaterial is used to replace a part of the body that is no longer functioning correctly. The second time frame focuses on the regeneration of tissues, where the biomaterial is used to stimulate the growth of new tissue in the body. Examples include skin grafts and bone regeneration. The third time frame involves the engineering of tissues, where the biomaterial is used to create an entirely new tissue or organ in the laboratory, which can then be transplanted into the body. Biomaterials play a crucial role in modern medicine, and their usage has significantly improved the quality of life for many individuals [1]. However, there are still challenges to overcome, such as ensuring biocompatibility and long-term safety and improving the materials' functionality. Nonetheless, the development of biomaterials is an exciting area of research that has the potential to revolutionize the field of medicine in the future [2, 3].

The past: removal of tissues
The present: replacement of tissues
The future: regeneration of tissues

Bioactive materials are designed to interact with the surrounding biological environment, promoting tissue growth and regeneration. These materials have chemical, mechanical, and biological properties that allow them to bond with living tissue, which can lead to improved tissue regeneration and healing. Some examples of bioactive materials include ceramics, glasses, and certain polymers.

One of the most well-known bioactive materials is hydroxyapatite, a calcium phosphate ceramic similar in composition to the mineral component of bone. Hydroxyapatite can promote the growth of new bone tissue and is often used in orthopedic implants, such as artificial joints and bone screws. Another example of bioactive material is bioglass, which is a type of glass that can bond with bone tissue. Bioglass has been used to repair bone defects and promote bone regeneration in a variety of applications.

In addition to promoting tissue growth and regeneration, bioactive materials can also be used for drug delivery, as they can release therapeutic agents at the site of tissue regeneration. This can help to further promote tissue growth and healing.

Overall, the development of bioactive materials has revolutionized the field of biomaterials and has opened up new avenues for tissue regeneration and repair.

The formation of an HCA layer is a key characteristic of bioactive materials, as it indicates that the material is interacting with the surrounding biological environment and promoting tissue growth and regeneration [4]. The HCA layer forms as a result of specific surface reactions between the material and the body's fluids, and it provides a surface that allows for the bonding of living tissue to the material. This ability to bond with living tissue is what sets bioactive materials apart from bioinert materials, which do not interact with the surrounding biological environment in the same way. While bioinert materials can provide mechanical strength and corrosion resistance, they do not promote tissue growth and regeneration in the same way as bioactive materials.

Overall, the ability to form an HCA layer and promote tissue growth and regeneration is a key characteristic of bioactive materials and has led to their use in a wide range of medical applications, from orthopedic implants to drug delivery systems.

9.2 Aim and Objective of the Work

The present study aims to improve the *in vitro* biomechanical properties of 45S5 bioactive glass substituted with rare earth oxide (CeO_2, La_2O_3, and Sm_2O_3) and Al_2O_3 up to 2 wt%. The motivation for selecting CeO_2, La_2O_3, Sm_2O_3, and Al_2O_3 is due to their good biological and mechanical properties with low toxicity.

- Preparation of bioactive glass by the melting route
- Preparation of bioactive glass ceramics by heat treatment
- Structural study of bioactive glass by DTA, FTIR, and XRD
- Physical behavior of bioactive glass and glass ceramics
- Assessment of *in vitro* bioactivity in SBF
- Biocompatibility with human blood cell
- Identification of HCA formation by pH, FTIR, and SEM
- Mechanical behavior of bioactive glass and glass ceramic (destructive test)
- Elastic moduli of bioactive glass and glass ceramic (non-destructive test)

Bioactive glasses were one of the first types of bioactive materials to be developed, and they have since been followed by a range of other materials, including hydroxyapatite, β-tricalcium phosphate, and bioactive glass ceramics. Silica-based bioactive glasses are particularly popular due to their ability to form an HCA layer quickly and effectively. They contain a range of elements, including silicon, calcium, sodium, and phosphorus, and they are able to dissolve in bodily fluids, releasing ions that can stimulate tissue growth and regeneration. Hydroxyapatite (HA) and β-tricalcium phosphate (TCP) are both forms of calcium phosphate, which is a major component of bone tissue. These materials can bond with surrounding bone tissue, promoting new bone growth and integration with existing bone. Bioactive glass ceramics, such as A-W glass ceramic, contain both glassy and crystalline phases and are designed to combine the mechanical strength of ceramics with the bioactivity of glasses. They can form a strong bond with bone tissue, making them particularly useful for orthopedic applications.

Overall, the development of a range of bioactive materials has opened up new possibilities for medical applications from bone repair to tissue engineering and drug delivery as shown in Figure 9.1.

In 1991, T. Kokubo developed a simulated body fluid (SBF) that could be used for the *in vitro* investigation of material bioactivity [7]. The SBF

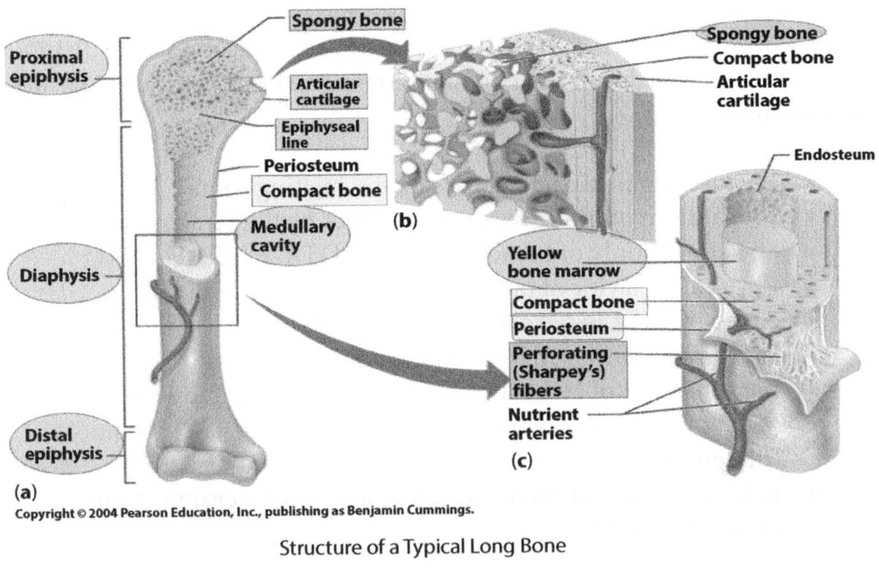

Figure 9.1 *In vitro* and *in vivo* performance (©Pearson Education).

solution was designed to closely mimic the composition of human blood plasma, and it has become the most widely used solution for this purpose. The use of SBF provides researchers with a fast, economical, and reliable way to test the bioactivity of any material. By immersing a material in SBF, researchers can simulate the conditions that the material would encounter in the human body. The SBF solution is designed to endorse the formation of a layer of hydroxyapatite on the material's surface, which is an indicator of its bioactivity. *In vitro* tests using SBF can provide valuable information about the bioactivity and biocompatibility of a material before it is tested *in vivo*, which can help reduce the number of animal studies needed. Moreover, SBF can also be used to evaluate the performance of different materials and compare their bioactivity levels.

Bioactive glasses possess unique characteristics that differentiate them from other bioactive materials. They can be customized by altering the oxide composition of the glass to regulate their properties and bonding rate with living tissues. Among them, the 45S5 bioactive glass and glass ceramic are extensively utilized due to their ability to bond with both hard and soft tissues, making them promising biomaterials for biomedical applications, especially as an artificial bone graft. Nevertheless, their mechanical properties are not comparable to that of cortical bone, rendering them unsuitable for bearing heavy loads. Researchers are currently exploring ways to enhance the mechanical strength of bioactive glasses by introducing reinforcing materials such as fibers or polymers. Overall, bioactive glasses hold great potential for various biomedical applications, particularly in tissue engineering and regenerative medicine, due to their bonding ability and customizable properties.

Bioactive glass 45S5 has a specific chemical composition consisting of SiO_2 as the network former and a Ca/P molar ratio of 5:1, which is responsible for its bioactivity. This glass is characterized by a low SiO_2 content, high Na_2O and CaO contents, and a high CaO/P_2O_5 ratio, which contribute to its bioactivity. The network former, SiO_2, plays a vital role in maintaining the three-dimensional non-periodic glass structure during the selective dissolution of cations such as Na^+ and Ca^{2+} while preventing the detachment of some other ions. The presence of SiO_2 also promotes the formation of a hydroxyapatite layer by facilitating the precipitation or surface reconstruction of the silica-rich layer, which is crucial for its bioactivity. Earlier studies have thoroughly examined the interactions between bioactive implants and bone tissues, including the reaction kinetics at the interfacial region and the sequence of reactions. Despite its low fracture toughness and mechanical weakness, the 45S5 bioactive glass remains a promising biomaterial for various biomedical applications due to its distinctive properties and bonding capabilities.

This study seeks to investigate the *in vitro* bioactivity and mechanical properties of bioactive glasses and their derivatives that incorporate oxides of rare earth elements (REEs), specifically CeO_2, La_2O_3, Gd_2O_3, and Sm_2O_3. These particular REEs were chosen due to their biocompatibility, ease of processing, high-temperature stability, and adjustable release kinetics, rendering them a promising substitute for growth factors and genetic methods in tissue engineering. The study aims to evaluate the potential of these materials as a viable option for biomedical applications [4, 5]. Based on the literature survey, Ce^{4+}, La^{3+}, Sm^{3+}, and Al^{3+} ions have been considered. Cerium and lanthanum are particularly attractive due to their good biological properties, and their incorporation is not expected to negatively impact bone formation. Incorporating CeO_2, La_2O_3, and Sm_2O_3 into 45S5 bioactive glass can potentially enhance its biological and mechanical properties. Recently, Sm_2O_3-doped bioactive glasses and their ceramic derivatives have been synthesized for the first time. To evaluate the physical, mechanical, and bioactive properties of these materials, additional investigations are required.

The addition of La, Ce, and Sm to bioactive glasses can affect their network connectivity, such as ion release rate and HA formation. However, at higher concentrations, these oxides can be cytotoxic. Studies have shown that bioactive scaffolds containing less than 4 wt% of these oxides are not cytotoxic. Obtaining structural information from a single total diffraction pattern is challenging due to the low concentration of REEs in these glasses. Therefore, multiple techniques in order to examine the structure and properties of these bioactive glasses, including X-ray diffraction (XRD), scanning electron microscopy (SEM), and Fourier transform infrared spectroscopy (FTIR), are employed.

9.3 Role of REEs (CeO_2, La_2O_3, and Sm_2O_3)

Researchers in the field of orthopedic tissue engineering have shown great interest in redox-active rare earth oxide-like cerium oxide nanoparticles due to their therapeutic potential. These nanoparticles have been reported to have oxygen-buffering capacity and act as neuroprotective agents by limiting the amount of oxygen required to kill cells. The radical-scavenging potential of cerium nanoparticles is dependent on the reversibility of the Ce^{3+}/Ce^{4+} redox couple, which results in the generation of surface defects. Partially reduced cerium oxides contain oxygen vacancies in their lattice structure due to the loss of oxygen during the reaction [Eq. (9.1)]. Although non-stoichiometric cerium oxide ($CeO_{2-\delta}$) can be readily reoxidized back to CeO_2 by oxygen, oxidized CeO_2 nanoparticles can be easily reduced when placed in water or coated with PEG.

$$CeO_2 \longleftrightarrow CeO_{2-\delta} + \delta/2\ O_2 \qquad \text{Eq. (9.1)}$$

Additionally, previous studies have shown that the introduction of cerium-based bioglass or its oxide into the glass network may affect bioactivity and, therefore, needs to be evaluated. These studies have also demonstrated that bioactivity decreases with increasing cerium content in bioglass. Specifically, the deposition of an apatite layer was delayed in these glasses forming after 7–14 days of incubation in simulated body fluid. Nonetheless, despite the potential impact of cerium on bioactivity, the incorporation of cerium oxide nanoparticles in bioactive glasses has shown to have a positive effect on the proliferation, differentiation, and mineralization of primary osteoblasts, as well as antibacterial activities.

9.4 Uses of Rare Earth Elements

Lanthanum is a rare earth element that has shown promising results in enhancing the properties of orthopedic bioimplants. It is commonly used to substitute calcium in hydroxyapatite or bioceramic materials such as bioglass, which can improve the physicochemical and biological properties of the substituted materials. The incorporation of La^{3+} in hydroxyapatite stabilizes its apatite structure, leading to improvements in tensile and bending strengths. Additionally, La-incorporated hydroxyapatite has been observed to facilitate the formation of fibrous tissue around the matrices without causing significant inflammation after implantation in the bones of rats. Lanthanum substitution for calcium in apatite also enhances the resistance of hard tissues to acid dissolution. These findings suggest that lanthanum has significant potential for improving the performance of orthopedic implants in biomedical applications.

Aluminum oxide (Al_2O_3) has been extensively studied as an additive in melt-quenched glasses since the 1990s to improve their mechanical properties. When added to alkali glasses or glazes, alumina acts as a network intermediate, providing strength, chemical resistance, and greater devitrification resistance. Additionally, the addition of alumina raises the softening and melting points of glass. Incorporating aluminum was observed to

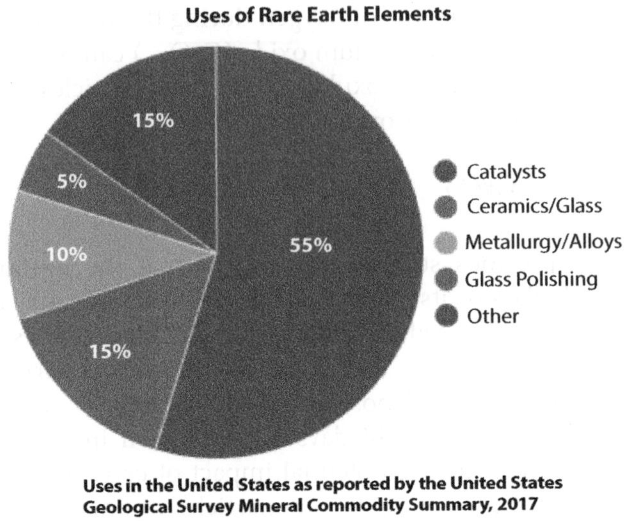

Figure 9.2 Classification of bioactive materials according to the United States Geological Survey Mineral Commodity, 2017 [14].

accelerate the formation of bioactive hydroxyapatite layers on the surface of phosphate-rich bioglasses. Al_2O_3 has been incorporated into bioactive glasses with the expectation of enhancing bone tissue defect regeneration potential and controlling the biodegradation of bioglass-based implants for long-term stability. M. A. K. Elfayoumi reported that incorporating Al^{3+} into bioglass at an appropriate concentration enhances the mechanical resistance of the bioglass for successful application as orthopedic implants. The incorporation of Al_2O_3 leads to a glass network, the P = O bonds break and are exchanged by Al–O–P linkages, while the Si–O–P linkages are also substituted, thus enhancing the resistance of bioglass to biodegradation by increasing its stability under physiological conditions (Figure 9.2).

9.5 Biomaterials

Biomaterials are materials, whether natural or synthetic, that are designed to interact with living tissues in the body. According to the definition by Williams, biomaterials are non-toxic materials that are expected to connect with biological systems to treat, replace, or augment any tissue, organ, or function of the body while evoking a minimal biological response. In other words, biomaterials are used to fabricate artificial organs, prostheses, and rehabilitation devices, as well as to replace natural tissues. The attachment of an implant to tissue is a critical factor in determining the success of the implantation process. It is important to note that no material implanted in the body is entirely inert, as all materials elicit a response from the host tissue. Therefore, biomaterials are typically categorized into four types based on the nature of the attachment between the implant and the surrounding tissue summarized in Table 9.1.

The compositions in region A have a fixed P_2O_5 content of 6 wt%, and the A/W glass has a higher P_2O_5 content, as illustrated in Figure 9.3. Region A can form hydroxyapatite both *in vitro* and *in vivo*. The content within the dashed line can also bind to soft tissues. Materials in region B are inert, while those in region C are restorable. Region D is non-glass forming and does not bond.

Biomaterials can achieve different types of tissue attachment. Nearly inert implants provide only morphological fixation, leading to a non-adherent fibrous capsule. Porous biomaterials provide biological fixation, allowing tissue ingrowth into the implant. Resorbable implants gradually degrade and are replaced by natural host tissues. Bioactive implants offer bioactive fixation, which involves a series of biophysical and biochemical reactions that result in a strong chemical interfacial bonding between the implant and tissue. The type of attachment depends on the biomaterial used, and its application and design should consider the desired attachment and tissue response.

Table 9.1 Types of tissue attachment for biomaterials based on the nature of the interface between the implanted material and the surrounding tissue [1].

Type of implant	Type of attachment	Example
Nearly inert	Mechanical interlock	Metals, alumina, zirconia, polyethylene
Porous	In the growth of tissues into pores	Hydroxyapatite, hydroxyapatite-coated porous metals
Bioactive	Interfacial bonding with tissues	Bioactive glasses, hydroxyapatite
Resorbable	Replacement with tissues	Tricalcium phosphate, polylactic acid

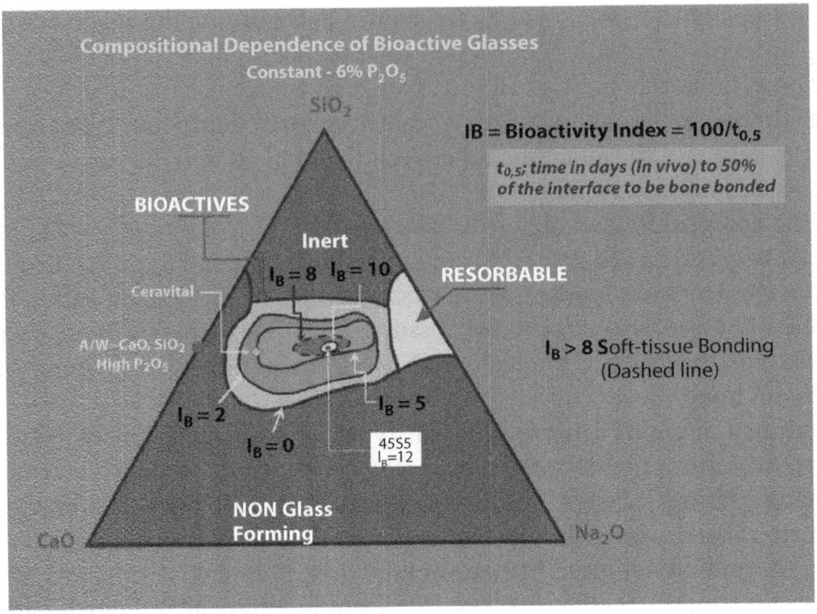

Figure 9.3 A kinetic diagram illustrating the bioactivity of SiO_2–Na_2O–CaO bioactive glasses is shown, along with a compositional dependence (in wt%) of the materials' bonding to bone and soft tissue [2].

9.6 Simulated Body Fluid

Simulated body fluid (SBF) is utilized as a solution to replicate the chemical environment of human blood plasma, and it serves as an effective tool for assessing the bioactivity of materials intended for biomedical applications. Kokubo first developed SBF in 1991, with the aim of producing a material that could bond with living tissues by generating hydroxyl-carbonate apatite (HCAp) on its surface after being implanted in the body [7]. Although the ion concentrations in SBF are comparable to those in human blood plasma, there are differences in some ions. Consequently, other researchers have formulated alternative SBF solutions. In 2006, Kokubo and colleagues developed a revised SBF (r-SBF) that included Cl^- and HCO^{3-} ions adjusted to the levels present in human blood plasma. However, r-SBF exhibited a propensity to precipitate calcium carbonate. A modified SBF (n-SBF) was later proposed, where only the Cl^- ion concentration was increased. n-SBF does not differ in terms of stability and reproducibility from the SBF developed by Kokubo. The use of SBF and its variants has greatly facilitated the development and assessment of biomaterials for bone regeneration and tissue repair.

9.7 Bioactive Glasses

To address the issue of poor mechanical properties of silica-based bioactive glasses, researchers have explored the use of other materials as alternatives or as composites with silica-based glasses. For example, bioactive glasses based on borate, phosphate, or fluoride have been investigated as potential alternatives to silica-based glasses. These glasses have shown promising results in terms of mechanical properties, but their bioactivity and biocompatibility are still being studied. Composites of bioactive glasses with other materials, such as polymers or ceramics, have also been explored as a way to improve the mechanical properties of bioactive glasses while maintaining their bioactivity and biocompatibility. For example, bioactive glass–polymer composites have been developed for bone tissue

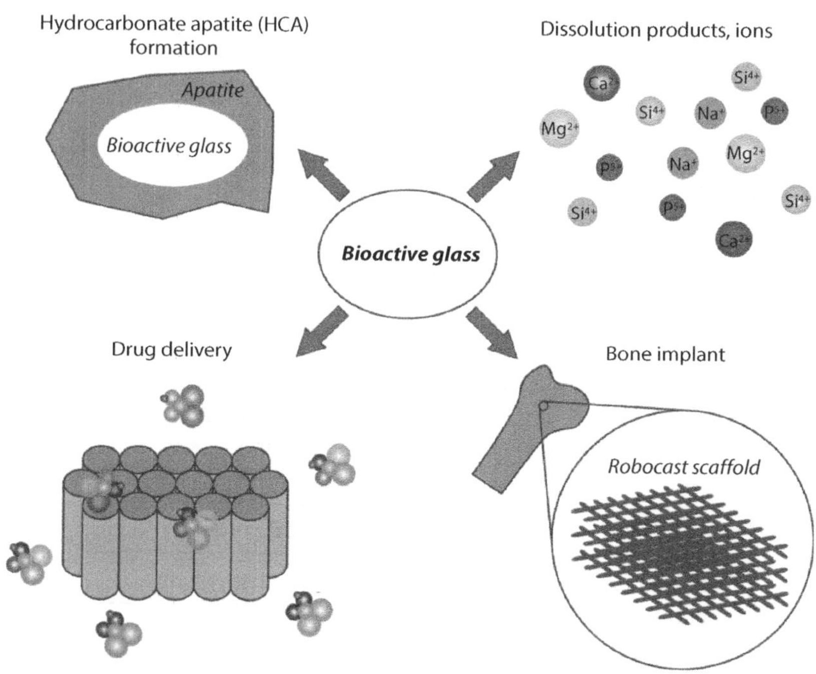

Figure 9.4 Specific biological activities can be induced by exploiting different aspects of bioactive glass (BG) materials [5].

engineering applications, and bioactive glass–ceramic composites have been explored for dental applications [8].

Overall, while silica-based bioactive glasses have been widely studied and used in various clinical applications, the issue of poor mechanical properties has led to the exploration of alternative materials and composites to expand their potential clinical applications shown in Table 9.2.

Silica-based glasses have an amorphous network structure with SiO_4^- tetrahedra as the structural unit. The open structure of silica-based glass is created by the presence of non-bridging oxygen ions, which are generated by network modifiers such as Na^+, K^+, and Ca^{2+}. The bioactivity of glass is dependent on the quantity of non-bridging oxygen ions in the silica tetrahedron., which should be greater than 2.6 for a silica-based glass to be bioactive. Traditional silica-based glasses contain over 65% SiO_2 by weight, less than 15% Na_2O by weight, and around 10% CaO by weight. Bioactive glasses (Figure 9.4), on the other hand, have a composition that is different

Table 9.2 Clinical applications of silica-based bioactive glasses [15].

Material form	Clinical application
Solid shapes	- Ossicle replacement in the middle ear
	- Cone-shaped devices for jaw defect filling
	- Curved plates for restoring eye orbit floor
	- Soft tissue sealing for transdermal implants
Particulates	- Bone tissue replacement in periodontal diseases
	- Soft tissue augmentation in the paralysis of vocal cords
Particulates and autologous bone	- Maxillofacial reconstructions
	- Spine
Particulates by injection	- Urological tissue augmentation

from traditional silica-based glasses, containing less than 60% SiO_2 by weight and more alkali and/or alkaline earth oxides.

In summary, bioactive glasses differ from traditional SiO_2–Na_2O–CaO glasses in their composition, with a SiO_2 content of 40%–60% by weight, high Na_2O and CaO content, and a high CaO/P_2O_5 ratio being the key features that make them bioactive. A SiO_2 content above 60% will reduce the network dissolution rate, while a content below 40% will result in totally dissolved monomeric SiO_4^{4-} units, making it unlikely to obtain a glass phase of this composition. P_2O_5 was previously thought to be required for bioactivity, but it was later found that it only aids in the nucleation of the calcium phosphate phase on the surface and is not a critical constituent [6]. The base components in most silica-based bioactive glasses are SiO_2, Na_2O, CaO, and P_2O_5.

The process by which silica-based bioactive glass bonds with tissues is believed to involve the creation of a layer of hydroxyl-carbonate apatite (HCA) on the surface of the glass upon contact with body fluids. The HCA layer is formed through a sequence of reactions that occur on the surface of the bioactive glass implant. Initially, the surface of the glass reacts with water molecules to form silanol (Si-OH) groups. These silanol groups then hydrolyze the ions in the body fluid, such as Ca^{2+} and PO_4^{3-}, to form a layer of amorphous calcium phosphate (ACP). The ACP layer then undergoes a transformation to form a crystalline HCA layer. The HCA layer provides a surface that is biocompatible and allows for the attachment and growth of

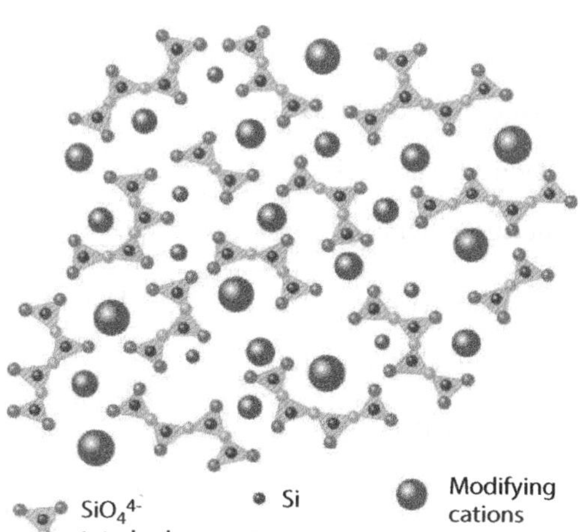

Figure 9.5 Two-dimensional presentation of a random glass network is a non-periodic arrangement of network modifiers and network formers [7].

bone tissue. The formation of the HCA layer is a critical step in the process of osseointegration, which is the integration of the bioactive glass implant with the surrounding bone tissue (Figure 9.5).

9.8 Bioactive Composites

Bioactive materials are useful in forming biochemical bonds with the bone and soft tissues, but their mechanical properties are often unsuitable for load-bearing applications. To address this, bioactive composites are created by combining bioactive materials with a fracture-tough material. There are two main types of composites: those in which a tough material reinforces a bioactive matrix and those in which bioactive materials reinforce an elastically compliant and biocompatible polymer matrix. The former type of composite uses metal fibers or tough ceramic particles to reinforce the bioactive material, while the latter type uses powders, particles, or fibers of bioactive materials to reinforce a polymer matrix. Composites derived from two bioactive materials have also been developed for applications without load-bearing requirements [9–13]. *In vivo*, tests have shown that these composites can provide an alternative for bone defect reconstruction, especially in the head and neck area.

9.9 Area of Biomaterials

✓ Glasses ✓ Tissue culture vessels ✓ Diagnostic devices ✓ Thermometers	✓ Bone and dental sector
✓ Filling materials ✓ Gold porcelain coating ✓ Prosthetic parts	✓ Dental sector

References

1. Hench, L.L., The story of bioglass TM. *J. Mater. Sci. Mater. Med.*, 17, 967–978, 2006.
2. Hench, L.L. and Polak, J.M., Third-generation biomaterials. *Science*, 295, 1014–1017, 2002.
3. Beherei, H.H., Fabrication and characterization of bioactive glass (45S5)/ titania biocomposites. *Ceram. Int.*, 35, 1991–1997, 2009.
4. Kim, C.Y., Early stages of calcium-phosphate layer formation in bioglasses. *J. Non-Cryst. Solids*, 113, 195–202, 1989.
5. Polini, A., Bai, H., Tomsia, A.P., Dental applications of nanostructured bioactive glass and its composites. *Wiley Interdiscip. Rev. Nanomed. Nanobiotechnol.*, 5, 4, 399–410, 2013.
6. Deliormanlı, A.M., Synthesis and characterization of cerium-and gallium-containing borate bioactive glass scaffolds for bone tissue engineering. *J. Mater. Sci. Mater. Med.*, 26, 67, 2015.
7. Kokubo, T., How useful is SBF in predicting *in vivo* bone bioactivity? *Biomaterials*, 27, 15, 2907–2915, 2006.
8. Benedetto, A., Effects of the rare elements lanthanum and cerium on the growth of colorectal and hepatic cancer cell lines. *Toxicol. In Vitro*, 46, 9–18, 2018.
9. Calahoo, C. and Wondraczek, L., Ionic glasses: Structure, properties, and classification. *J. Non-Cryst. Solids X*, 8, 100054, 2020.
10. Ali, A., Studies on the effect of CuO addition on mechanical properties and *in vitro* cytocompatibility in 1393 bioactive glass scaffold. *Mater. Sci. Eng. C*, 93, 341–355, 2018.
11. Ershad, M., Mechanical and biological response of (CeO2+ L2O3)-substituted 45S5 bioactive glasses for biomedical application. *J. Aust. Ceram. Soc.*, 56, 1243–1252, 2020.
12. El Batal, F.H., Preparation and characterization of some substituted bio-glasses and their ceramic derivatives from the system SiO2–Na2O–CaO–P2O5 and effect of gamma irradiation. *Mater. Chem. Phys.*, 110, 352–362, 2008.
13. Greenlee Jr., T.K., Glass-ceramic bone implants. A light microscopic study. *J. Biomed. Mater. Res.*, 6, 235–244, 1972.
14. U.S. Geological Survey, Mineral commodity summaries 2017: U.S. Geological Survey, 202p., 2017. https://doi.org/10.3133/70180197
15. Al-Harbi, N. *et al.*, Silica-based bioactive glasses and their applications in hard tissue regeneration: A review. *Pharmaceuticals*, 14.2, 75, 2021.

10

Biomedical Applications of Composite Materials

Mulugundam Siva Surya*, Atla Sridhar and Maddula Satya Prasad

GITAM School of Technology, Hyderabad, Telangana, India

Abstract

Materials with two or more unique constituent phases on a scale more significant than the atomic level are called composites. The real benefit of composites over conventional homogeneous materials like metals, ceramics, and polymers is that their mechanical, biological, and other physical properties may be tuned to the needs of particular applications. As a result, during the past 40 years, composites have grown steadily; now, composite materials are used in various industries, including aerospace, automotive, marine, and more. As a result, numerous composite biomaterials have recently undergone research and testing for use in medicine. Due to their benefits over conventional materials, some are currently being commercialized. Most human tissues, including bones, tendons, skin, ligaments, and teeth, comprise single constituents. How much, where it is distributed, how it is shaped, and how it behaves ultimately decide how the tissue or organ will behave. Prosthetics that closely resemble these biological tissues, match their mechanical behavior, and restore the automated functions of the injured tissue can be created using synthetic composites.

***Keywords*:** Composite, biomedical application, matrix, fiber

10.1 Introduction

A composite material comprises two or more components with very different physical and chemical properties. The matrix phase and the reinforcement phase are the two main phases of a composite. The development

**Corresponding author*: sivasurya.mtech@gmail.com

Jay Prakash Srivastava, Dražan Kozak, Vinayak Ranjan, Pankaj Kumar, Ranjan Kumar and Shubham Tayal (eds.) Mechanical Engineering in Biomedical Applications: Bio-3D Printing, Biofluid Mechanics, Implant Design, Biomaterials, Computational Biomechanics, Tissue Mechanics, (277–290) © 2024 Scrivener Publishing LLC

of the material properties is independently aided by both phases, which do not mix and keep their characteristics [1, 2]. The automotive sector is interested in using bio-based composites in the future. Investigating new materials with desirable features is a response to the challenge of making automotive goods lighter, safer, and more affordable. Additionally, the growing awareness of environmental problems compels researchers and producers to focus on the domains of composite and biocomposite materials [3–5]. Ecological concerns and issues, including sustainability and cost, have lately been addressed worldwide. Biomaterial technology, which has great promise, has been the focus of manufacturing research. Many researchers have been drawn to biocomposites because of how easily they can be disposed of and how they can be composted beyond their use-by date, which is typically not achievable with standard synthetic materials. Additionally, biofibers are increasingly desirable due to the advantages of natural fibers over synthetic counterparts, including low weight, corrosion resistance, and high strength. However, biofibers have significant disadvantages, such as anisotropy and increased moisture absorption [6, 7]. Due to their ease of disposal and capacity for composting after expiry, which is often not achievable with conventional synthetic materials, biocomposites have attracted the attention of many researchers. Additionally, natural fibers are preferable to synthetic equivalents due to their low weight, high strength, and corrosion resistance, elevating biofibers' appeal. However, anisotropy and higher moisture absorption are the drawbacks of biofibers. Natural or artificial materials have been employed since the dawn of medicine in prehistoric societies to heal tissue, replace or integrate biological functions or organs destroyed by trauma or disease, or correct abnormalities [8–10]. Higher patient survival rates, a considerable gain in life expectancy, and improvements in quality of life are all made possible by the development of advanced medical technologies and therapeutic advances. To create a composite with specified physical, chemical, and mechanical properties, two or more materials with various compositions, morphologies, and general physical properties are blended heterogeneously on a macroscopic scale. Multiple chemicals are combined to produce a substance that maximizes specific properties. The advantage of composites is that they often demonstrate features that the individual parts do not and the finest qualities of each of its constituents. As a result of their ability to adjust their structure and properties to specific applications, composite materials also enable flexible design.

Numerous elements, including the matrix material, the fiber, the orientation of the fiber, the type of use for the fiber, the production process, and others, influence the final properties of composite materials. The matrix

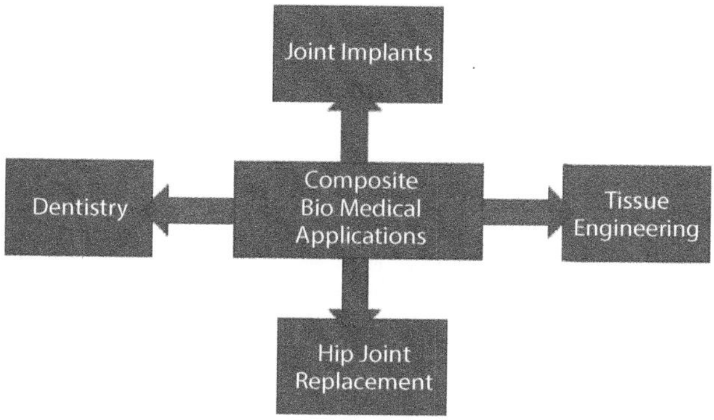

Figure 10.1 Biomedical applications of composites.

material is a crucial factor in determining a composite material's mechanical and chemical properties, and most matrix materials are derived from petrochemicals. A significant problem for composites is the depletion of fossil fuel sources. As a result, several researchers are researching plant-derived natural resins to provide a matrix material for composites. Figure 10.1 shows the biomedical applications of composites.

10.2 Different Types of Composites Used in Biomedical Applications

Composite materials, which are made by combining two or more different materials, have found extensive applications in the field of biomedical engineering due to their unique properties and performance characteristics. Here are some examples of composite materials commonly used in biomedical applications as shown in Figure 10.2.

Carbon Fiber-Reinforced Polymers (CFRP): Carbon fiber-reinforced polymer composites, or CFRP composites, are renowned for their high strength, low weight, and great biocompatibility. For usage in orthopedic implants, prosthetics, and other load-bearing applications, CFRP composites are appropriate because carbon fibers have exceptional mechanical qualities, such as high tensile strength and stiffness. Additionally, CFRP composites can be utilized in non-load-bearing applications including tissue engineering scaffolds and medication delivery systems.

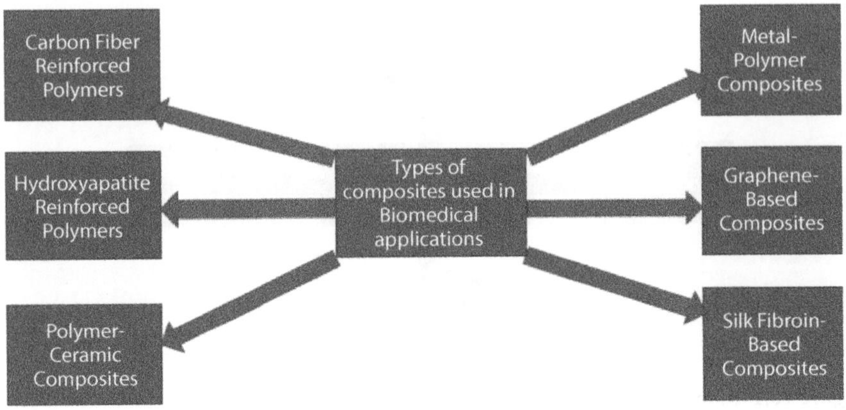

Figure 10.2 Composites used in biomedical applications.

Hydroxyapatite (HA)-Reinforced Polymers: HA is a biocompatible ceramic substance that mirrors the mineral makeup of human bones very closely. HA-reinforced composites with polymers show great bioactivity and biocompatibility, making them appropriate for bone tissue engineering applications. The increased mechanical strength and osteoconductivity that HA offers encourage the formation and regeneration of new bone. The application of HA-reinforced polymer composites in bone implants, bone graft alternatives, and tissue engineering scaffolds is possible.

Polymer–Ceramic Composites: Polymer–ceramic composites blend ceramic particles like calcium phosphates or bioactive glass with biocompatible polymers like PLA or PCL. Depending on the composition and processing methods employed, these composites may have a variety of qualities, including mechanical strength, biodegradability, and bioactivity. Dental materials, drug delivery systems, and bone tissue engineering are just a few biomedical uses for polymer–ceramic composites.

Metal–Polymer Composites: Metal–polymer composites combine biocompatible metals with polymers to produce materials with distinctive characteristics, such as titanium or stainless steel. While the polymer component enhances biocompatibility and lessens the effects of stress shielding, the metal component offers mechanical strength and can be used in load-bearing applications. Orthopedic implants including joint replacements and fixation devices frequently use metal–polymer composites.

Graphene-Based Composites: Polymer matrices can be combined with graphene, a two-dimensional nanomaterial with extraordinary characteristics, to produce composites with improved qualities for biomedical applications. Due to their distinct mechanical, electrical, and thermal properties, graphene-based composites have shown promise in applications like drug delivery systems, biosensors, and scaffolds for tissue engineering. Graphene is useful for a variety of biological applications due to its large surface area and biocompatibility.

Silk Fibroin-Based Composites: A natural protein derived from silkworm cocoons called silk fibroin can be mixed with ceramics or polymers to make composites that have favorable qualities for biological applications. The mechanical property, biocompatibility, and biodegradability of silk fibroin-based composites make them useful for tissue engineering, dressings for wounds, and drug delivery systems.

These are only a few examples of the broad spectrum of composite materials utilized in biological applications. Composites are extremely adaptable for a variety of biomedical applications because the mixing of multiple materials allows for the customization of properties to fit certain requirements. In order to develop enhanced composite materials with better performance for healthcare applications, research and development is ongoing in the field of composite materials in biomedical engineering.

10.3 Application of Composites in Tissues

Composites have shown promise as materials in tissue engineering, which tries to create functioning replacement tissues or organs for regenerative medicine. To assist in cell development, proliferation, and tissue regeneration, composites used in tissue engineering are often designed to match the structural and functional characteristics of the natural tissue. Here are some instances of how tissue engineering makes use of composites:

Scaffold Materials: Composites are frequently employed as scaffolding materials in tissue engineering. Scaffolding materials give cells a place to attach, develop, and differentiate into valuable tissues. Biodegradable polymers like poly(lactic acid), poly(glycolic acid), and their copolymers (PLGA), as well as other materials like ceramics, metals, or organic polymers like collagen or chitosan, are frequently combined to create composite scaffolds. To encourage tissue regeneration, the composite scaffold's composition, structure, and qualities can be adjusted to match the demands of the target tissue, such as mechanical strength, porosity, and degradation rate.

Bone Tissue Engineering: In bone tissue engineering, composites are frequently utilized to create scaffolds that replicate genuine bone's mechanical and biological characteristics. The mechanical strength and osteoconductivity required for bone regeneration, for instance, can be provided by composites consisting of biodegradable polymers and ceramics, such as hydroxyapatite (HA) or tricalcium phosphate (TCP). These composites can facilitate bone tissue regeneration by encouraging cell adhesion, proliferation, and differentiation into bone cells.

Cartilage Tissue Engineering: Using composites in cartilage tissue engineering allows for replicating the unique characteristics of cartilage, such as its high load-bearing capacity and minimal friction. Composites containing biodegradable polymers, like poly(caprolactone) (PCL) or poly(ethylene glycol) (PEG), as well as organic or inorganic substances, like collagen or hyaluronic acid, can offer the necessary mechanical properties and encourage the differentiation of cells into cartilage tissues, or chondrogenesis.

Skin Tissue Engineering: To create scaffolds that can encourage skin cell adhesion, proliferation, and differentiation, composites are used in skin tissue engineering. Composite scaffolds that imitate the mechanical characteristics and structure of the skin may be created using natural materials like collagen or elastin and biodegradable polymers like polycaprolactone (PCL). These composites can speed up skin tissue regeneration for skin restoration or wound-healing applications.

Vascular Tissue Engineering: In vascular tissue engineering, composites create scaffolds with mechanical characteristics and a structure resembling blood vessels. Composites made of biodegradable polymers like poly(lactic acid) (PLA) or poly(caprolactone) (PCL), combined with other substances like hydrogels or elastin, can offer good mechanical properties and support the growth and organization of vascular cells, resulting in the formation of functional blood vessels.

Neural Tissue Engineering: In neural tissue engineering, composites are also used to create scaffolds that encourage nerve cell adhesion, development, and regeneration. To support nerve cell growth and renewal for applications in spinal cord repair or nerve regeneration, composites made from biodegradable polymers, such as poly(lactic-co-glycolic acid) (PLGA), and natural or synthetic materials, such as laminin or poly(ethylene glycol) (PEG), can provide the necessary mechanical properties and bioactivity [11–13].

10.4 Application of Composites in Dentistry

Composites have found numerous applications in dentistry due to their unique properties, including aesthetics, biocompatibility, and versatility. Here are some examples of how composites are used in dentistry:

Dental Fillings: As tooth-colored dental fillings, composite resins are frequently utilized to repair teeth that have suffered from dental caries or tooth decay. Due to their natural look and capacity to blend in with neighboring teeth, composite fillings are favored by many patients. They are often made of a combination of biocompatible fillers, such as silica or glass particles, and inorganic fillers, like bisphenol-A-glycidyl methacrylate (Bis-GMA), which offer the appropriate mechanical qualities and aesthetics.

Dental Bonding: In dental bonding operations, composites are used to fix cracked, chipped, or discolored teeth. Applying a composite resin material directly to the tooth surface, shaping, polishing, and attaching the substance to the tooth using a curing light are all steps in the dental bonding process. Without considerable tooth reduction, composite bonding is a conservative and less invasive surgery that can restore the natural appearance and function of teeth.

Veneers: Dental veneers are thin shells glued to the front surface of the teeth to enhance their appearance. Veneers are made of composite materials. Composite veneers offer a cosmetic fix for discolored, out-of-place, or broken teeth since they may be manufactured specifically to match the original color, shape, and size of the teeth.

Dental Inlays and Onlays: Composites can be used to fabricate dental inlays and onlays, which are indirect restorations used to repair larger cavities or damaged tooth surfaces. Inlays and onlays are custom-made restorations fabricated in a dental laboratory and bonded to the tooth using dental adhesives. Composites used for inlays and onlays can be color-matched to natural teeth and provide durable and aesthetic restorations.

Orthodontic Appliances: Composites are used in orthodontics to fabricate various appliances, such as orthodontic brackets, retainers, and aligners. Composite brackets are tooth-colored and provide an aesthetic alternative to traditional metal brackets. Composite retainers and aligners are also commonly used as removable appliances to maintain tooth alignment after orthodontic treatment.

Dental Implants: Dental inlays and onlays are indirect restorations covering bigger cavities or repairing chipped or broken tooth surfaces. Composites can be utilized to create these restorations. Inlays and onlays are individualized restorative materials created in a dental lab and then attached to the tooth with dental adhesives. Inlays and onlays made of composite material can be color-matched to natural teeth to create long-lasting and aesthetically pleasing repairs.

Endodontics: Composite materials are employed in endodontics, the area of dentistry that deals with root canal therapy. Composites can be employed as filler materials or root canal sealers to seal the root canal and stop reinfection. Their superior biocompatibility and sealability provide effective root canal treatments [14–16].

10.5 Application of Composites in Total Joint Replacements

Composites can fabricate total joint replacement implants like hip, knee, shoulder, and elbow implants. In these applications, composites are often used as the bearing surface material, which comes into direct contact with the opposing joint surface. Composites used as bearing materials in joint implants consist of ultra-high molecular weight polyethylene (UHMWPE) reinforced with various forms of reinforcement, such as fibers or particles, to improve their mechanical properties and wear resistance. Composites in total joint replacements provide low friction, durability, and long-term performance.

Spinal Implants: Composites, such as interbody fusion cages and spinal rods, are also used in spinal implants. Interbody fusion cages provide stability and promote fusion in spinal fusion surgeries. Composites used in interbody fusion cages are typically made of biocompatible polymers, such as polyetheretherketone (PEEK), reinforced with fibers, such as carbon or glass, to provide mechanical strength and stability. Spinal rods of composites stabilize the spine during fusion surgeries and can provide excellent biocompatibility, mechanical strength, and radiolucency for better visualization during postoperative imaging.

Dental Implants: Composites are used in dental implants as well, where they are used for the fabrication of implant abutments, which are the components that connect the implant fixture to the prosthetic crown or bridge. Composites used in dental implant abutments are typically made of biocompatible polymers, such as PEEK, and can provide high strength, durability, and aesthetics. Composite implant abutments can be customized to match the patient's natural tooth color and shape, providing a natural-looking and functional replacement for missing teeth.

Custom Joint Implants: Composites are also used to fabricate custom collaborative implants, specifically designed and manufactured for individual patients with unique anatomical requirements. Custom joint implants can be made using additive manufacturing (3D printing) techniques, where composites are used to create patient-specific implants with precise geometry, fit, and mechanical properties. Custom joint implants made of composites can improve patient outcomes, improve implant fit, and reduce surgical complications.

Revision Joint Surgeries: Composites can be used in revision joint surgeries, which are performed to replace failed or worn-out joint implants. In these cases, composites can be used as modular components, such as revision femoral heads or acetabular cups, to replace the worn-out parts of the existing implant while preserving the intact portions. Composites used in revision joint surgeries can provide mechanical strength, wear resistance, and ease of modification [17–21].

10.6 Application of Composites in Hip Joint Replacement

Composites are commonly used in hip joint replacement applications due to their excellent mechanical properties, biocompatibility, and wear resistance. Here are some examples of how composites are used in hip joint replacement:

Acetabular Cups: The acetabular cup is the socket component of the hip joint replacement that replaces the damaged or diseased acetabulum (hip socket). Composites are used to fabricate acetabular cups to provide durability, wear resistance, and long-term performance. Typically, acetabular cups are made of metal-backed or ceramic-backed composites. Metal-backed composites consist of a metal shell with a composite liner, usually made of ultra-high molecular weight polyethylene (UHMWPE) reinforced with fibers, such as carbon or ceramic, to improve their mechanical properties. Ceramic-backed composites, on the other hand, consist of a ceramic shell with a composite liner, providing excellent wear resistance and biocompatibility.

Femoral Heads: The femoral head is the ball component of the hip joint replacement that replaces the damaged or diseased femoral head. Composites are used to fabricate femoral heads to provide high strength, wear resistance, and low friction against the acetabular cup. Composite femoral heads are typically made of ceramic composites, such as zirconia-toughened alumina or ceramic-coated composites, providing superior wear performance to traditional metal femoral heads.

Revision Components: Composites are also used in revision hip joint replacement surgeries, which are performed to replace failed or worn-out hip joint implants. In these cases, composites can be used as revision components, such as revision acetabular cups or femoral heads, to replace the failed components of the existing implant while preserving the intact portions. Revision components made of composites provide mechanical strength, wear resistance, and ease of revision.

Custom Implants: Composites are used to fabricate custom hip joint implants, specifically designed and manufactured for individual patients with unique anatomical requirements. Custom hip joint implants can be made using additive manufacturing (3D printing) techniques, where composites are used to create patient-specific implants with precise geometry, fit, and mechanical properties. Custom hip joint implants made of composites can improve patient outcomes, improve implant fit, and reduce surgical complications [22–25].

Conclusions

This work shows composite material applications in the biomedical implant field due to their inherent properties. The combination of different materials in composites allows for the customization of properties to meet specific requirements, making them highly versatile for various biomedical applications. Composites have a wide range of applications in tissue engineering, providing versatile and tunable materials for developing scaffolds that can mimic the properties of native tissues. Composites have many applications in dentistry, providing versatile and aesthetic materials for various restorative, cosmetic, and orthodontic procedures. Their unique properties make them a popular choice for dental applications, offering functional and aesthetic solutions for patients. They are used in various joint implant types, including total joint replacements, spinal implants, dental implants, custom joint implants, and revision joint surgeries, providing functional and durable solutions for patients with joint conditions or injuries. They are used in various components of hip joint replacements, such as acetabular cups, femoral heads, revision components, and custom implants, offering advantages such as high strength, wear resistance, and customization to meet the unique needs of patients.

References

1. Surya, M.S., Effect of SiC weight percentage and sintering duration on microstructural and mechanical behaviour of Al6061/SiC composites produced by powder metallurgy technique. *Silicon*, 14, 6, 2731–2739, 2022.
2. Surya, M.S., Kumar, V.N., Sridhar, A., To study the effect of varying sintering temperature and reinforcement on physical and mechanical characteristics of AA6061/SiC composites. *Silicon*, 15, 7, 3003–3009, 2022.
3. Yildizhan, Ş., Calik, A., Ozcanli, M., Serin, H., Bio-composite materials: A short review of recent trends, mechanical and chemical properties, and applications. *Eur. Mech. Sci.*, 2, 3, 83–91, 2018.
4. Salernitano, E. and Migliaresi, C., Composite materials for biomedical applications: A review. *J. Appl. Biomater. Biomech.*, 1, 1, 3–18, 2003.
5. Migliaresi, C. and Nicolais, L., Composite materials for biomedical applications. *Int. J. Artif. Organs*, 3, 2, 114–118, 1980.
6. Ramakrishna, S., Mayer, J., Wintermantel, E., Leong, K.W., Biomedical applications of polymer-composite materials: A review. *Compos. Sci. Technol.*, 61, 9, 1189–1224, 2001.

7. Silva, M., Pinho, I.S., Covas, J.A., Alves, N.M., Paiva, M.C., 3D printing of graphene-based polymeric nanocomposites for biomedical applications. *Funct. Compos. Mater.*, 2, 1, 1–21, 2021.
8. Biswal, T., BadJena, S.K., Pradhan, D., Synthesis of polymer composite materials and their biomedical applications. *Mater. Today Proc.*, 30, 305–315, 2020.
9. Zhang, S., Li, F., Yu, J.Y., Novel cellulose/SPI blend bio-fibers prepared via direct dissolving approach. *J. Eng. Fibers Fabr.*, 6, 1, 155892501100600105, 2011.
10. Tavares, T.D., Antunes, J.C., Ferreira, F., Felgueiras, H.P., Biofunctionalization of natural fiber-reinforced biocomposites for biomedical applications. *Biomolecules*, 10, 1, 148, 2020.
11. Salernitano, E. and Migliaresi, C., Composite materials for biomedical applications: A review. *J. Appl. Biomater. Biomech.*, 1, 1, 3–18, 2003.
12. Boccaccini, R. and Blaker, J.J., Bioactive composite materials for tissue engineering scaffolds. *Expert Rev. Med. Devices*, 2, 3, 303–317, 2005.
13. Aslam Khan, M.U., Abd Razak, S., II, Al Arjan, W.S., Nazir, S., Sahaya Anand, T.J., Mehboob, H., Amin, R., Recent advances in biopolymeric composite materials for tissue engineering and regenerative medicines: A review. *Molecules*, 26, 3, 619, 2021.
14. Ensaff, H., O'Doherty, D.M., Jacobsen, P.H., The influence of the restoration–Tooth interface in light cured composite restorations: A finite element analysis. *Biomaterials*, 22, 23, 3097–3103, 2001.
15. Fujihara, K., Teo, K., Gopal, R., Loh, P.L., Ganesh, V.K., Ramakrishna, S. *et al.*, Fibrous composite materials in dentistry and orthopaedics: Review and applications. *Compos. Sci. Technol.*, 64, 6, 775–788, 2004.
16. Matinlinna, J.P., Lassila, L.V., Özcan, M., Yli-Urpo, A., Vallittu, P.K., An introduction to silanes and their clinical applications in dentistry. *Int. J. Prosthodont.*, 17, 2, 155–164, 2004.
17. Costa, R.R.C.D., Almeida, F.R.B.D., Silva, A.A.X.D., Domiciano, S.M., Vieira, A.F.C., Design of a polymeric composite material femoral stem for hip joint implant. *Polímeros*, 29, 1–8, 2020.
18. Song, J., Liao, Z., Shi, H., Xiang, D., Liu, Y., Liu, W., Peng, Z., Fretting wear study of PEEK-based composites for bio-implant application. *Tribol. Lett.*, 65, 1–11, 2017.
19. Katti, K.S., Biomaterials in total joint replacement. *Colloids Surf. B Biointerfaces*, 39, 3, 133–142, 2004.
20. Affatato, S., Ruggiero, A., Merola, M., Advanced biomaterials in hip joint arthroplasty. A review on polymer and ceramics composites as alternative bearings. *Compos. B Eng.*, 83, 276–283, 2015.
21. Cho, S.M., Park, J.W., Han, H.S., Seok, H.K., Moon, M.W., Kim, Y.C., Multifunctional composite coating as a wear-resistant layer for the bearing in total hip joint replacement. *ACS Appl. Mater. Interfaces*, 5, 2, 395–403, 2013.

22. Ghalme, S.G., Mankar, A., Bhalerao, Y., Biomaterials in hip joint replacement. *Int. J. Mater. Sci. Eng.*, 4, 2, 113–125, 2016.
23. Skinner, H.B., Composite technology for total hip arthroplasty. *Clin. Orthop. Relat. Res.*, 235, 224–236, 1988.
24. Goswami, C., Patnaik, A., Bhat, I.K., Singh, T., Mechanical physical and wear properties of some oxide ceramics for hip joint application: A short review. *Mater. Today Proc.*, 44, 4913–4918, 2021.
25. Rahaman, M.N., Yao, A., Bal, B.S., Garino, J.P., Ries, M.D., Ceramics for prosthetic hip and knee joint replacement. *J. Am. Ceram. Soc.*, 90, 7, 1965–1988, 2007.

Part III

BIOFLUID MECHANICS

11

Materials Advancement, Biomaterials, and Biosensors

Ashish Kumar Bhui[1]*, Priyanka Singh[2], Yunus Raza Baig[2], Sanvedna Shukla[2], Satish Sen[3], Amar Dey[4] and Rajmani Patel[5]

[1]*Govt. Shahid Koushal Yadav College, Gunderdehi, Balod, Chhattisgarh, India*
[2]*Govt. Digvijay College, Rajnandgaon, Chhattisgarh, India*
[3]*Govt. V.Y.T. College, Durg, Chhattisgarh, India*
[4]*Bhilai/Institute of Technology, Durg, Chhattisgarh, India*
[5]*Govt. Ghanshyam Singh Gupt P.G. College Balod, Chhattisgarh, India*

Abstract

The development of material science research and its association with biomaterials has resulted in its technical improvisation toward wider biomedical applications. Biomaterials are bioactive analytes for advanced interaction applications and are non-toxic. These materials can enhance, replace, or replicate a live natural ingredient device. Biosensors simultaneously provide a convenient scientific zone for various biomedical and industrial multidisciplinary applications in an ever-expanding manner through specified biochemical reaction processes mediated by isolated organelles, tissues, immune systems, enzymes, or whole cells for monitoring cell optical, electrical, or thermal signals. Therefore, biomaterials and biosensors provide a sensitive diagnosis for various pharmacological species. They decrease the use of toxic or hazardous chemicals and solvent systems during fabrication, reducing chemical waste. This review article discusses various aspects of biomaterials and biosensors, from their origin to applications in the present context. Biomaterial and biosensor research has a vivid opportunity when shared with other elements like supramolecular chemistry, chemoreceptors, and protein engineering.

***Keywords*:** Biomaterial, biosensor, green sensing platform, supramolecular chemistry, chemoreceptor, protein engineering

**Corresponding author*: bhuiashishkumar@gmail.com

Jay Prakash Srivastava, Dražan Kozak, Vinayak Ranjan, Pankaj Kumar, Ranjan Kumar and Shubham Tayal (eds.) Mechanical Engineering in Biomedical Applications: Bio-3D Printing, Biofluid Mechanics, Implant Design, Biomaterials, Computational Biomechanics, Tissue Mechanics, (293–326) © 2024 Scrivener Publishing LLC

11.1 Introduction

Global network expansion has upended ordered equity, and individualized insane practices have resulted in natural disasters. Restoring missing or injured body parts has traditionally been viewed as a technique to relieve pain. The employment of entrenched materials by the Celts, Mayans, and Egyptians demonstrates such evolution [64]. Because of this degeneration, harmless biological intervention with their allusion is gaining popularity. As a kind of treatment, replacements of entirely destroyed organs or tissues are implanted [7, 34]. Despite unnecessary bodily reactions due to unfavorable failure in these medical alternatives, whether due to perceptions or bodily infections, human intrusion frequently forecasts further research in the concerned field of medical diagnosis in biomaterials and biosensors [9, 29]. Nanomaterials research and molecular cell biology changed biomaterials science in the early 20th century. Biomaterials nowadays mimic many biochemical and biophysical features of native tissues while triggering little foreign body response. In high-risk procedures, it employs sophisticated medical equipment to replace injured tissues with dental implants, artificial joints, ocular lenses, or vascular stents. Biomaterials and biosensors have substantially improved the lives of patients.

Dr. Jonathan Cohen [16] defined biomaterial as any material used to mechanize biomedical implants. In joint replacement by Sheffield, biomaterials are employed for heart valves and biocoating hydroxyapatite in hip implants known as furlong hip, which last over considerable years.

11.2 Design of Biomaterials

Biomaterials are impractical, sophisticated enough to function efficiently after implanting, and capable of incorporating many biological designs. Biomaterials' dexterity and persistence make their employment in a corporeal medium possible [89]. Combining these critical features with physical, mechanical, synthetic, and organic aspects creates traditional designs [85]. Polymers, metals, composite textiles, and ceramics were designed for fabrics. These fabric classes have different nuclear compositions that have basic, detectable, synthetic, and machine-like properties, allowing a wide range of frame usage.

11.3 Polymers

Polymers, sensitive biological substances, replace and produce malleable tissues like cardiovascular device tissues. The implant technique uses polymeric polymers. Contact and intraocular lenses, vascular grafts, dental materials, breast prostheses, cardiac valves, artificial hearts, plasmapheresis fixtures, medical devices, surgical materials with coatings, dialysis applications, tissue adhesives, and other polymeric materials are commonly utilized [51]. Macromolecule composition, structure, and organization determine polymer properties [89]. The polymers' flexibility requires their manufacture in various forms and designs to satisfy particular objectives with acceptable physicochemical, interfacial, and biomimetic properties. Polymeric biomaterials are the best at being made and reworked, available with the correct mechanical and physical properties, and affordable. Synthetic and natural polymers dominate biological uses. Acrylics, polyamides, polyesters, polyethylene, polysiloxanes, and polyurethane are artificial polymeric systems. However, they are not biocompatible and readily combustible. Natural polymers can solve these issues [2]. Biomedical polymers include alginate, chitosan, and carrageenan. Drug delivery methods and tissue regeneration utilize them [2].

11.4 Metals as Biomaterials

In the medical field, metallic implant materials have consistently been used for their therapeutic relevance. Metal–metal alloys used for medical applications include cobalt chromium alloys, aluminum alloys, stainless steel, titanium and aluminum alloys (cpTi, Ti6Al4V), zirconium, niobium, and heavy tungsten alloys. Numerous metallic medical products have been developed to keep up with the recent rapid advancements in medicine, including pacemakers, clips, balloon catheters, valves, dental materials, radiation shielding, bone fixation devices, prosthetic and orthodontic devices, craniofacial plates and screws, dental implants, and parts of artificial hearts [89]. Table 11.1 shows the lengths of time that metals, polymers, and ceramics have been used in biomedical applications.

Table 11.1 An overview of the development of metals, plastics, and ceramics in biomedical applications from the onset of the Industrial Revolution till present (reproduced from Gobbi *et al.* [24]).

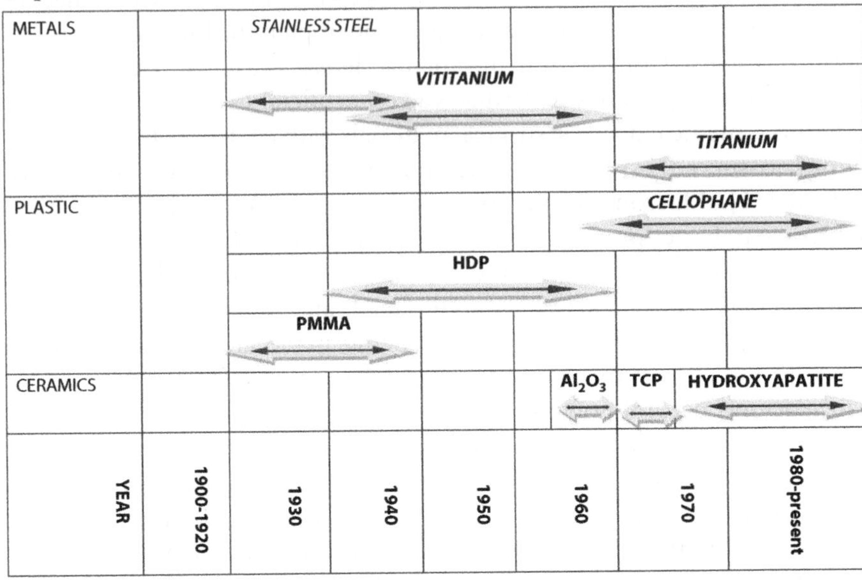

11.5 Bioactive Material and Concept of Bioactivity

From this perspective, it is significant to comprehend the role of bioactive material. The significance of the term bioactive material came into light by Hench *et al.* [31] who defined the term "bioactive" as one which reduces an explicit biological response in the interface resulting in the development of linkage between the biomaterial and living tissue. This response is a direct result of the chemical reactivity of the bioactive material at the interface due to its partial dissolution *in vivo* and *in vitro*. Calcium phosphate is layered out due to precipitation which in turn forms a stable interface by preventing its rapid dissolution. A tough bonding interface between the host tissue and the biomaterial implanted results due to the interaction of this layer with collagen and biomolecules [23, 31, 77]. A list of ceramics and glass as bioactive materials provoking satisfaction for bone bonding for the system SiO_2–CaO–Na_2O interaction with minimal rejection competence to bioactive bonding is given in Table 11.2.

Thus, biomaterials find common usage in dentistry applications, surgery, and drug delivery even though existing conventional tissue engineering procedures run parallel. Due to the non-availability of the majority

Table 11.2 List of a few ceramics and glass as bioactive materials.

Glass/ ceramic type	Bioactive component	Batch no.	(%) Chemical composition												Reference
			SiO_2	P_2O_5	CaO	$Ca(PO_3)_2$	CaF_2	MgO	MgF2	Na2O	K_2O	Al_2O_3	B_2O_3	Ta_2O_5/TiO_2	
Glass	Bioglass	45S5	45	6	24.5	0	0	0	0	24.5	0	0	0	0	[30]
Glass	Bioglass	45S5F	45	6	12.25	0	12.25	0	0	24.5	0	0	0	0	[30]
Glass	Bioglass	45S5.4F	45	6	14.7	0	9.8	0	0	24.5	0	0	0	0	[30]
Glass	Bioglass	40S5B5	40	6	24.5	0	0	0	0	24.5	0	0	5	0	[30]
Glass	Bioglass	52S4.6	52	6	21	0	0	0	0	21	0	0	0	0	[30]
Glass	Bioglass	55S4.3	55	6	19.5	0	0	0	0	19.5	0	0	0	0	[30]
Ceramic	Ceravital	KGC	46.2	0	20.2	25.5	0	2.9	0	4.8	0.4	0	0	0	[25]
Ceramic	Ceravital	KGC	46	0	33	16	0	0	0	5	0	0	0	0	[25]
Ceramic	Ceravital	KGy213	38	0	31	13.5	0	0	0	4	0	7	0	6.5	[25]
Ceramic	Ceravital	A-W GC	34.2	16.3	44.9	0	0.5	4.6	0	0	0	0	0	0	[37]

of material-based products coupled with cells and/or growth factors to the patient's stakeholders despite ambiguous large-scale traditional tissue engineering development and limited clinical translation technology partly due to limited performance, negative side effects, and partly due to partial cumbersome regulatory processes and cost factor, the creation of intelligent, instructive biomaterials that can draw and accumulate pertinent cells and growth factors *in situ* and commence and stimulate the process of regeneration without requiring the addition of cells or biologics before surgery is the current field of interest for research. After playing a vital role in the human body, with the advent of time, these materials, which are polymers, ceramics, or their mixtures, normally decay. Table 11.3 lists some important contributors in the field of biomaterials and biosensors.

Table 11.3 Few noteworthy contributors from the mid-16th century to the 21st century when the onset of elaborate R&D blasted in the field of biomaterials and biosensors.

1628:	*William Harvey promoted the modern concept of heart functioning as "One of the prime role of heart is the transmission of the blood and its propulsion by means of the arteries to the extremities everywhere."* [28]
1812:	*Physiologist Le Gallo pioneered the concept that biological organs could be kept alive by a constant flow of blood through them.*
1829:	*Henry S. Levert evaluated the efficacy as well as safety of different metallic sutures on dogs.*
1870:	*Dr. Joseph Lester introduced aseptic surgical techniques free from harmful bacteria, viruses, or other microorganisms.*
1886:	*Dr. H. Hansmann used metal plates for internal fixation.*
1931:	*Dr. Smith Peterson developed a metal cup for partial hip implants.*
1939 to 1945:	*During World War II, many materials and orthopedic surgical techniques were introduced involving metals mostly because at that time there were very few plastics.*
1947:	*The first paper on polyethylene was published as a synthetic implant material.*
1949:	*A paper was published about plastics "sweating out" additives such as cellophane, Lucite, and nylon, resulting in a strong biological reaction.*

11.6 Biocompatibility of the Titanium Binder Element

Due to their exceptional corrosion resistance, titanium alloys have relatively high biocompatibility than stainless steel and cobalt alloys [4, 13, 42, 71]. Due to its appreciable corrosion rate, 316L stainless steel has good biocompatibility although lower than Co–Cr–Mo and titanium alloys. The Ti–6Al–4V alloy, representing the first generation of titanium alloys, has been linked to human allergy reactions [46, 58]. Pure titanium (cpTi, grade 2) and Ti–6Al–4V (grade 5) are frequently used as substitutes for hard tissue in artificial bones, joints, and dental implants [53]. Since pure titanium and its alloys have relatively low modulus, they are commercially preferred over other metals or alloys during the replacement of hard-bone tissue, as a low modulus of elasticity minimizes the effect of voltage shielding. Compared with vanadium and aluminum, metals such as zirconium, tantalum, and niobium are utilized as alloying elements and are considered safe for biomedical applications [24, 53, 75]. The Ti–6Al–4V alloy is frequently used in situations with high mechanical properties, such as hip or knee implants, screws, and plates [72, 73]. One of the most popular uses for titanium alloys includes artificial hip joints, which consist of a joint (head and femoral cup) and stem bearing. Titanium is a composite component with a metal cup and hip stem. They are also frequently utilized to replace knee joints, including a polyethylene joint surface and titanium femoral and tibial components. In recent years, cardiovascular devices like heart valves, heart pumps, and vascular stents have also utilized titanium and its alloys. Titanium is a thrombogenic substance [87]. Titanium surface-induced coagulation has been the subject of extensive investigation, and numerous techniques have been used to enhance *in vitro* biocompatibility. The exact treatment procedure involving luminescent discharge and air-furnace processing of Ti and Ti–6Al–4V alloy produces a thicker and tougher protective oxide layer, which improves the biocompatibility of human umbilical vein endothelial cells (HUVECs) [10]. Diamond-like carbon-coated titanium (DLC) may enhance cytocompatibility for HUVECs *in vitro*, according to Kumari *et al.* [38]. A native extracellular matrix (ECM) was produced on the surface of pure titanium after HUVEC cells were cultured to create modified characters for cardiovascular implants. This indicates that the ECM produced by HUVECs enhances endothelial cells' ability to adhere and increase, inhibiting the adhesion and activation of the HUVECs [88].

11.7 Classification

Biomaterials are classified broadly based on a series of historical artifacts. Several bioactive, polymeric matrix composite materials like tar were used for biomedical operations before being created as the first inert ceramic; it was used in procedures like ocular implants [70]. Classification is done based on the following

- Chemical composition: metal, polymer, ceramic, or composite.
- Degree of interaction with biological surrounding tissues or host bioactive or inert property with more or slight host reaction, bioresorbable or biodegradable, which resorbs in the body over time.
- Based on origin: synthetic or natural.
- Another classification is based on energy: the amount of energy a biomaterial consumes to attain its final form. Due to missing links of any technological criteria inputs, the classification failed. Newer methods of the classification of biomaterials include the following three main categories:
  - Primitive: naturally occurring and reshaped using hand tools like stones and bone.
  - Handicraft: cultured using human skills like weaving natural fibers, linen threads, etc.
  - Transformed: materials altered through different processes, physical or chemical, using technologies of applied nature like metallurgical operations.

While ceramics and polymers might provide bioactivity or resorption features, metals are typically chosen for their inert properties. Austenitic stainless steels, cobalt–chromium alloys, and titanium are the three most frequently utilized metallic components. Because of their strength, elastic modulus, and corrosion resistance, titanium alloys, particularly, Ti–6A1–4V, have recently been developed.

11.8 Interaction Between Biomaterials and Biological Systems

The uses of biomaterials have evolved from the clinical treatment of damaged organs and tissues, where they were used to submissively override the functions of damaged tissues, to the activation and exploitation of the body's in-place autoregenerative capacity, followed by the mortification of the effects of regenerative medicine. To meet such criteria, understanding the nature of the interaction of biomaterials with biological systems is noteworthy for understanding its regenerative capabilities supporting their design with improved functionalities such as biocompatibility and bioactivity. The role of protein in interacting with biomaterials and cells or tissues is significant. It includes adsorption at the material surface followed by subsequent processes of cells cultured on materials, cellular growth, cellular differentiation, and extracellular matrix formation. Investigations of the interactions between materials and biological systems using mass spectrometry imaging, or proteomics, provide a rich platform for such interface studies.

11.9 Biomaterials: Protein Surface Interactions

The biocompatibility of biomaterials applied to various medical devices depends primarily on the protein–surface interactions it sustains. When a fluid containing soluble proteins (such as blood, interstitial fluid, or cell culture media) comes into contact with a solid material, proteins quickly adsorb onto the material's surface. The adsorbed molecular structure of the biomaterial surface interacts with the molecular structure of the protein layer that has been adsorbed. The cells bind to these adsorbed proteins through membrane-bound receptors having specific bioactive characteristics. These receptor–protein binding processes are subsequently transmitted in the cell membrane through synchronized molecular mechanisms, stimulating particular intracellular activities that ultimately characterize a cell response to such interaction.

Different factors affect the extent of protein adsorption on the biomaterial, which include chemical and physical properties of the biomaterial, the nature of the solvent, pH, temperature, the concentration of different protein species in the buffer solution, and ionic strength of the surrounding solvent [35]. Proteins with diversified nature, like size or structure, also influence the adsorption behavior. For instance, small-sized proteins like lysozyme, α-chymotrypsin, or β-lactoglobulin have a relatively small affinity for structural modification on adsorption. In contrast, proteins of intermediate size like transferrin, immunoglobulin, and albumin have a relatively high relationship to undergo conformational reorientation on coming in contact with the surface with additional characteristics to expel preadsorbed proteins when required. Thus, protein adsorption is a diverse and dynamic process often consummated with an overlap of transport and repulsion affected by many internal and external factors [61]. The nature of bioactive sites in the adsorbed layer of proteins ultimately regulates the overall key to regulating cellular response. Thus, controlling the quantities and the type of proteins adsorbed, their orientation, conformation, and packing arrangement on the biomaterial's surface regulate the cellular response. While soluble proteins present in physiological fluids represent a broad category with several variations and the complicated structural makeup development of our understanding of these processes is critical, studies on the topic are not fully conclusive without totally understanding the submolecular-level mechanisms concerning surface chemistry which affects not only the organization of adsorbed proteins but also its orientation as well as its conformation.

The practice of substitution/replacement of damaged or injured/wounded tissues whether they are bone or cartilage by adequate natural or synthetic biomaterials has been effectively performed in the past years, and it has been remolded in the modern era in the form of science of tissue engineering. The pooled purpose of tissue engineering is to instill the progression of regeneration of tissue that is injured or fully damaged through either cell grown in culture, systemic and/or local hormones/mediators, biomaterial-based scaffold, or genetic modulators.

Biomaterials used for various products of tissue engineering have characteristic structural variations and find their usage in musculoskeletal diseases and disorders for long decades. Various elemental species as part of biomaterials like metallic elements such as titanium in pure or alloy form, cobalt–chrome, ceramics of calcium phosphate and hydroxyapatite (HA), and polymethyl-methacrylate polymers find application as replacements of cartilage, bone, and joints. The implantation procedure damages vascularized tissue, which causes a provisional matrix to form quickly at the implant site. From the perspective of wound healing, blood protein deposition on a biomaterial surface is referred to as "provisional matrix formation." The provisional matrix is composed of fibrin, which results from the activation of coagulated thrombosis systems and inflammatory mediators. It is expected to be released by the complement system, activated platelets, inflammatory cells, and endothelial cells [8]. Resolution, reorganization, and repair are all aided by fibrin, which also triggers the recruitment of fibroblasts and inflammatory cells. Recruitment of inflammatory cells and fibroblasts is one of the repair, reorganization, and resolution processes that fibrin starts [15]. The development of neovascularization, or angiogenesis, has also been demonstrated to be significantly influenced by fibrin. Within 4 days, fibrin-filled porous surfaces that have been implanted are anticipated to produce new vessel growth. The provisional matrix may be seen as a naturally occurring, biodegradable, sustained release system that controls the release of mitogens, chemoattractants, cytokines, and other growth factors during the implantation of biomaterial devices.

The degree of biocompatibility depends on the *in vivo* functioning of the implant having apt mechanical properties, viz., lubrication in joints, load-bearing capacity of the bone, etc. with a minimal tolerable response at the local or systemic level. These responses in turn depend on other factors like physical and chemical properties, surface topography, and heterogeneity. To evaluate the behavior of the biomaterial to be implanted, the substance's physical qualities must match those of the organism, ideally being identical or at the very least compatible. Therefore, these materials are created in accordance with the tissue characteristics of the intended application. Materials are modified for the best acclimatization to the biological environment based on the tissue type for which the biomaterial is used, physical strengths (tensile, compressive, shear, and elasticity modulus), thermal properties, photoreactivity or translucency, color, calcification potency, surface structure, chemical features, or degradation resistance. Prior to biologic behavioral testing, such characteristics are evaluated in a lab setting [20].

Many biomaterials, including MMA (acrylic), silicone used in intraocular lenses, and UHMWPE used in the implantation of artificial hip joints, have properties that make them biocompatible with materials with optical properties like stainless steel, titanium, and its alloys. In order to avoid blood clotting and maintain mechanical veracity and long-term blood compatibility, vascular grafts made of polyurethane, Teflon, and Dacron, Indian Chitra heart valves, and substitute heart valves are widely used having such challenging attributes. Hip surgery additionally employs a dual tapered stem design made of titanium to support a physiologic proximal load transfer. This helps to reduce the chances of calcar resorption along with distal hypertrophy. The overall impact of the system is to furnish an anatomic fit together with designing a straight stem supporting system. In addition, a major role of cartilage is displayed containing polyethylene as a biomaterial.

A brief classification of biomaterials on conventional pattern and their advantages as well as disadvantages are given in Table 11.4.

Table 11.4 Classification of biomaterials based on conventional patterns with their applications, advantages, and disadvantages.

Type	Advantages	Disadvantages	Applications
Metals and metal alloys, e.g., gold, platinum, titanium, steel, chromium, cobalt	High-strength material	Corrosive	Orthopedic implants, screws, pins, and plates
	Easy to fabricate and sterilize	Aseptic loosening	
		Excessive elastic modulus	
Ceramics and carbon compounds, e.g., calcium phosphate salts (HA), glass, oxides of aluminum and titanium	High-strength material	Difficult to mold	Bioactive orthopedic implants
	Biocompatibility	Excessive elastic modulus	Dental implants
	Corrosion resistance		Artificial hearing aids
Polymers, e.g., PMMA*, polycaprolactone (PCL), PLA, polycarbonates, polyurethanes	Biodegradable	Leachable in body fluids	Orthopedic and dental implants
	Biocompatible	Hard to sterilize	Prostheses
	Easily moldable and readily available		Tissue engineering scaffolds
	Suitable mechanical strength		Drug delivery systems
Composites, e.g., dental filling composites, carbon fiber-reinforced methyl methacrylate bone cement + ultra-high molecular weight polyethylene	Excellent mechanical properties	Expensive	Porous orthopedic implants
	Corrosion-resistant	Laborious manufacturing methods	Dental fillings
			Rubber catheters and gloves

11.10 Dental Material Cavity Fillers

Silver–mercury amalgams were formerly a popular component of dental fillers, although they had limitations in that they could only produce brittle and corrodible intermetallic Sn_7Hg in the second phase. A composite alloy of silver, tin, and copper that possesses a copper content of around 12%, high tensile strength, and corrosion-resistant properties and is more biocompatible than earlier dental amalgams is among the phase suppressors used in modern amalgam. Because of the formation of the Cu_6Sn_5 (rf) phase, adding Pd may further reduce corrosion. In subsequent attempts, pastes composed of gallium, indium, silver, tin, and copper were used to replace the mercury amalgam. The development of a methacrylate monomer-filled resin paste incorporating ceramic particles with silane coatings at the micron scale was pursued; however, none of these methods proved successful. Glass cement for cavities is being developed at the moment, which is a chemical silicate cement formed when phosphoric acid displaces metal ions from an alumina–silica glass containing metal oxides and fluorides. When aluminum phosphate precipitates between the glass particles, this cement hardens. To increase longevity, additional changes were made utilizing different resin alternatives, particularly carboxylate groups to produce suitable bonding to the tooth and dimethacrylate as composite resin.

11.11 Bridges, Crowns, and Dentures

Removal partial dentures (RPDs) with cast metal frameworks with Ni–Cr or Co–Cr alloy holding prosthetic teeth are commonly utilized in tooth replacement. Dental porcelains are vitrified feldspar with metallic oxide pigments that are used to replicate real tooth enamel. Initially, methyl methacrylate was employed as the basic material for denture replacement, but in recent years, other materials such as vinyl addition silicone and polyether elastomers have taken its place.

11.12 Bone Fractures

While bones are composed of a collagen (polymer) matrix reinforced with approximately 50% volume fraction of hydroxyapatite (ceramic) nanometer-scale crystals, metallic alloys have been universally used as plates, pins, screws, and so on in fracture joints due to their inherent properties

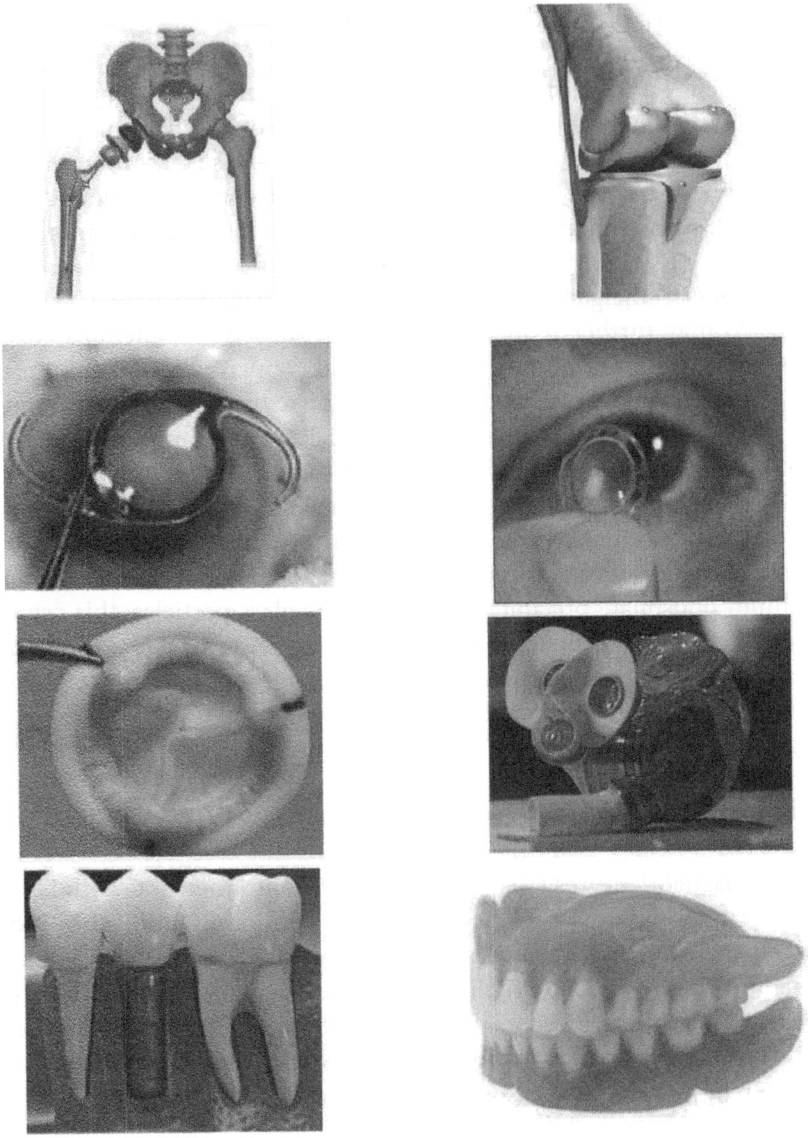

Figure 11.1 Visualization of the biomedical applications of biomaterials: orthopedic applications (artificial hip/artificial knee), ophthalmologic applications (artificial cornea hip/intraocular lenses), cardiovascular applications (heart valve/artificial heart), and dental applications (endosteal root form dental implant/biomaterial tooth gums) (reproduced from Raghavendra *et al.* [62]).

that provide strength and rigidity as well as corrosion resistance. Titanium and its alloys are the most suitable metals and are progressively being employed. Figure 11.1 gives a brief visualization of the biomedical application of biomaterials.

11.13 Biosensors

Enzyme electrodes were introduced by Leland C. Clark, who initially proposed the biosensor concept, signaling the beginning of a new branch of biomaterials, i.e., biosensors [14]. Nevertheless, Updike and Hicks found a working enzyme electrode for glucose detection, which is recognized as the first biosensor [79]. Since then, research groups from various disciplines, including VLSI, physics, chemistry, and material science, have collaborated to develop more advanced, reliable, and mature biosensing devices for biotechnological applications, medicine, agriculture, and military applications such as bioterrorism detection and prevention [47]. Due to its potential uses in the monitoring and detection of opportunistic infections, the biosensor bacteriophage has been widely employed. For the detection, identification, and monitoring of pathogenic microbes as well as hazardous and carcinogenic substances, a number of bacteriophage-based biosensors have been developed and used [26].

A biosensor is a device made up of two main parts: a transducer and a bioreceptor. Thus, a biosensor is a functionally integrated device with a biological recognition element that can analyze particular quantitative or semiquantitative data [54]. The bioreceptor has features that interact (bind or recognize) with a specific bioanalyte, are obtained from biological sources, and are biomimetic. The transducer changes the signal generated by a particular bioanalyte's interaction into another signal more amenable to measurement and quantification. The transducer frequently relies on one or more mechanisms, such as electrochemical, piezoelectric, optical, or physicochemical effects.

Typically, the primary category standards for biosensors can be the type of signal transducer or the type of biorecognition components.

The working efficiency of biosensors depends on the biochemical specificity of the biologically active substance. This includes the specificity, storage, and operational and environmental stability of the biological material [55, 79]. At the same time, there are numerous biosensor applications, including biomedical, commercial, and military uses. Among them, the one having huge market potential at present is the blood glucose sensors [6, 47]. As biological sensing components, biomolecules like enzymes, antibodies, receptors, organelles, microbes, and animal and plant cells or tissues have been used [19]. For biosensor devices, microorganisms were used in combination with a range of transducers in various analytical instrumentation like amperometric, potentiometric, calorimetric, conductimetric, colorimetric, luminescence, and fluorescence [44, 50, 78]. Figure 11.2 illustrates the main parts of a biosensor, which typically combines the capabilities of a bioreceptor and a transducer.

11.14 Biosensor Classification

11.14.1 Resonant Biosensor

This biosensor combines an antibody and an acoustic wave transducer (bioelement). The membrane mass changes when the analyte molecule (or antigen) binds. The consequent change in mass then alters the transducer's resonance frequency. The subsequent frequency shift is measured [32]. Figure 11.2 shows the rear view of the components of biosensors. An alternative classification of biosensors is also proposed based on signal transduction and biorecognition mode [27] (Table 11.5).

11.14.2 Optical Biosensors

In these biosensors, the output transduced signal is light which is measured. This biosensor functions by linking changes in the properties of light to variations in concentration, mass, or the number of molecules.

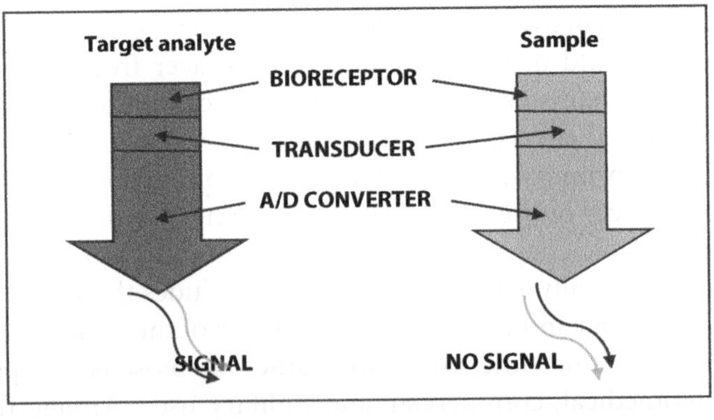

Figure 11.2 Diagram depicting the components of biosensors (reproduced from Perumal *et al.* [55]).

Table 11.5 Classification of biosensors based on signal transduction and biorecognition mode (reproduced from Hani A. Alhadrami [27]).

Recognizing biomolecule	Example	Signal transduction	Example
Antibodies (immunosensors)	Monoclonal Polyclonal	Electrochemical	Amperometric Conductimetric Impedimetric Potentiometric
Protein receptors	Metabotropic receptors lonotropic receptors	Optical	Absorbance Fluorescence Phosphorescence Bio/ chemiluminescence Reflectance Raman scattering Refractive index
Enzymes	B-glucuronidase (encoded by the gusA marker gene) B-galactosidase (encoded by the lacZ marker gene)	Mass sensitive	Surface acoustic wave biosensor Cantilever biosensors
Whole cells	Microbial sensors of mammalian cell tissue	Optical	*E. coli* K12C600 *E. coli* HB101 *E. coli* DPD1718
Nucleic acids	Hybridization DNA-aptamer-based biosensors	Optical (fluorescence)	Fluorescence-based aptasensors

These biosensors make use of the principles of electrochemiluminescence or optical diffraction. Optical transducers can be readily engaged for use in (label-free) bacterial detection. When cells connect to receptors fixed on the transducer surface, tiny changes in the refractive index or thickness occur, which these sensors can detect. When cells bind to receptors set on the transducer surface, slight differences in the refractive index or thickness are detected by the sensors. Surface plasmon resonance (SPR), ellipsometry, the resonant mirror, the interferometer, and other optical techniques are generally used for the detection of bacterial infections [39, 74, 83].

11.14.3 Surface Plasmon Resonance (SPR) Biosensor

These sensors are based on an evanescent field using a thin gold sheet for visual sensation. The presence of reflection minima on photodetector array sensors allows for the investigation of the interaction between analytes flowing through an immobilized gold surface. SPR has effectively identified pathogenic bacteria by immunoreactions [33, 74].

11.14.4 Piezoelectric Biosensors

The piezoelectric (PZ) biosensor functions by coating the PZ sensor's surface with a material that preferentially binds antibodies to bacteria when immersed in a bacterial test solution. Due to this binding, the resultant crystal's bulk is increased. The resonant frequency of oscillation drops causes the crystal mass to rise [39, 74].

11.14.5 Thermal Biosensors

Based on sensing the temperature disparity of the reaction medium owing to the heat absorbed or released in explicit biological reactions, the sensor utilizes immobilized enzyme molecules and temperature sensors for the purpose. The measured heat of the enzymatic reaction is calibrated against the analyte concentration coming into contact with each other. Pesticide detection and pathogenic bacterial detection are the two general applications of this type [74].

11.14.6 Electrochemical Biosensors

These detect hybridized DNA, DNA-binding substances, glucose levels, etc. Applying the principles of electrical parameters, i.e., conductimetric, amperometric, and potentiometric, as against optical methods, it is vaguely mediocre in terms of selectivity and sensitivity [39, 80]. Clark oxygen electrode is frequently used in amperometric biosensors [39, 74].

11.14.7 Bioluminescence Sensors

Bioanalytical sensors are based on the capacity of some enzymes to release photons in a reaction. The concept of luciferase reporter phages sparked interest in the possible application of bioluminescence for bacterial detection. In studies using the bacterial luminescence lux gene in an inducible or constitutive manner, the reporter lux gene is combined with a promoter controlled by the concentration of the concerned molecule in an inducible way using bioluminescence intensity. The compound concentration is further quantified through this technique. Various bacteria have been detected using bioluminescence devices [19, 74].

11.14.8 Nucleic Acid-Based Biosensors

This analytical biosensor tool combines an oligonucleotide with a signal transducer. The nucleic acid probe is the biorecognition molecule to identify DNA/RNA fragments and is immobilized on the transducer [74].

11.14.9 Nanobiosensors

These sensors follow the conceptual principle of nanotechnology using silver and other noble metal nanoparticles in biolabeling, drug delivery systems, filters, and antimicrobial medicines for analysis [63].

11.14.10 Microbial Biosensors

They are utilized for biological sensing using microbes as a significant constituent which notably assimilate various chemical substances and are diversified. Microorganisms are incredibly adaptable and can readily break down novel compounds with time. Microbes are a valuable source of intracellular enzymes and compliant with genetic variation due to either mutagenic alternation or technologies about the recombinant DNA process [41]. These biosensors emphasize viable cells as numerous organic substances are metabolized by possible bacterial processes aerobically or aerobically. A number of organic compounds are metabolized aerobically or anaerobically by bacterial activities, resulting in the formation of products such as carbon dioxide, ammonia, and acids, which may be detected using transducers. Since microbes possess the ability to assimilate diverse substrates as a measure of their respiratory and metabolic activity, it can animatedly serve as a tool for various estimations such as biological oxygen demand (BOD) or metabolically related nutrients such as sugars, vitamins, nitrogenous compounds, organic acids, and viable cells such that the respiratory and metabolic activity is considered by and large the total substrate assimilation capacity of these microbial species. The other fields of analysis of interest by viable microbial biosensors include those of environmental contaminants, inhibiting microbial respiration [18, 66]. For such analysis, a microbial biosensor consisting of an oxygen microelectrode with microbial cells bound in polyvinyl alcohol is used. BOD sensors use microorganisms interfaced with electrodes to measure oxygen depletion rates owing to their high reaction rates [18, 53].

11.14.11 Bioreceptor-Based Category

This bioreceptor or biological recognition element includes a biological ingredient like an enzyme, DNA probe, or antibody, which identifies analytes like an enzyme substrate, complementary DNA, or antigen. To avoid interference from other signal sources or compounds, it includes a sample matrix enhancing the fact that a bioreceptor has a unique sensitivity to the target analyte [6, 45].

Introduced in 1962 by Clark and Lyons, an electrochemical sensor was developed based on the enzyme glucose oxidase in blood plasma to detect glucose [14]. Effective biocatalysts include enzymes like peroxidase, urease, and redoxase, which serve to identify analytes accurately [32, 50, 78]. A mechanism for analyte recognition involves an enzymatic transformation of the analyte into a sensor-detectable product. In contrast, an alternative method consists of detecting enzymes blocked or activated by the analyte or observing modification of enzyme characteristics. Since enzymes are stable, enzyme-based biosensors can be used for considerable periods. Apart from other biosensors, an electrochemical immunosensor (EI) has considerably drawn significant attention for clinical diagnosis of a variety of illnesses and their early identification, such as malignancies, cardiovascular diseases, and hepatitis A [33, 63, 66, 74, 80, 83]. For building up a stable antigen–antibody complex, all immunosensors depend on the sensitivity of molecular recognition. In contrast, the nature of the site of binding of an antibody to an antigen depends on its amino acid components [49, 67]. The biological recognition element in nucleic acid-based biosensors uses the complementary strands of the nucleic acids; double-stranded DNA is created when two single-strand DNA (ssDNA) chains link together (dsDNA). Creating, tracking, and ultimately immobilizing on the sensor complementary nucleic acid sequences to a recognized target nucleic acid sequence are possible. Cell-based biosensors monitor internal and extracellular microenvironment conditions and physiological factors using living cells as biospecific sensors. They subsequently produce a response through the interaction of the cell with the stimulus [44]. Biologically produced materials can also be used to create cell-based sensors. The significant development in cell-based sensors for prior detection of infections and oral cancer in recent years is a significant achievement in the medical field [65]. Cell vitality, which can be drastically inclined by factors including cellular lifetime, sterilization, and the device issue of biocompatibility, appears to be the main constraint of cell-based sensors. Synthetic sensors that imitate the features or methods by which natural systems detect outside inputs are known as biomimetic sensors. An aptasensor is a typical biomimetic biosensor that is based on the theory of antibodies and antigens chemically connected to nucleic acid probes [11]. Compared with antibody-based biosensors, an aptamer has advantages like high-binding effectiveness, is simpler and smaller, and does not involve the input of any animal ingredient. However, the characteristic issues of nucleic acid, like structural pleomorphism and chemical simplicity of aptasensors, often result in the shrinkage of the efficiency of its assay.

11.14.12 Transducer-Based Category

A transducer functions to generate an electrical signal by transforming different physical, chemical, or biological inputs and thus plays a crucial role in the signal detection route. Electrodes used as a transducer by the electrochemical sensor bear a reference electrode, a working electrode, and a counter electrode which characteristically make up its substrate.

The transfer of electrons or the potential difference between two layers is produced when the reaction occurs on the electrode surface with the target analyte, which results in the production of current or voltage monitored. In electrochemical biosensors, the voltage or current generated by oxidation and reduction processes is determined. Based on electrochemical alteration features of the biorecognition process, electrochemical biosensors are classed into potentiometric, amperometric, impedance, and conductometric biosensors. Optical biosensors may assess many analyte characteristics and provide label-free, parallel concurrent detection coupled with optical spectroscopy techniques, like fluorescence, phosphorescence, absorption spectroscopy, refraction, Raman spectroscopy, and dispersion spectroscopy. Since a piezoelectrical crystal undergoes elastic deformation when subjected to the applied electrical potential, it can effectively behave like a transducer to primarily detect electrical or mechanical inputs. Depending upon the elastic qualities of a crystal, a standing wave at a specific frequency instantly develops an electrical potential. A crystal covered with a biological recognition element allows target analyte–receptor interaction causing resonating variation in frequency, ultimately leading to the typical current value, which is the outcome of sensation for the sensor.

11.15 Biosensors: Precursors of Contemporary Biomaterial Succession

11.15.1 Carbon-Based Nanomaterials

Due to their exceptional chemical, thermal, electrical, and optical capabilities, carbon nanomaterials, carbon nanotubes, and graphene are employed in biosensing devices. Dopamine and serotonin are two examples of biomolecules that carbon nanostructures can identify and analyze [36, 81]. Since carbon nanostructures have high electron mobilities, these biosensors react fast. Preliminary molecular biomolecule and gas molecule identification is made feasible by the monolayer structure and substantial surface area of 2D carbon nanomaterials like graphene [5, 69].

11.15.2 Carbon Nanotubes

Carbon nanotubes, which have a high mode of conductivity more than 100 times that of copper, can be used to improve signal transduction for recognition and functional properties of other functional materials as a platform and functionalized through conjugation, such as conductive polymer [43] and nanoparticles [56], supporting the positive efficacy of analysis. Detection is also improved. Carbon nanotubes may penetrate cell membranes, making *in vivo* imaging and diagnostics more likely [12].

11.15.3 Electrochemical Biosensors Based on Carbon Nanotubes

Carbon nanotube-based electrochemical sensors are widely used for enzymatic reaction-based biosensing due to their high conductivity, electrocatalytic effect, and electron mobility. Carbon nanotube-based biosensors may also detect nucleic acids like RNA, DNA, epinephrine, glucose, nitric oxide, and other chemicals. Lin *et al.* constructed a glucose biosensor utilizing NEEs [52]. Glucose oxidase was covalently bound on carbon nanotube nanoelectrode ensembles using amide linkages in amine residues–carboxylic acid groups at the ends. Once the carbon nanotube nanoelectrode ensembles collect glucose molecules, glucose oxidase starts an enzymatic activity that reduces hydrogen peroxide catalytically, allowing glucose detection at 0.08 mM. Consequently, glucose sensors are widely used in medical diagnostics to detect glucose in biological fluids like perspiration, urine, and saliva. H_2O_2 sensing may minimize interference in carbon nanotube nanoelectrode ensembles. This sensor detects choline, xanthine, hypoxanthine, lactate, alcohol, acetylcholine, and cholesterol by immobilizing oxidases on carbon nanotube nanoelectrode ensembles. Fu *et al.* [22] created a non-enzymatic glucose biosensor using a nanocomposite Au/LDH-CNTs-G (ternary layer of nickel and aluminum double hydroxide/single-walled carbon nanotubes/graphene). This changed electrode performed well in glucose oxidation and glucose detection in human serum with a low detection limit (1.0 mM) and high sensitivity of 1.989 × 103 A/mM/cm^2. In carbon nanotube-based electrochemical biosensors, N-hydroxyphenyl maleimide functionality was combined with multiple carbon nanotube walls comprising monomeric units bound to terminal sites for chain propagation and simultaneous imprinting of epinephrine [59]. It detects epinephrine with great sensitivity and selectivity on graphite electrode tips at 0.02 ng/ml. Ali *et al.* [1] developed a poly(aniline boronic acid)/carbon nanotube composite that detects dopamine better than boronic acid. This gadget electrochemically detects dopamine at 1 nM instead of oxidatively.

11.15.4 Carbon Nanotube-Based Immunosensors

Vertically aligned single-walled carbon nanotube (silver nanoparticle) forests were produced for amperometric enzyme-linked immunodetection of human serum albumin [90]. These silver nanoparticle forests facilitate electron exchange between enzymes connected to their tips and carbon nanotubes, lowering the detection limit below 1 nM and increasing sensitivity. Carbon nanotube-based immunosensors can detect gonadotrophin, interleukin, fetoprotein, and other antigens by modifying recombinant antibodies or antibody fragments, making them useful in medical diagnosis and the development of cancer biomarkers like the multifunctional cytokine interleukin-6 (IL-6), which regulates immunological and inflammatory responses. Silver nanoparticle forests detect IL-6 electrochemically using antibodies (Ab1) [48]. This silver nanoparticle immunosensor detects IL-6 at 0.5 pg/ml. Consequently, silver nanoparticle immunosensors may detect IL-8, HER2, and prostate-specific antigen biomarkers for early cancer detection (PSA).

11.15.5 Optical Sensors Composed of Carbon Nanotubes

Using surface plasmon resonance, waveguides, and resonant mirrors, optical fibers detect and monitor target analytes. Carbon nanotubes are attractive optical sensors for biological samples and live tissue that seldom absorb NIR luminescence. By adsorbing biomolecules, the near-infrared sensor's silver nanoparticle approach can drastically change its emission [3]. It may control nanotube fluorescent characteristics using electron-withdrawing or fluorescence-quenching chemicals, making a groundbreaking biological research contribution.

11.15.6 Graphene-Based Biosensors

Graphene's biosensor buoyant break is unique. It has inherent crystalline properties, band structures with a shallow noise level, and intrinsic ability to enhance sensitivity and lower detection limit to detect chemicals like CO, NH_3, and NO_2.

11.15.7 Electrochemical Biosensors Based on Graphene

Kang developed a glucose-detecting nanocomposite electrochemical biosensor utilizing glucose oxidase/graphene/chitosan [86]. Stability and enzyme loading capacity (1.12 mol/cm^2) distinguish this graphene-based platform. Its 37.93 A/mM/cm^2 sensitivity allows it to detect 0.02 mM glucose. Graphene's conductivity and surface/volume ratio promote enzyme uptake and electron transfer between redox enzymes and electrode surfaces. Wu *et al.* made 0.6 M glucose biosensors [84]. Glucose oxidase-Pt-functional graphene sheets and chitosan minimized uric acid and ascorbic acid interference in the nanocomposite screen. Graphene compounds with different enzyme combinations may detect H_2O_2 and cholesterol utilizing enzymatic electrochemical methods. Wu *et al.* [84] created porphyrin-functionalized graphene, an enzymatic biosensor that can detect dopamine up to 0.01 M in hydrochloride-injected samples, blood serum, and urine [84]. Porphyrin-modified graphene and dopamine electrostatically energize it. Xue *et al.* [88] electrochemically measured uric acid utilizing poly(diallyldimethylammonium chloride)-graphene nanosheets (PDDA-GNs) at 103.08 A/M/cm^2 (0.1 M). The adrenaline biosensor can detect uric acid. The adrenaline-induced acid detection PDDA-GN electrode resolved the anodic peaks of uric acid and adrenaline.

11.15.8 Graphene-Based Immunosensors

Graphene-based immunosensors can detect hormones, proteins, and germs as biomarkers. A recent advancement toward familiar and wearable immunosensors like impedimetric immunosensor [82] for detecting particular strains of *Escherichia coli* (*E. coli* O157:H7) is a modified free-standing graphene paper gold nanoparticle-coated electrode. Biotin–streptavidin binds anti-*E. coli* O157:H7 antibodies to an altered electrode. This biosensor has a large linear range (1.53102–1.53107 CFU/ml) and a minute detection limit (1.53102 CFU/ml) and is outstanding explicitly. Graphene paper-based sensors showed excellent lenience to mechanical deformation, enabling new prospects for flexible device design—long-lasting immunosensors. Teixeira *et al.* [76] created an epitaxial graphene immunosensor to detect "human chorionic gonadotropin," a pregnancy marker and potential biomarker for prostate, ovarian, and bladder cancer. Graphene-bound antibody binds human chorionic gonadotropin at 0.62 ng/ml. The graphene biosensor's channel current is nearly 30 times more sensitive than an ELISA test under identical conditions.

11.15.9 Graphene-Modulated Gene Biosensors

Graphene rapidly interacts with proteins, protecting other biological functions while producing a favorable environment for DNA fixing and greatly speeding the transport of electrons straight from DNA to an electrode. Chen *et al.* created a nitrogen-doped gold nanoparticle-functionalized graphene nanosheet DNA biosensor [13]. Graphene nanosheets doped with nitrogen can boost electrode surface property and electrical conductivity, enabling accurate multidrug resistance gene detection in humans at 3.12 mM. Gold nanoparticles immobilize DNA probes and hybridize extensively. Hence, a label-free/enzyme-free biosensor might be a potential portable DNA analyzer.

Au nanoclusters and an exonuclease III stabilized by graphene-modified glassy carbon electrode by Wang *et al.* [88] are the other target recycling amplification biosensors for HIV AIDS virus activities. The graphene-modified electrode is a conductive platform with a raised area and has large preset locations for collecting cytosine-rich probes. This electrical biosensor provides ultrasensitive signal-off selectivity to HIV RNA at 3.31 $\times 10^6$ mM. The cytosine-rich capture probe's signal allows early HIV detection in blood and AIDS diagnosis.

11.15.10 Conductive Polymers

Conductive polymers are commonly employed in bioanalytical applications due to their excellent biocompatibility, inherent charge transfer, and sensitivity to system reactivity to tiny molecules.

11.15.11 Polypyrrole

Biocompatible thin polypyrrole films on anodes were intensively studied for immobilizing nucleic acids, antibodies, and enzymes to find target analytes. Wang *et al.* [81] combined CNTs and glucose oxidase into an electropolymerized polypyrrole screen. A "polypyrrole/glucose oxidase film" employing an "oxidized" carbon nanotube to balance charge was produced. Amperometric enzyme electrodes can accurately detect glucose at 2.33 nA/mM. In DNA detection, mercapatoacetic acid-coated magnetite nanoparticles (Fe_3O_4) were immobilized in polypyrrole. The collector comprised Fe_3O_4 nanoparticles, and a modified polypyrrole electrode detected specific sequence DNA at a low limit (2.33 mol/L) with good selectivity (correlation coefficient 50.99). Enhanced DNA probes can also find more complementary or non-complementary sequences. A new micropotentiometric hemoglobin immunosensor uses an electrochemically produced polypyrrole Au nanoparticle composite. A polypyrrole Au nanoparticle layer produced electrochemically on gold microelectrode (HbA1c) can be employed as an immunosensor for clinical hemoglobin (Hb) and hemoglobin A1c detection. Mounting antibodies on both functional electrodes on an immunogen allows whole blood hemoglobin (Hb) and hemoglobin A1c detection.

11.15.12 Polythiophene

Polythiophene is chemically and thermally stable and very conductive. Functionalization may also improve electrocatalysis and sensing selectivity.

Direct polymerization of monomers yields polythiophene backbones for biosensors. The latter technique immobilizes enzymes or aptamers in polythiophene layers. Electroactive, cationic, water-soluble polythiophene with complementary oligonucleotide coupled to DNA probes as a negatively charged backbone can interact strongly with the transducer. Room-temperature hybridization might be transformed into an electrical signal to identify unlabeled targeted DNA. PTEBS, a derivative of anionic polythiophenes, showed good thermal and environmental stability, high catalytic efficiency, and fast reactivity to horseradish peroxidase. Linking glucose oxidase to PTEBS might provide a simple, inexpensive, sensitive, and discriminating colorimetric glucose biosensor. Le Floch *et al.* [40] invented a transducer.

11.15.13 Polyaniline and Its Byproducts

Polyaniline exhibits redox conductivity, chemical explicitness, and electrode deposition ease. Enzymatic transformations and reagent and transducer bioaffinity must be determined to produce polyaniline-based biosensors. Polyaniline acts as a perfect mediator in enzymatic biosensors by directly linking the enzyme at its active site, where the redox reaction occurs, to the electrode surface. For example, a polyaniline-grafted silver nanoparticle electrode for peroxide detection with a relatively high sensitivity of 1 nM detects H_2O_2 preferentially while filtering out potential interferences like acetaminophen, ascorbic acid, uric acid, and cholesterol esterase, oxidase, and peroxidase immobilized on polyaniline films [68]. These bioelectrode sensors have an optimal detection range of 25 mg/dl to 0.042 A mg/dl, can survive temperature drops up to 4°C for 40–45 days, and are thermally stable up to 48°C. Polyaniline-based sensors for DNA quantification employ bioaffinity reagents. An oxidized nanomaterial graphene and polyaniline nanowire DNA biosensor was demonstrated. The synergistic interaction between graphene and polyaniline gives this DNA sensor remarkable sensitivity and selectivity to complementary sequences with a minimum detection limit of 3.2531 mol/L. For biosensing, polyaniline derivatives like poly(o-phenylenediamine) have high selectivity. Cosimino *et al.* [17] pioneered electropolymerized polymers (o-phenylenediamine). A glucose oxidase film detected glucose. This sensor can be used in HPLC and FIA chromatography since its response time is faster than enzyme electrodes. Based on these beliefs, brain extracellular fluid glucose sensors were constructed [21]. A biosensor based on glucose oxidase fixed on a Pt electrode electropolymerized with poly(o-phenylenediamine) was used to study rats' reactions to fluctuating glucose levels in the brain's extracellular fluid. Ascorbic acid and oxygen concentrations varied randomly, indicating precise glucose level monitoring *in vivo* in peripatetic animals. Poly(o-phenylenediamine) and silver nanoparticles formed a sensitive sensor for elemental identification to measure brucine in human serum [57]. Poly(o-phenylenediamine)-modified glassy carbon electrode complexes detected non-enzymatic hydrogen peroxide and kanamycin.

11.15.14 Polyacetylene

Long carbon atom sequences in conjugated polymer polyacetylene bear alternated single and double bonds. A disubstituted polyacetylene polymeric chemosensor can withstand heat breakdown and emit blue light efficiently. The study found copper-ion-containing α-amino acids using disubstituted polyacetylene [60]. Copper ions (7.031 mol/L) completely quenched the strong green fluorescence of post-functional disubstituted polyacetylenes. As copper ions showed a strong affinity for imidazole moieties, they were thought to transfer the absorbed energy to them, suppressing fluorescence.

References

1. Ali, S.R., Ma, Y., Parajuli, R.R., Balogun, Y., Lai, W.Y.C., He, H., A nonoxidative sensor based on a self-doped polyaniline/carbon nanotube composite for sensitive and selective detection of the neurotransmitter dopamine. *Anal. Chem.*, 79, 6, 2583, 2007.
2. Anderson, J.M., Biological responses to materials. *Annu. Rev. Mater. Res.*, 31, 1, 81–110, 2001.
3. Barone, P., Stranges, F., Barberio, M., Renzelli, D., Bonanno, A., Xu, F., Study of band gap of silver nanoparticles—Titanium dioxide nanocomposites. AI Hindawi. *J. Chem.*, 1–7, Article ID 589707, 2014. https://doi.org/10.1155/2014/589707.
4. Bermejo, R. and Danzer, R., High failure resistance layered ceramics using crack bifurcation and interface delamination as reinforcement mechanisms. *Eng. Fract. Mech.*, 77, 11, 2126–35, 2010.
5. Besteman, K., Lee, J.-O., Wiertz, F.G., Heering, H.A., Dekker, C., Enzyme-coated carbon nanotubes as single-molecule biosensors. *Nano Lett.*, 3, 6, 727–730, 2003.
6. Bilitewski, U. and Turner, A.P.F. (Eds.), *Biosensors for Environmental Monitoring*, Harwood Academic, Amsterdam, 2000.
7. Black, J., The education of the biomaterialist: Report of a survey, 1980-81. *J. Biomed. Mater. Res.*, 16, 2, 159–167, 1982.
8. Broadley, K.N., Aquino, A.M., Woodward, S.C., Buckley-Sturrock, A., Sato, Y., Rifkin, D.B., Monospecific antibodies implicate basic fibroblast growth factor in normal wound repair, *Lab. Invest.*, 61, 571–5, 1989.
9. Bronzino, J.D., *The Biomedical Engineering Handbook 2*, 1, 15,1–12, Boca Raton, FL, CRC Press, 1st ed., 1995; 2nd ed., 2000.
10. Bruni, S., Martinesi, M., Stio, M., Effects Of surface treatment Of Ti–6al–4v titanium alloy on biocompatibility in cultured human umbilical vein endothelial cell. *Acta Mater.*, 1, 23–234, 2005.

11. Brys, E., Tombelli, S., Minunni, M.E., *Approaches to Allergy Detection Using Aptasensors*, pp. 539–565, Taylor & Francis, 2007.
12. Chen, D., Luque, M.D., Castro, D.E., Valcarcel, M., Flow-through sensor for the fluorimetric determination of cyanide, *Talanta*, 37, 11, 1049, 1990.
13. Chen, Q. and Thouas, G.A., Metallic implant biomaterials. *Mater. Sci. Eng. R*, 87, 1–57, 2015.
14. Clark, L. and Lyons, C., Electrode systems for continuous monitoring in cardiovascular surgery, *Ann. N.Y. Acad. Sci.*, 102, 29–45, 1962.
15. Clark, R.A., Lanigan, J.M., DellePelle, P., Manseau, E., Dvorak, H.F., Colvin, R.B., Fibronectin and fibrin provide a provisional matrix for epidermal cell migration during wound reepithelialization. *J. Invest. Dermatol.*, 79, 264–9, 1982.
16. Cohen, J., Biomaterials in orthopedic surgery. *Am. J. Surg.*, 114, 31–41, 1967.
17. Malitesta, C., Palmisano, F., Torsi, L., Zambonin, P.G., Glucose fast-response amperometric sensor based on glucose oxidase immobilized in an electropolymerized poly(o-phenylenediamine) film. *Anal. Chem.*, 62, 24, 2735–2740, 1990. https://doi.org/10.1021/ac00223a016.
18. D'Souza, S.F., Immobilized enzymes in bioprocess. *Curr. Sci.*, 77, 69–79, 1999.
19. D'Souza, S.F., Microbial biosensors. *Biosens. Bioelectron.*, 16, 337–353, 2001.
20. Davis, J.R., *Handbook of Materials for Medical Devices*, ASM International, 1–317, 2003.
21. Brown, F.O., Finnerty, N.J., Lowry, J.P., Nitric oxide monitoring in brain extracellular fluid: Characterisation of Nafion-modified Pt electrodes *in vitro* and *in vivo*. *Analyst*, 134, 10, 2012–20, Oct. 2009.
22. Fu, Y., Liang, F., Tian, H., Hu, J., Nonenzymatic glucose sensor based on ITO electrode modified with gold nanoparticles by ion implantation. *Electrochim. Acta*, 120, 314–318, 2014.
23. Gerhardt, L.C. and Boccaccini, A.R., Bioactive glass and glass-ceramic scaffolds for bone tissue engineering. *Materials*, 3, 3867–3910, 2010.
24. Gobbi, S.J., Gobbi, V.J., Reinke, G., Rocha, Y., Orthopedic implants: Coating with tin. *Biomed. J. Sci. Technol. Res.*, 16, 1, 2019.
25. Gross, U., Kinn, R., Schmitz, H.J., Strunz, V., The Response of bone to surface active glasses/glass-ceramic. critical reviews in Biocompatibility, CRC, 4, 2, 155–179, 1988. https://www.researchgate.net/publication/292531507.
26. Goodridge, L., Chen, J., Griffiths, M., Development and characterization of a fluorescent-bacteriophage assay for detection of Escherichia coli O157: H7, *Appl. Environ. Microbiol.*, 65, 1397–1404, 1999.
27. Alhadrami, H.A., Biosensors: Classifications, medical applications, and future prospective. *Biotechnol. Appl. Biochem.*, 65, 3, 497–508, 2017.
28. Harvey, W., *Exercitatio Anatomica de Motu Cordis et Sanguinis in Animalibus*, Sumptibus Guilielmi Fitzeri, vol. 17, pp. 1–72, 1628.

29. Helmus, M.N., Gibbons, D.F., Cebon, D., Biocompatibility: Meeting a key functional requirement of next-generation medical devices. *Toxicol. Pathol.*, 36, 1, 70–80, 2008.
30. Hench, L.L. and West, J., The sol-gel process. *Chem. Rev.*, 90, 33–72, 1990.
31. Hench, L.L. and Ethridge, E.C., *Biomaterials: An Interfacial Approach*, p. 385, Academic Press, New York, 1982.
32. Higson, S.P.J., Reddy, S.M., Vadgama, P.M., Enzyme and other biosensors: Evolution of a technology. *Eng. Sci. Educ. J.*, 3, 41–48, 1994.
33. Homola, J., Yee, S.S., Gauglite, G., Surface plasmon resonance sensors: Review. *Sens. Actuators B Chem.*, 54, 3–15, 1994.
34. Kalita, S.J., Nanostructured biomaterials, in: *Nanostructure Science and Technology*, pp. 168–219, 2008.
35. Kastantin, M., Langdon, B.B., Schwartz, D.K., A bottom-up approach to understanding protein layer formation at solid-liquid interfaces. *Adv. Colloid Interface. Sci.*, 207, 240–252, 2014.
36. Kim, S.K., Kim, D., Jeon, S., Electrochemical determination of serotonin on glassy carbon electrode modified with various graphene nanomaterials. *Sens. Actuators B. Chem.*, 174, 285–291, 2012.
37. Kokubo, T. and Takadama, H., How useful is SBF in predicting *in vivo* bone bioactivity? *Biomaterials*, 27, 2907–2915, 2006.
38. Kumari, T.V., Kumar, P.R., Muraleedharan, C.V., *In vitro* cytocompatibility studies of diamond like carbon coatings on titanium. *Bio-Med. Mater. Eng.*, 12, 329–338, 2002.
39. Lazcka, O., Del Campo, F.J., Munoz, F.X., Pathogen detection: A perspective of traditional methods and biosensors. *Biosens. Bioelectron.*, 22, 1205–1217, 2007.
40. Le Floch, F., Simonato, J.P., Bidan, G., Electrochemical signature of the grafting of diazonium salts. A probing parameter for monitoring the electro-addressed functionalization of devices. *Electrochim. Acta*, 54, 3078, 2009. http://dx. doi.org/10.1016/j.electacta.2008.11.063.
41. Lei, Y., Chen, W., Mulchandani, A., Microbial biosensors. *Anal. Chim. Acta*, 568, 1-2, 200–210, Epub 2006.
42. Lemons, J.E., Niemann, K.M., Weiss, A.B., Biocompatibility studies on surgical-grade titanium-, cobalt-, and iron-base alloys. *J. Biomed. Mater. Res.*, 10, 549–553, 1976.
43. Linares, P., Luque, M.D., Castro, D.E., Valcarcel, M., Integrated reaction and photometric detection with enzymes immobilized in the flow cell of a flow-injection system, *Anal. Chim. Acta*, 230, 199–202, 1990.
44. Liu, Q. and Wang, P., *Cell-Based Biosensors: Principles and Applications*, United States Artech House, 1–286, 2009.
45. Lowe, C.R., Overview of biosensor and bioarray technologies, in: *Handbook of Biosensors and Biochips*, 2008.

46. Madri, J.A., Pratt, B.M., Tucker, A.M., Phenotypic modulation of endothelial cells by transforming growth factor-beta depends upon the composition and organization of the extracellular matrix. *J. Cell Biol.*, 106, 1375–84, 1988.
47. Malhotra, B.D., Singhal, R., Chaubey, A., Sharma, S.K., Kumar, A., Recent trends in biosensors. *Curr. Appl. Phys.*, 5, 92–97, 2005.
48. Malhotra, R., Patel, V., Vaque, J.P., Gutkind, J.S., Rusling, J.F., Ultrasensitive electrochemical immunosensor for oral cancer biomarker IL-6 using carbon nanotube forest electrodes and multilabel amplification. *Anal. Chem.*, 82, 8, 3118–3123, 2010.
49. Marty, J.L., Olive, D., Asano, Y., Measurement of BOD-correlation between 5-day BOD and commercial BOD biosensor values. *Environ. Technol.*, 18, 333–337, 1997.
50. Matrubutham, U., Sayler, G.S., Mulchandani, A., Rogers, K.R., *Enzyme and Microbial Biosensors: Techniques and Protocols*, p. 249, Humana Press, Totowa, NJ, 1998.
51. Mehdizadeh, M. and Yang, J., Design strategies and applications of tissue bioadhesives. *Macromol. Biosci.*, 13, 3, 271–88, 2013.
52. Moretto, L.M., Panero, S., Scrosati, B., Ugo, P., Lin, Y., Nalwa, H.S., Templated ensembles of nanoelectrodes, in: *Handbook of Electrochemical Nanotechnology*, vol. 1, pp. 87–105, American Scientific, Stevenson Ranch, CA, USA, 2009.
53. Niinomi, M., Recent metallic materials for biomedical applications. *Metall. Mater. Trans. A*, 33a, 477–486, 2002.
54. Palchetti, I., New trends in the design of enzyme-based biosensors for medical applications. *Mini Rev. Med. Chem.*, 16, 14, 1125, 2016.
55. Perumal, V. and Hashim, U., Advances in biosensors: Principle, architecture and applications. *J. Appl. Biomed.*, 12, 1, 115, 2014.
56. Petersson, B.A., Andersen, H.B., Hansen, E.H., Determination of urea in undiluted blood samples by flow injection analysis using optosensin, *Anal. Lett.*, 20, 1977–1994, 1987.
57. Liu, P., Zhang, X., Xu, W., Guo, C., Wang, S., Electrochemical sensor for the determination of brucine in human serum based on molecularly imprinted poly-o-phenylenediamine/silver nano particles composite film. *Sens. Actuators B Chem.*, 163, 1, 84–89, 2012. https://doi.org/10.1016/j.snb.2012.01.011.
58. Pramanik, S., Agarwal, A.K., Rai, K.N., Chronology of total hip joint replacement and materials development. *Trends Biomater. Artif. Organs*, 19, 1, 15–26, 2005.
59. Prasad, B.B., Prasad, A., Tiwari, M.P., Madhur, R., Multiwalled carbon nanotubes bearing 'terminal monomeric unit' for the fabrication of epinephrine imprinted polymer-based electrochemical sensor. *Biosens. Bioelectron.*, 45, 114–122, 2013. http://dx.doi.org/10.1016/j.bios.2013.01.042.
60. Zeng, Q., Jim, C.K.W., Lam, J.W.Y., Dong, Y., Li, Z., Qin, J., Tang, B.Z., A new disubstituted polyacetylene for the detection of α-amino acids. *Macromol.*

Rapid Commun., Wiley, 30, 3, 170–175, 2009. https://doi.org/10.1002/marc.200800616.

61. Rabe, M., Verdes, D., Seeger, S., Understanding protein adsorption phenomena at solid surfaces. *Adv. Colloid Interface Sci.*, 162, 87–106, 2011.
62. Raghavendra, G.M., Varaprasad, K., Jayaramudu, T., Biomaterials, design, development and bio-medical applications. Elsevier, 2015.
63. Rai, M., Gade, A., Gaikwad, S., Marcato, P.D., Duran, N., Biomedical applications of nanobiosensors, the state-of-the-art. *J. Braz. Chem. Soc.*, 23, 14–24, 2012.
64. Ratner, B.D. and Hoffman, A.S., A history of biomaterials. Biomaterials science: A multidisciplinary endeavor, in: *Biomaterial Science: An Introduction to Materials in Medicine*, Seconded, B.D. Ratner, A.S. Hoffman, F.J. Schoen, J.E. Lemons, (Eds.), pp. 1–9, Elsevier's Science & Technology, Oxford, 2004.
65. Rider, T.H., Petrovick, M.S., Nargi, F.E., Harper, J.D., Schwoebel, E.D., Mathews, R.H., A B cell-based sensor for rapid identification of pathogens. *Science*, 301, 5630, 213, 2003.
66. Riedel, K., Mulchandani, A., Rogers, K.R. (Eds.), *Enzyme and Microbial Biosensors: Techniques and Protocols*, p. 199, Humana Press, Totowa, 1998.
67. Rogers, K.R. and Gerlach, C.L., Update on environmental biosensors. *Environ. Sci. Technol.*, 33, 500–506, 1999.
68. Santhosh, P., Gopalan, A., Lee, K.-P., Gold nanoparticles dispersed polyaniline grafted multiwall carbon nanotubes as newer electrocatalysts: Preparation and performances for methanol oxidation. *J. Catal.*, 238, 177–185, 2006.
69. Schedin, F., Geim, A., Morozov, S., Hill, E., Blake, P., Katsnelson, M., Detection of individual gas molecules adsorbed on graphene. *Nat. Mater.*, 6, 9, 652, 2007.
70. Seymour, R.B. and Kauffman, G.B., Polyurethanes: A class of modern versatile materials. *J. Chem. Educ.*, 69, 11, 909, https://doi.org/10.1021/ed069p909.
71. Sporn, M.B. and Roberts, A.B., Peptide growth factors are multifunctional. *Nature*, 332, 217–219, 1988.
72. Stadlinger, B., Ferguson, S.J., Eckelt, U., Biomechanical evaluation of a titanium implant surface conditioned by a hydroxide ion solution. *Br. J. Oral. Maxillofac. Surg.*, 50, 74–79, 2012.
73. Subramani, K. and Mathew, R., Titanium surface modification techniques for dental implants from microscale to nanoscale, in: *Emerging Nanotechnologies in Dentistry*, pp. 85–102, 2012.
74. Syam, R., Justin Davis, K., Pratheesh, M.D., Anoopraj, R., Surej Joseph, B., Biosensors: A novel approach for pathogen detection. *VETSCAN*, 7, 14–18, 2012.
75. Taddei, E.B., Henriques, V.A.R., Silva, C.R.M., Ensaio de citotoxicidade e influência do tratamento de solubilização na microestrutura da liga Ti-35Nb-7Zr-5Ta para potenciais aplicações ortopédicas. *Rev. Matér.*, 12, 1, 120–127, 2007.

76. Teixeira, S., Burwell, G., Castaing, A., Gonzalez, D., Conlan, R., Guy, O., Epitaxial graphene immunosensorfor human chorionic gonadotropin. *Sens. Actuators B Chem.*, 190, 723–729, 2014.
77. Tilocca, A., Cormack, A.N., de Leeuw, N.H., The structure of bioactive silicate glasses: New insight from molecular dynamics simulations. *Chem. Mater.*, 19, 1, 95–103, 2007. https://doi.org/10.1021/cm061631g.
78. Turner, A.P.F., Karube, I., Wilson, G.S. (Eds.), *Biosensors: Fundamentals and Applications*, Mir Publishers, Moscow, 1992.
79. Updike, S. and Hicks, G., The enzyme electrode. *Nature*, 214, 5092, 986, 1967.
80. Wang, J., Rivas, G., Cai, X., Palecek, E., Nielsen, P., Shiraishi, H., Dontha, N., Luo, D., Parrado, C., Chicharro, M., Farias, P., Valera, F.S., DNA electrochemical biosensors for environmental monitoring: A review. *Anal. Chim. Acta*, 347, 1–8, 1997.
81. Wang, Y., Li, Y., Tang, L., Lu, J., Li, J., Application of graphene-modified electrode for selective detection of dopamine. *Electrochem. Commun.*, 11, 4, 889–892, 2009.
82. Wang, Y., Ping, J., Ye, Z., Wu, J., Ying, Y., Impedimetric immunosensor based on gold nanoparticles modified graphene paper for label-free detection of escherichia coli O157: H7. *Biosens. Bioelectron.*, 49, 492–498, 2013.
83. Watts, H.J., Lowe, C.R., Pollard-Knight, D.V., Optical biosensor for monitoring microbial cells. *Anal. Chem.*, 66, 2465–2470, 1994.
84. He, W., Yixuan, H., Wu, J., Enzyme-free glucose biosensors based on MoS2 nanocomposites. *Nanoscale Res. Lett.*, 15, 60, 2020. https://doi.org/10.1186/s11671-020-3285-3.
85. Williams, D.F., Review: Tissue biomaterial interactions. *J. Mater. Sci.*, 22, 10, 3421–45, 1987.
86. Kang, X., Wang, J., Wu, H., Aksay, I.A., Liu, J., Lin, Y., Glucose Oxidase–graphene–chitosan modified electrode for direct electrochemistry and glucose sensing. *Biosens. Bioelectron.*, 25, 901–905, 2009.
87. Xue, X., Ang, J., Zhu, Y., Biocompatibility of pure titanium modified by human endothelial cell-derived extracellular matrix. *Appl. Surf. Sci.*, 256, 3866–3873, 2010.
88. Xue, Y.H., Zhou, W.J., Zhang, L., Li, M., Chan, S.H., Poly (diallyldimethylammonium chloride)-functionalized reduced graphene oxide supported palladium nanoparticles for enhanced methanol oxidation. *RCS Adv.*, 42, 5, 32983–32989, 2015.
89. Yoruc, A.B.H. and Sener, B.C., Biomaterials, in: *A Roadmap of Biomedical Engineers and Milestones*, 2012.
90. Yu, J., Zhang, Y., Liu, S., Enzymatic reactivity of glucose oxidase confined in nanochannels. *Biosens. Bioelectron.*, 55, 307–312, 2014.

12

Blockage Study in Carotid Arteries

Bushra Khatoon and M. Siraj Alam*

Chemical Engineering Department, MNNIT-Allahabad, Prayagraj, U.P., India

Abstract

The brain is one of the most important organs of the human body that commands the human nervous system, it is made of 100 billion nerves, and the supply of blood to these nerves is possible through the carotid arteries. In the human body, carotid arteries are located on both sides of the neck and present as a pair of blood vessels. If the supply of blood is hindered through the carotid arteries, then the human brain would stop working. The reason for this hindrance could be the blocking of arteries, and blockage appeared due to the presence of plaques. These plaques are made of different materials like wax/fats and cholesterol. In the present study, a simulation was performed for the blood flow through the carotid artery using the CFD tool, and the hydrodynamics of blood flow in the arteries with and without plaques was studied. This study would help to understand the presence of stress, pressure, velocity, and viscosity of the blood in the carotid arteries, and all the results will be generated in the form of different contours (velocity, pressure, and stress) and graphs (mass distribution and viscosity) using ANSYS Fluent.

Keywords: Blood, non-Newtonian flow, artery, plaque, CFD, brain

12.1 Introduction

As we know, the human body is a complicated structure, and this complex body has many organs, tissues, arteries, veins, blood circulation activities, and many more. The human body works smoothly when all the organs work properly, and that would be possible when the supply of blood and oxygen [1] goes through the arteries at the required rate and with the

**Corresponding author*: msalam@mnnit.ac.in

Jay Prakash Srivastava, Dražan Kozak, Vinayak Ranjan, Pankaj Kumar, Ranjan Kumar and Shubham Tayal (eds.) Mechanical Engineering in Biomedical Applications: Bio-3D Printing, Biofluid Mechanics, Implant Design, Biomaterials, Computational Biomechanics, Tissue Mechanics, (327–342) © 2024 Scrivener Publishing LLC

definite composition of blood. The supply of blood from the heart to the different organs goes through the arteries, and if we want to take out the web of arteries from the human body and check its length, it would reach 60,000 miles which is a huge value, and if some problem arises, obviously, it cannot be easily detected and it will disturb the supply of blood and oxygen to the human organs. Ultimately, it will disrupt or create a number of problems for the human body. The most common diseases related to the arteries that can be frequently seen in the human body include aneurysms, blood clotting/plaque formation, Buerger's disease, heart attack, brain strokes, atherosclerosis, abdominal aortic aneurysm, thoracic aortic aneurysm, thoracic outlet syndrome, subclavian steal syndrome, and cardiovascular diseases [2] such as coronary artery disease, carotid artery disease, vertebrobasilar disease, renal vascular disease, and peripheral artery disease. Cardiovascular disease is one of the most dangerous and leading problems in the world, and this is the main reason for heart attacks, brain strokes, and deaths in the world [3]. The current study is related to the occurrence of stenosis in carotid arteries located on both sides of the neck and present as a pair of blood vessels. If the supply of blood is hindered through the carotid arteries, then the human brain would stop working causing strokes in body organs [4]. The reason for this hindrance could be the blocking of arteries, and blockage appears due to the presence of plaques. These plaques are made of different materials like wax/fats, calcium, and cholesterol which leads to atherosclerosis disease, and arteries lose their properties, e.g., elasticity and structure.

People are showing their continuous interest, doing efforts in cardiovascular disease, and trying to find the solution using different techniques, methods, and simulations [5]. In their work, Gao *et al.* [6] monitored the number of ECG reports of patients after their heart attacks. The studies of Nitish *et al.* [7] and Tan *et al.* [8] were related to the clotting/blockage issues in the arteries using carotid angiography and digital subtraction catheter angiography for the nature of blockage (due to clots/plaques). Kotalczyk *et al.* [9] observed the 1-year data for those patients who have problems/issues in the arteries. If you look at the studies of Edwards *et al.* [10], Gaziano *et al.* [11], Hoorweg *et al.* [12], and others, it can be easily seen that the researchers were collecting and analyzing data by observing medical reports, but it was found that the solution to this coronary health problem is lacking. As per the WHO data, 17.9 million people die each year, and according to the Centers for Disease Control and Prevention (CDC), one person dies every 36 s in the USA due to this disease. Prabhakaran *et al.* [13] have provided a detailed review of cardiovascular diseases and presented all the reasons and outcomes in their work (Figure 12.1).

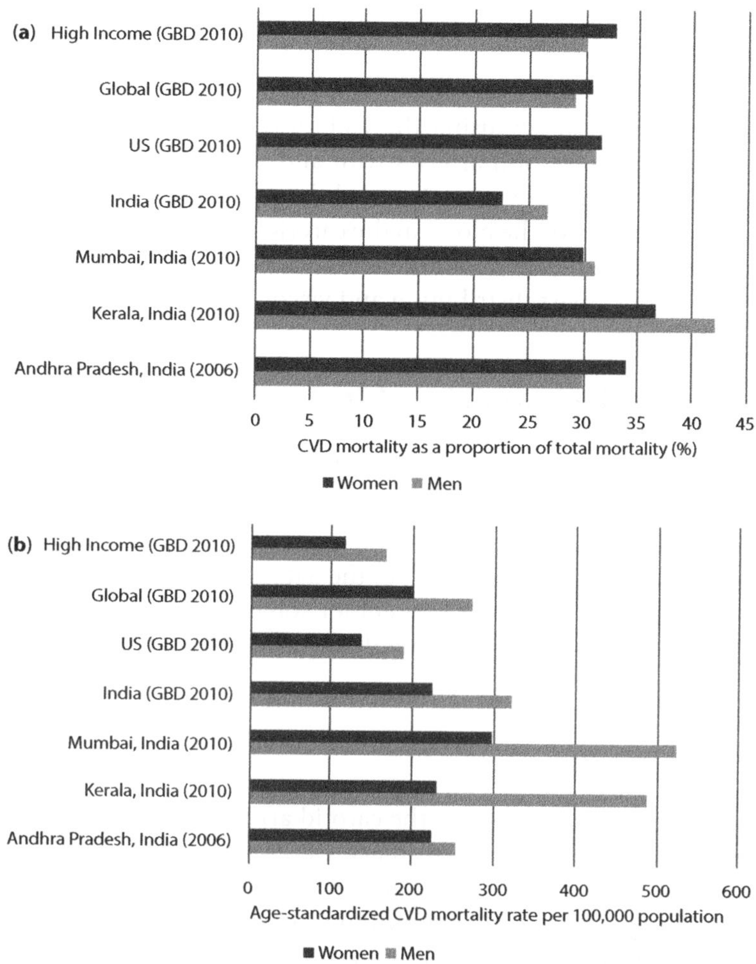

Figure 12.1 (a) Proportion of cardiovascular disease mortality in India based on the data from available prospective studies and the Global Burden of Disease estimates [13]. (b) Age-standardized rate of cardiovascular disease mortality in India based on the Global Burden of Disease estimates and data from available prospective studies. CVD, cardiovascular disease; GBD, Global Burden of Disease [13].

Doctors often judge the severity based on the patient's physical symptoms as well as growth rate, constriction size, and pressure drop often putting the patient's life at high risk as stated by many researchers [14–16]. However, the results of the numerical simulation can be better than MR, CT, and ultrasound techniques in analyzing the severity of stenosis. Apart from this, an attempt is made in this present study to demonstrate the significant changes in flow behavior for 65% and 85% severities of stenosis.

In the present study, a simulation was performed for the blood flow through the carotid artery using the CFD tool, and the hydrodynamics of blood flow in the carotid arteries with and without plaques was studied. This study would help us understand the presence of stress, pressure, velocity, and viscosity of the blood in the carotid arteries, and all the results will be generated in the form of different contours (velocity, pressure, and stress) and graphs (mass distribution and viscosity) using ANSYS Fluent.

12.2 Numerical Model and Its Implementation

Two-dimensional geometry of the carotid artery is intended in the ANSYS design modular (DM), the designed geometry has meshed in the ICEM meshing tool and the grid independency test (GIT) is performed to assure the repeatability and reasonableness of the data recorded, and then meshed geometry is uploaded in ANSYS Fluent for solving all mathematical equations. All steps that belong to the simulation are represented in the form of a flow diagram (Figure 12.2).

12.2.1 Geometry

Angiographic and MRI image of the carotid artery was collected from the literature and developed as a 2D image using a commercial tool. In this work, normal, 50%, and 75% blocked carotid arteries were selected which have 12 mm inlet, 8 mm, and 6.4 mm internal and external carotid artery outlets for each case (Figure 12.3). A flexible wall of the artery is considered for the normal artery as well as stenosis in the arteries.

12.2.2 Meshing and GIT

2D Two-dimensional triangular elements were employed to discretize the geometry of the carotid artery as mentioned in Figure 12.4. Tables 12.1–12.3 depict the outcomes of the grid independence test (GIT) being done to assure that grid size has no impact on the result of the simulation. The GIT test is used for four different mesh sizes ranging from 0.15 to 0.35 m to obtain the constant blood flow velocity concerning mesh size at the outlet of the internal carotid artery (ICA) for all the three cases.

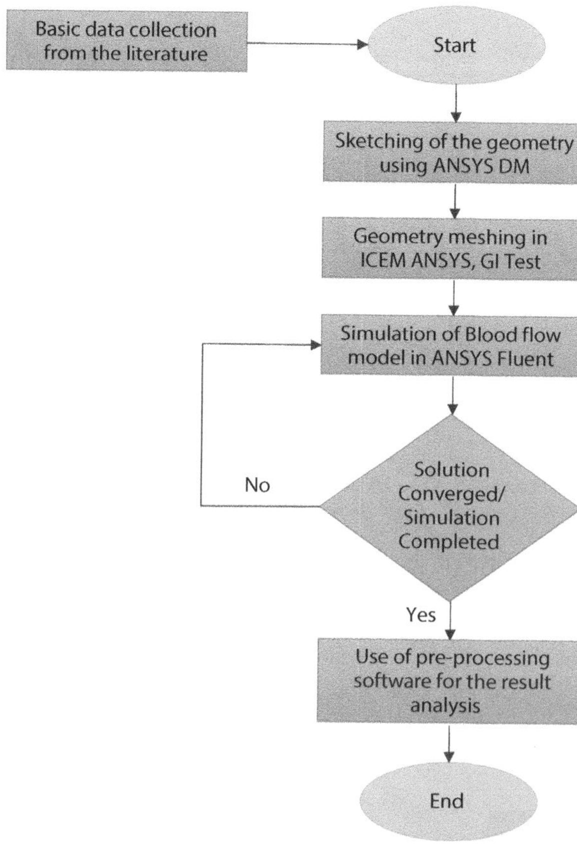

Figure 12.2 Flow diagram for the simulation steps.

12.2.3 Governing Equations

Considering the blood as a single-phase non-Newtonian liquid and using the shear rate-dependent Carreau model to obtain viscosity, model equations [17] with boundary conditions are defined according to the system. A representative set of the equations for the study of velocity (u) and pressure (P) for single-phase blood flow (incompressible) with the help of the Carreau model (Eq. 12.4) [18] in the carotid artery can be explained as follows:

$$\nabla.\rho A u = 0 \tag{12.1}$$

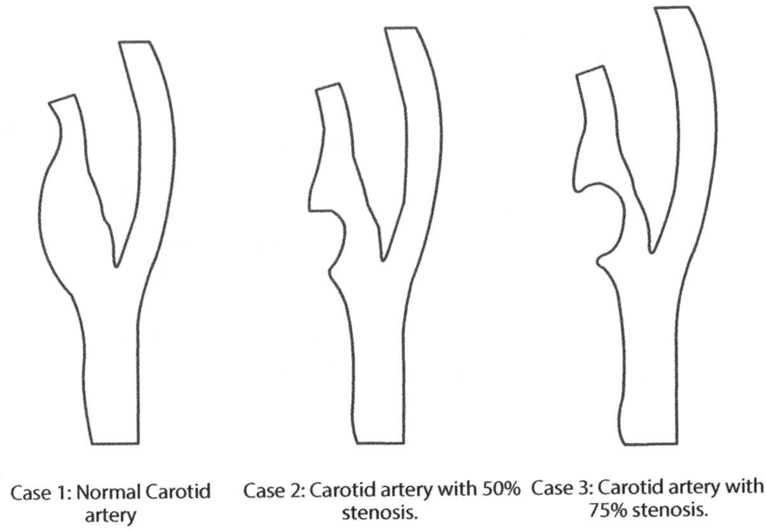

Figure 12.3 Geometry used in the design modular.

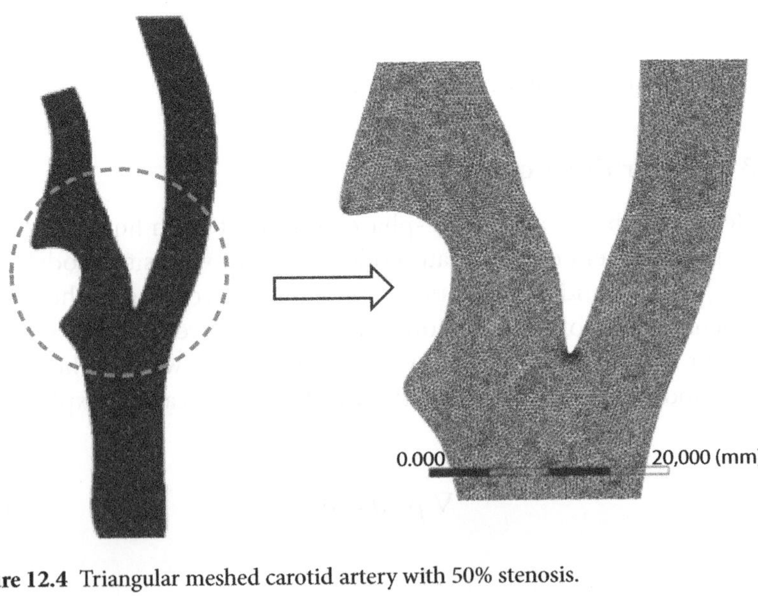

Figure 12.4 Triangular meshed carotid artery with 50% stenosis.

Table 12.1 Grid independency test for case 1, i.e., normal artery.

Sr. no.	Mesh size (mm)	No. of elements	Maximum velocity (m s^{-1})
1.	0.15	11,532	0.086
2.	0.20	23,561	0.125
3.	0.25	34,925	0.143
4.	*0.30*	*41,375*	*0.150*
5.	0.35	46,621	0.152

Table 12.2 Grid independency test for case 2, i.e., 50% blocked carotid artery.

Sr. no.	Mesh size (mm)	No. of elements	Maximum velocity (m s^{-1})
1.	0.15	9,444	0.092
2.	0.20	14,527	0.127
3.	0.25	25,454	0.141
4.	*0.30*	37,489	*0.164*
5.	0.35	42,295	0.165

Table 12.3 Grid independency test for case 3, i.e., 75% blocked carotid artery.

Sr. no.	Mesh size (mm)	No. of elements	Maximum velocity (m s^{-1})
1.	0.15	7,752	0.092
2.	0.20	12,505	0.127
3.	0.25	21,890	0.141
4.	*0.30*	*34,621*	*0.174*
5.	0.35	39,124	0.176

$$Re = \frac{\rho u D}{\mu} \tag{12.2}$$

$$\frac{\partial u}{\partial t} + \nabla.(uu) = -\nabla P + \nabla.\left[\mu\left(\nabla\vec{u} + \nabla\vec{u}^{T}\right)\right] + \rho g + F \tag{12.3}$$

Table 12.4 lists the Carreau model coefficient values that were used to simulate blood flow in arteries, and Figure 12.5 illustrates how strain rate affects blood flow viscosity with the Carreau model [19].

Table 12.4 Coefficients obtained from ANSYS Fluent for the Carreau model [19].

Sr. no.	Carreau model coefficients	Values
1.	Time constant (λ)	3.1300 (s)
2.	Power law index (n)	0.3568
3.	Zero shear viscosity (μ_0)	0.0560 (kg m^{-1} s^{-1})
4.	Infinite shear viscosity (μ_∞)	0.0350 (kg m^{-1} s^{-1})

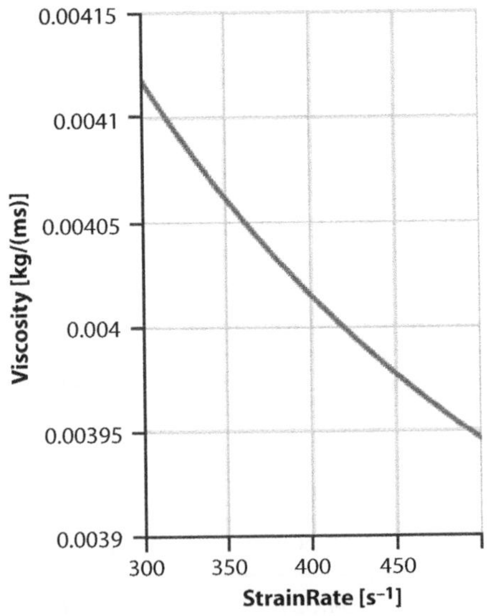

Figure 12.5 Viscosity behavior of blood flow using the Carreau model (data obtained from ANSYS Fluent) [19].

$$\mu = \mu_\infty + \frac{(\mu_0 - \mu_\infty)}{[1+(\lambda\gamma)^2]^{\frac{1-n}{2}}} \tag{12.4}$$

Where ρ, A, D, g, F, μ, μ_0, μ_∞, γ, n, and λ are the blood density (1,080 kg m^{-3}), area, equivalent diameter, gravity, surface tension force, non-Newtonian viscosity, zero shear viscosity, infinite shear viscosity, shear strain rate, power law index, and time constant, respectively.

12.2.4 Boundary Conditions

Table 12.5 provides the simulation conditions employed in this research work. For the continuity equation and velocities, convergence criteria are considered as 10^{-5}.

Table 12.5 Model, methods, and initialization conditions for the present work.

Model		
Viscous model		*Laminar*
Methods		
Pressure–velocity coupling		*Coupled*
	Flux type	Rhie–Chow: distance based
Spatial discretization		
	Gradient	Least squares cell based
	Pressure	Second order
	Momentum	Second-order upwind
	Transient formulation	First-order upwind
Initialization		
Initialization methods		*Standard*
	Reference frame	Relative to cell zone
	Gauge pressure	0 Pa
	Velocity	0.1000003 m s^{-1} (obtained from UDF)

12.3 Results and Discussion

In the present work, the simulation started with a normal carotid artery, and then the same model was employed for both the 50% and 75% stenosis arteries.

12.3.1 Effect of Blockage on Blood Flow Velocity

Figure 12.6 represents the velocity contour in the carotid artery for each case. From these contours, it can be seen that as the percentage of blockage changes from 0 to 75, the magnitude of velocities increases, which gives justification to the continuity and the Reynold's equation (Eqs. 12.1 and 12.2).

Figures 12.7–12.9 justify the results of Figure 12.6 through calculation of velocity magnitude (black line in the above diagram), and a maximum velocity in the 75% stenosis passage of the carotid artery was found.

12.3.2 Effect of Blood Flow Velocities on Wall Stress

Increasing stenosis from 0% to 75% increases shear stress on the wall (Figure 12.10) because of the velocity hike, and wall shear stress (WSS) can damage the internal walls of the arteries.

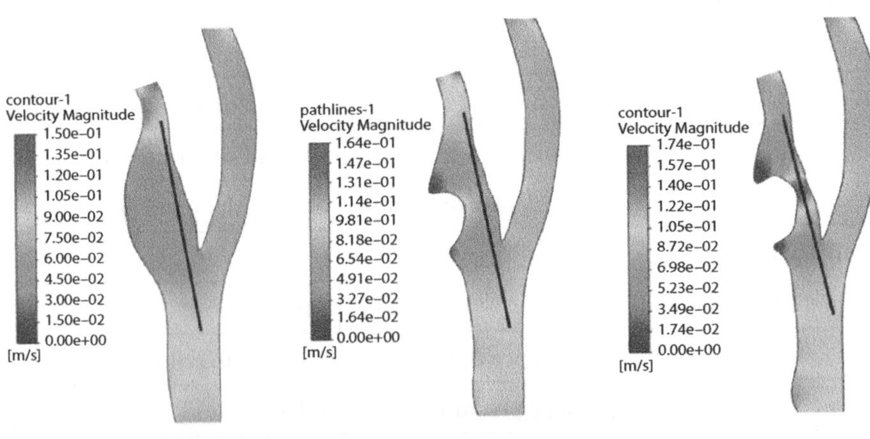

Figure 12.6 Velocity contour in the healthy, 50%, and 75% blocked carotid arteries.

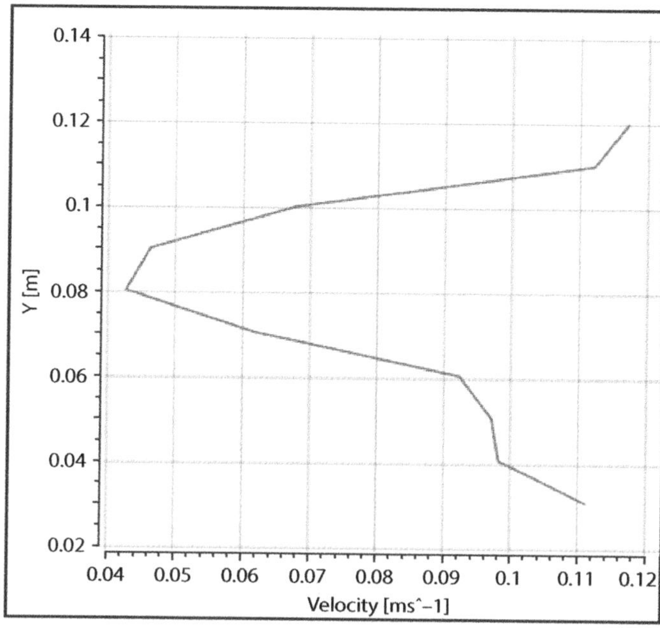

Figure 12.7 Velocity fluctuation with artery length in the normal carotid artery.

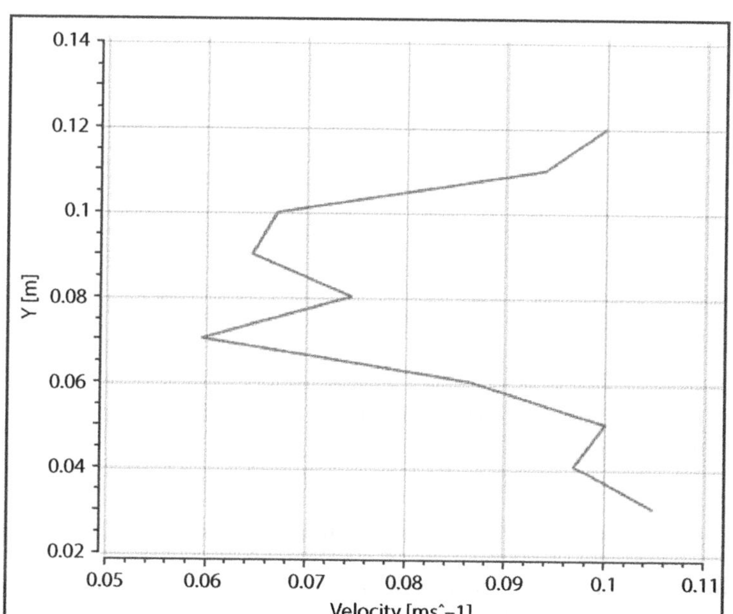

Figure 12.8 Velocity fluctuation with artery length in 50% stenosis carotid artery.

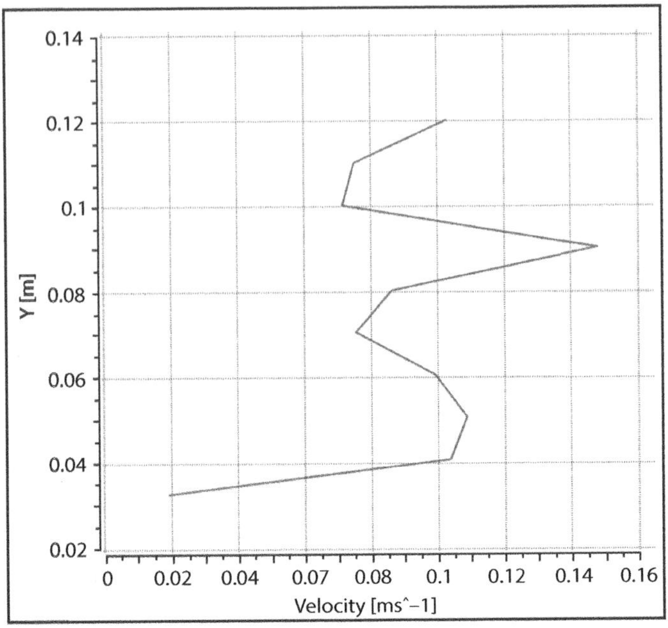

Figure 12.9 Velocity fluctuation with artery length in 75% stenosis carotid artery.

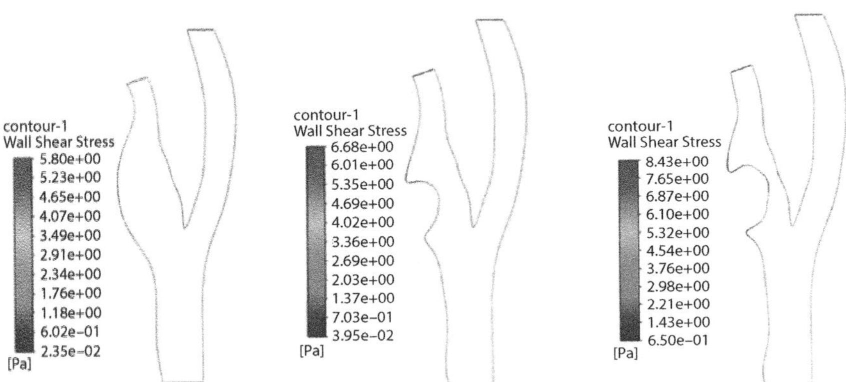

Figure 12.10 Wall shear stress (WSS) distribution contour in the healthy, 50%, and 75% blocked carotid arteries.

12.3.3 Effect of Stenosis on Dynamic Pressure Distribution

With the help of Figure 12.11, uniform pressure was obtained throughout the arteries, and similar behavior and pressure magnitude were found at the inlet, outlet 1, and outlet 2 for every three cases. A change in the value

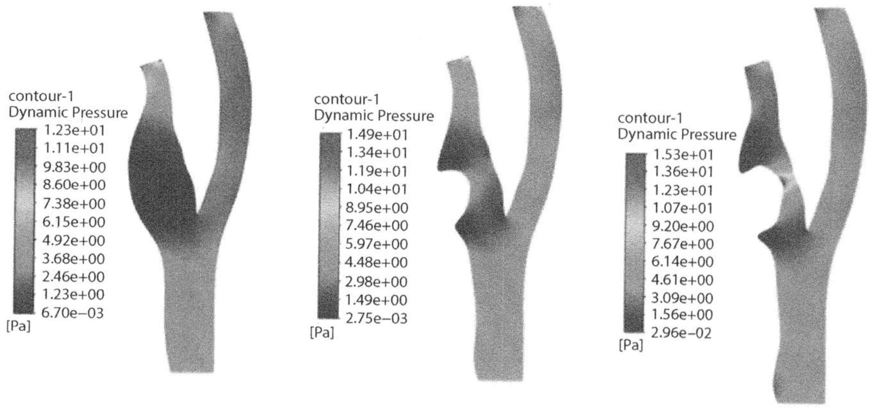

Figure 12.11 Pressure distribution contour in the healthy, 50%, and 75% blocked carotid arteries.

of dynamic pressure from low to high can be visualized from normal to stenosis arteries in Figure 12.11. In 75% of a blocked artery, the obtained dynamic pressure is much higher than in cases 1 and 2.

12.3.4 Effect of Stenosis on Viscosity and Mass Imbalance

Figures 12.12 and 12.13 show similar viscosity behavior and mass distribution during blood flow in carotid arteries for each case, i.e., normal, 50%, and 75% blocked carotid arteries at time (t) = 50 s.

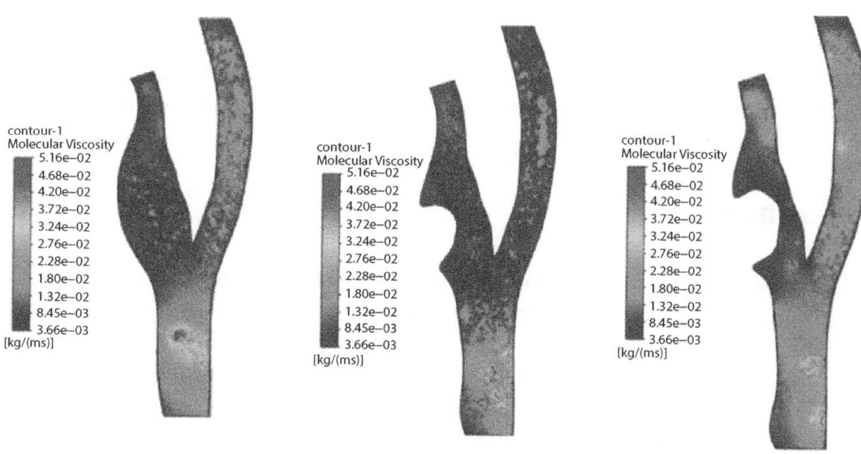

Figure 12.12 Molecular viscosity contour in the healthy, 50%, and 75% blocked carotid arteries.

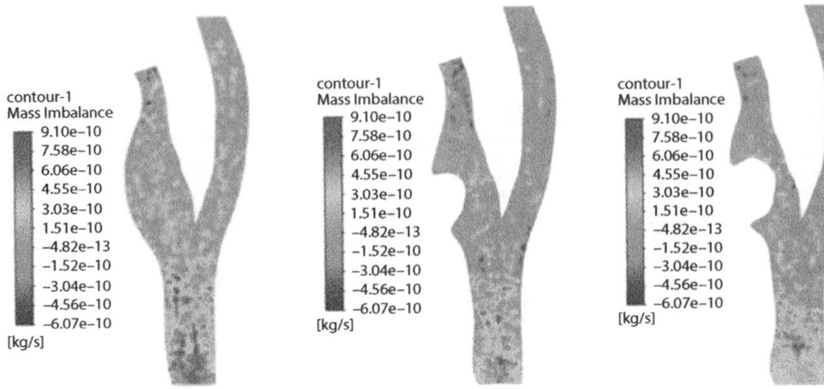

Figure 12.13 Mass distribution contour in the healthy, 50%, and 75% blocked carotid arteries.

12.4 Conclusion

A CFD simulation was performed for a detailed analysis of stenosis arteries in comparison to normal arteries. Modeling was performed by considering the blood as a non-Newtonian single-phase fluid with the Carreau model. Based on the outcomes of the CFD simulations, it can be concluded that stenosis of varying degrees has an impact on the carotid artery's blood flow characteristics, and the influence of 0% to 75% blockage on the velocity magnitude, wall shear stress, dynamic pressure distribution, blood molecular viscosity, and mass distribution was investigated. High velocity, high pressure, and high wall shear stress are all characteristics of stenosis.

Although blood is made up of red blood cells, white blood cells, and plasma, in this work, blood is assumed as a single-phase fluid and considered to have average blood properties for the computational study.

References

1. Bashar, M.S., Hossain, R., Bittu, R.S., Islam, M.S., CFD analysis of bifurcated human carotid artery: Newtonian approach. *Int. J. Eng. Res.*, 9, 03, 83–87, 2020.
2. Hameed, M.S., Shah, A.A., Khan, M., II, Ali, A., Hussain, I., Bukhari, M.D., Comparison of blood flow analysis in stenosed and stented carotid artery bifurcation models. *Cogent Eng.*, 10, 1, 2158624, 2023.
3. Ramdan, S.A. *et al.*, Blood flow acoustics in carotid artery. *J. Adv. Res. Fluid Mech. Therm. Sci.*, 94, 1, 28–44, 2022.

4. Romero, G., Martinez, M.L., Maroto, J., Felez, J., Blood clot simulation model by using the bond-graph technique. *Sci. World J.*, 519047, 2013.
5. Quanyu, W., Xiaojie, L., Lingjiao, P., Weige, T., Chunqi, Q., Simulation analysis of blood flow in arteries of the human arm. *Biomed. Eng.-Appl. Basis Commun.*, 29, 4, 1–17, 2017.
6. Gao, L. *et al.*, Cost-effectiveness of monitoring patients post-stroke with mobile ECG during the hospital stay. *J. Am. Heart Assoc.*, 11, 8, 022735, 2022.
7. Mittal, N., Rao, S., Ansari, M.M., The use of curved sheath technique to facilitate difficult internal carotid artery occlusions. *Southwest Respir. Crit. Care Chron.*, 9, 37, 74–77, 2021.
8. Tan, Q. *et al.*, Inner diameters of the normal carotid arteries measured using three-dimensional digital subtraction catheter angiography: A retrospective analysis. *BMC Neurol.*, 21, 1, 1–10, 2021.
9. Kotalczyk, A., Guo, Y., Stefil, M., Wang, Y., Lip, G.Y.H., Effects of the atrial fibrillation better care pathway on outcomes among clinically complex chinese patients with atrial fibrillation with multimorbidity and polypharmacy: A report from the ChiOTEAF registry. *J. Am. Heart Assoc.*, 11, 7, 024319, 2022.
10. Edwards, J.D., Kapral, M.K., Fang, J., Saposnik, G., Gladstone, D.J., Underutilization of ambulatory ECG monitoring after stroke and transient ischemic attack: Missed opportunities for atrial fibrillation detection. *Stroke*, 47, 8, 1982–1989, 2016.
11. Gaziano, T.A., Bitton, A., Anand, S., Abrahams-Gessel, S., Murphy, A., Growing epidemic of coronary heart disease in low- and middle-income countries. *Curr. Probl. Cardiol.*, 35, 2, 72–115, 2010.
12. Hoorweg, B.B.N. *et al.*, Frequency of chest pain in primary care, diagnostic tests performed and final diagnoses. *Heart*, 103, 21, 1727–1732, 2017.
13. Prabhakaran, D., Jeemon, P., Roy, A., Cardiovascular diseases in India: Current epidemiology and future directions. *Circulation*, 133, 16, 1605–1620, 2016.
14. Piskin, S. and Serdar Celebi, M., Analysis of the effects of different pulsatile inlet profiles on the hemodynamical properties of blood flow in patient specific carotid artery with stenosis. *Comput. Biol. Med.*, 43, 6, 717–728, 2013.
15. Lopes, D., Puga, H., Teixeira, J., Lima, R., Blood flow simulations in patient-specific geometries of the carotid artery: A systematic review. *J. Biomech.*, 111, 110019, 2020.
16. Savabi, R., Nabaei, M., Farajollahi, S., Fatouraee, N., Fluid structure interaction modeling of aortic arch and carotid bifurcation as the location of baroreceptors. *Int. J. Mech. Sci.*, 165, 105222, 2020.
17. Moradicheghamahi, J., Jahangiri, M., Mousaviraad, M., Sadeghi, M.R., Computational studies of comparative and cumulative effects of turbulence, fluid–structure interactions, and uniform magnetic fields on pulsatile non-Newtonian flow in a patient-specific carotid artery. *J. Braz. Soc. Mech. Sci. Eng.*, 42, 10, 1–22, 2020.

18. Akbar, N.S. and Nadeem, S., Carreau fluid model for blood flow through a tapered artery with a stenosis. *Ain Shams Eng. J.*, 5, 4, 1307–1316, 2014.
19. Krystian, J., Łukasz, M. and Wojciech, O., Model of blood rheology including hemolysis based on population balance. *Commun. Nonlinear Sci. Numer. Simul.*, 116, 106802, 2023.

13

Mechanical Properties of Human Synovial Fluid: An Approach for Osteoarthritis Treatment

Sunil More[1,2], K. L. Vasudev[1], N.N. Krishnadas[1] and Ankit Kotia[3]*

[1]Department of Mechanical Engineering, Amrita School of Engineering, Amrita Vishwa Vidyapeetham, Chennai, Tamil Nadu, India
[2]Department of Mechanical Engineering, JSPM's Rajarshi Shahu College of Engineering, Pune, Maharashtra, India
[3]Department of Mechanical Engineering, School of Engineering, University of Petroleum and Energy Studies, Dehradun, India

Abstract

The rheological properties of synovia are of interest because of their importance in the lubrication of knee joints. The synovial fluid has excellent lubrication properties at both low and high loads as well as shear (the friction coefficient is in the order of 0.01–0.02). The synovial fluid (SF) serves as a joint lubricant, a shock absorber, and a cause of sustenance for articular cartilage. The most prevalent kind of arthritis, osteoarthritis (OA), is a progressive joint condition categorized by the loss of articular cartilage. Changes in the rheological characteristics of the synovial fluid might result from damage to the articular cartilage in an osteoarthritic joint. The major aims of osteoarthritis (OA) treatment are to alleviate pain and preserve joint motion. Viscosupplementation, or intra-articular injections of hyaluronic acid (HA), is one of the non-operative actions for osteoarthritis (OA). It functions as a shock absorber for joint loads and a lubricant to allow bones to glide over one another. Polyvinylpyrrolidone (PVP) hyaluronate and moderate molecular weight PVP were found to be the closest to natural synovial fluid in artificial synovial fluid (ASF). Artificial articular lubricants were used in this treatment to help the joints function better. The present work proposes the usefulness of the PVP solution as a synovial mimic fluid and highlights the need to realistically assess the rheological characteristics of synovial fluid. Using viscosupplements that mimic

**Corresponding author*: ankitkotia@gmail.com

Jay Prakash Srivastava, Dražan Kozak, Vinayak Ranjan, Pankaj Kumar, Ranjan Kumar and Shubham Tayal (eds.) Mechanical Engineering in Biomedical Applications: Bio-3D Printing, Biofluid Mechanics, Implant Design, Biomaterials, Computational Biomechanics, Tissue Mechanics, (343–354) © 2024 Scrivener Publishing LLC

healthy synovial fluid is a sensible strategy since several viscosupplements used in osteoarthritis treatment help restore the synovial fluid to its natural composition and feature.

Keywords: Synovial joint, viscosupplement, HA, artificial synovial fluid, PVP

13.1 Introduction

The category of joint diseases known as osteoarthritis (OA) is diverse. The main effects of OA include loss of articular cartilage, acute and chronic synovial inflammation, modification of subchondral bone, the development of osteophytes, and changes in synovial fluid (SF). OA is a chronic, severe illness with an unknown cause that is also considered as synovitis, cartilage deterioration, osteophyte production, and subchondral bone sclerosis. OA is the most common joint condition, with a frequency of 10% in men and 13% in women among persons aged 60 and up [1]. Arthritis is thought to afflict 15% of India's population or approximately 180 million individuals. The medications and agents now used to treat OA have certain unwanted side effects and have been found to be ineffective in preventing joint degeneration over time. Several initiatives have targeted the defining aspects of OA in order to develop therapeutic alternatives. Hyaluronic acid (HA) is depolymerized in osteoarthritis, which improves the synovial fluid's elastic and viscous properties. Degenerated SF due to OA has lower viscosity than healthy SF. This drop in SF lubricating characteristics is attributed to a decrease in HA molecule size and concentration [2]. This worrying condition necessitates a more thorough study in the field, as well as more accurate diagnosis and more efficient patient care. There are more than 100 distinct forms of arthritis [3]. As the most common form of arthritis, OA is a progressive joint disorder marked by the loss of articular cartilage. Damage to the articular cartilage in an osteoarthritic joint can cause variations in the rheological properties of the synovial fluid. Due to its importance in lubricating joints, synovial fluid's rheological qualities are of interest. Since the synovial fluid (SF) is enormously elastic at high shear rates and very viscous at low shear rates, it may effectively protect the articular cartilage in a healthy joint from damage [4]. As people age, osteoarthritis becomes more prevalent, especially in the knee and hip joints from prolonged use. The characteristics and content of synovial fluid are significantly altered by this condition. This alteration is connected to the molecular characteristics of HA and a drop in its concentration. Viscosupplementation and non-steroidal anti-inflammatory medications are frequently used to restore it. Even if these procedures guarantee improved outcomes, a lasting cure has never been certain. The first step in this approach is to conduct studies looking at polyvinylpyrrolidone (PVP) as a synthetic joint lubricant.

13.1.1 Synovial Fluid

Joint fluid, commonly known as synovia, is synovial fluid. It is a viscid, non-Newtonian fluid that originates in the space between the bones where synovial joints are found. This fluid resembles an egg white fluid and contains extracellular fluid that is transcellular. The primary function of synovial fluid when synovial joints move is to reduce friction between the articular cartilages. The gaps in the joint cavities are filled by SF. Its purpose is to lubricate and moisturize joints. It acts as a conduit for the entry of nourishing elements into the cartilage [5]. A plasma dialysate called SF contains a significant amount of hyaluronic acid as well as other compounds produced by the joint tissues. Theoretically, this hyaluronic acid, also known as hyaluronan, is secreted by type-B synovial liner cells and forms a viscous lubricant inside the joint when it polymerizes with proteins in the SF [6]. A typical SF has high viscosity and is translucent and yellowish in color. Joint function deteriorates progressively as synovial fluid becomes less viscous and less efficient in lubricating joints. The source of joint discomfort and inflammation can be determined with the use of a synovial fluid investigation. The body's reaction to an injury or an illness is inflammation. The afflicted region may experience loss of function, discomfort, edema, and redness. Among the causes of joint issues are as follows:

- The most prevalent kind of arthritis is osteoarthritis. Joint cartilage degeneration is a long-lasting, degenerative condition. It may hurt and cause loss of movement and functionality.
- Gout is a kind of arthritis that often affects the big toe and produces inflammation in one or more joints.
- Rheumatoid arthritis is a sickness where the invulnerable system of the body assaults healthy joint cells.
- Joint effusion is a condition brought on by the accumulation of fluid around a joint. The knee is frequently affected. Knee effusion or fluid on the knee may be used to describe it when it affects the knee.

SF analysis has long been recommended as a routine test for an arthritis diagnosis. If signs of a joint problem are seen, this test is required. These include joint discomfort, swelling, redness of the joint, and a warm feeling to the touch.

13.1.2 Structure and Composition of Synovial Fluid

In the joint cavity, synovial fluid is secreted by the synovial membrane, which is a component of the inner synovial membrane. This SF serves as an ultrafilter for blood plasma, which has cells from joint tissues that create a protein derivative. The substance hyaluronic, which is abundant in synovial fluid, is generated by synovial membrane cells that resemble fibroblasts. Synovial fluid, which is filtered blood plasma, contains proteoglycan 4, an articular cartilage lubricant produced by chondrocytes. A thin 50-µm-thick layer of interstitial fluid is left on the cartilage's surface. In addition, it permeates the surface cracks and small fissures of the articular cartilage, filling each opening with fluid. Fluid from the articular cartilage is used as a synovial fluid reserve. During joint movement, the synovial fluid inside the cartilage is mechanically driven out in order to maintain the fluid layer on the cartilage surface. It is believed that synovial fluid has two main functions. By lubricating the articular surfaces, one contributes to the mechanical functionality of the joint, and the other serves to nourish the articular cartilage by serving as a transporter for nutrients like glucose [7]. Hyaluronate, a mucoprotein consisting of polymerized hyaluronan and proteins, is responsible for its high viscosity. The quantity of normal SF can be found up to 3.5 ml in large joints. Age-related decreases in hyaluronate content are between 3 and 4 g/l, glucose attentiveness is 0.55 mmol/l lesser than in plasma, and enzyme actions that move from the synovial lining cells, leukocytes, and surrounding tissue into SF from the plasma are lower in SF than in plasma [8]. Joint function progressively degrades as the synovial fluid loses viscosity and effectiveness as a lubricant. Pain management and maintaining joint mobility are the major objectives of treating OA. A variety of articular illnesses may have proteins associated with the disease that can be found in SF and utilized as biomarkers. The study of SF has attracted more attention in orthopedics and rheumatology in recent years in an effort to better understand the pathogenesis and improve the treatment of numerous joint diseases, including rheumatoid arthritis (RA), OA, osteochondrosis, articular infection, and periprosthetic joint infection [9].

13.2 Osteoarthritis and Its Treatments

"Arthritis" refers to about 100 distinct forms of inflammatory illnesses that affect the joints and connective tissues. Pain, edema, and stiffness in and around one or more joints are the main symptoms, which might appear gradually or all of a sudden. Osteoarthritis is the most prevalent type of arthritis. Degeneration of articular cartilage is a characteristic of OA, sometimes referred to as deteriorating joint and "wear-and-tear" sickness [10]. Joint pain and mobility loss are the frequent side effects of articular cartilage disintegration, both of which can result in long-term impairment. Articular cartilage and synovial fluid collaborate to lubricate, cushion, and form a protective obstruction between the ends of the bones [11]. Hence, articular cartilage degradation can cause the synovial fluid's rheological qualities to be inadequate. The performance of the joint will be impacted by changes in either one. Current OA treatments aim to manage pain and preserve joint function rather than change the disease course naturally [12]. The initial non-operative treatment for OA includes physical therapy, weight loss, physical therapy, patient education, assistive equipment, occupational therapy, and medication [13]. Although NSAIDs are frequently used to treat OA, they have serious gastrointestinal side effects and do not appear to change how the disease develops [14]. Glucosamine sulfate is a different OA therapy option. In healthy cartilage and synovial fluid, glucosamine is a component of glycosaminoglycans (GAGs) and proteoglycans. Additionally, it aids in the production of mucous secretions that lubricate synovial joints [15]. Recent revisions to the American College of Rheumatology's (ACR) treatment guidelines for OA of the hand, hip,

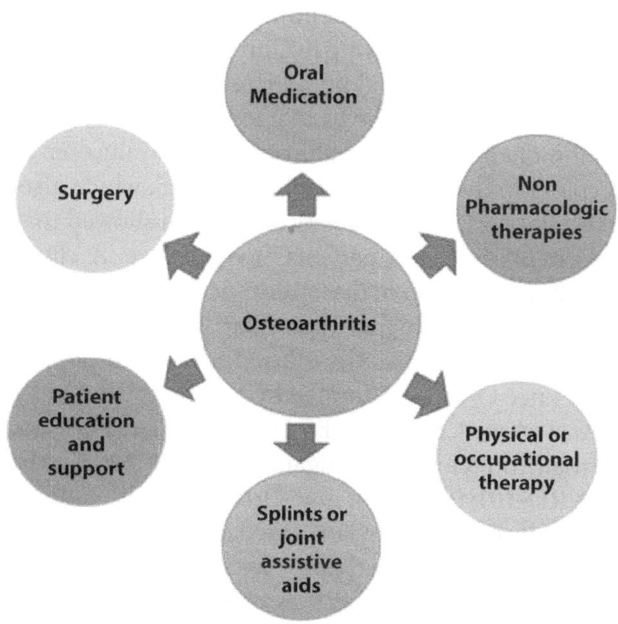

Figure 13.1 Six types of interventions considered for the treatment of OA.

and knee have been made. The ACR recommends the following for the treatment of arthritis [28]. The six different types of therapies that should be taken into account for the treatment of OA are depicted in Figure 13.1. Patients need to understand that the majority of these therapies only have a little to moderate effect on OA pain. It is advised to employ many modalities to treat OA because no single modality completely relieves symptoms.

13.3 Viscosupplements

Intra-articular hyaluronic acid injections, commonly known as viscosupplementation, are another non-surgical therapy for OA. Balazs *et al.* were the first to propose the notion of viscosupplementation for OA [16]. Viscosupplements are available in a variety of commercial forms. Viscosupplements can be obtained via streptococcal biological fermentation or from rooster combs [17]. The purpose of viscosupplementation is to restore the OA joint's physiological equilibrium. *In vitro* studies have shown that hyaluronic acid in viscosupplements increases hyaluronic acid synthesis [18]. The longest residence time for hyaluronic acid injections in the joint is between 5 and 30 days [19], with a half-life spanning from 24 h to 2 weeks. After the residence period, there is no longer a significant amount of artificial fluid in the joint. On the other hand, clinical studies have shown that the benefits of viscosupplements can extend for up to a year following the final injection [20]. At 3 months following hyaluronic acid injection, it is claimed that the content of HA in synovial fluid increased by 13%, and the complex shear modulus increased. This shows that viscosupplements might encourage the formation of endogenous hyaluronic acid. Hyaluronic acid interacts with pain receptors in the joint on a molecular level, which also contributes to the analgesic effect [21]. The majority of research showed that using hyaluronic acid, particularly in OA patients, was effective. A clinical benefit is often quantified as a reduction in discomfort or an improvement in function [22]. Hyaluronic acids with higher molecular weights are considered to be more beneficial in treating OA. The efficiency of viscosupplementation in the treatment of knee OA was examined by N. Bellamy *et al.* in 2006. Adant, Arthrum H, Artz (Artzal, Supartz), BioHy (Arthrease, Euflexxa, Nuflexxa), Durolane, Fermathron, Go-On, Hyalgan, hylan G-F 20 (Synvisc hylan G-F 20), Hyruan, NRD-101 (Suvenyl), and Orthovisc were among the brands that produced hyaluronan and hylan. The trials listed above show that viscosupplementation is an effective therapy for knee OA, with positive benefits on pain, function, and patient overall assessment at various post-injection intervals, but only at the 5- to 13-week post-injection durations. The clinical impact size for certain medications vs. placebo on specific variables at

Table 13.1 International knee OA guidelines and recommendations for intra-articular HA injections [24].

Scientific society		HA guidelines
EULAR (2003)	European League Against Rheumatism	Recognized effectiveness as a slow-acting therapy for OA symptoms
ACR (2014)	American College of Rheumatology	No general recommendation. Conditional recommendation: treatment resistance, elderly subject
AAOS (2013)	American Association of Orthopaedic Surgeons	Not recommended
NICE (2014)	National Institute for Health and Clinical Excellence	Not recommended
OARSI (2014)	Osteoarthritis Research Society International	Uncertain recommendation in isolated knee OA. Not appropriate in polyarthritis OA

various time periods ranges from modest to large [23]. Nevertheless, with modifications to their treatment guidelines for knee OA, major global scientific groups recently questioned the effectiveness of viscosupplementation. Table 13.1 shows the International knee OA guidelines and recommendations for intra-articular HA injections.

To treat knee osteoarthritis, a variety of hyaluronic acid derivatives, including sodium hyaluronate, hyaluronan, and cross-linked HA, such as hylan G-F 20, are now being used.

13.3.1 Hylan G-F 20

A cross-linked substance with a greater molecular weight made from rooster comb HA is called hylan G-F 20 (6,000 kDa). Under the trade names Synvisc and Synvisc-One, hylan G-F 20 is sold. Due to its higher molecular weight, Wobig *et al.* claim that hylan G-F 20 is more elastoviscous than other HA arrangements with lesser molecular weights [27]. According to Kearey *et al.*, a single injection of hylan G-F 20 significantly reduced pain and resulted in improvements; the outcomes were most noticeable at the 3- and 6-month points. Hylan G-F 20 is a successful interventional therapy for knee OA that may postpone complete knee replacement surgery.

13.3.2 Sodium Hyaluronate

Sodium hyaluronate is a glycosaminoglycan that is made up of lengthy-sequence disaccharide Na-glucuronate-N-acetyl glucosamine units and is a salt version of hyaluronic acid. Since it acts as a lubricant and shock absorber, it improves symptomatology when used as a viscosupplement in the treatment of knee OA. The arrangements include Euflexxa, Gelsyn, Durolane, and Hyalgan as supports. Intra-articular injections of sodium hyaluronate seem to be a secure and successful method for treating OA symptoms. These medications provide patients who have tried various non-operative treatments without success an affordable alternative therapy that has been shown to be safe and effective in otherwise healthy individuals.

13.3.3 Hyaluronan

Hyaluronan (HA) is a high molecular weight extracellular matrix (ECM) polysaccharide and hyaluronic acid derivative that is commonly found in soft connective tissues. Custom has traditionally attributed hyaluronan's distinctive ability to act as a viscosupplementation agent primarily due to its viscoelastic properties linked to a higher molecular weight and cream preparation. Hyaluronan was initially used as a viscosupplement due to its alleged viscoelastic properties. For intra-articular injection, a variety of hyaluronan brand name formulations are used to treat knee osteoarthritis. Orthovisc, Monovisc, and Hymovis are a few of them. There are several variations among the formulas. Although there is hope that HA may also offer therapeutic benefits to people of working age, most research examining hyaluronan has been carried out on older patients.

13.4 Synovial Mimic Fluid/PVP

The greatest resemblance to synovial fluid was seen in polyvinylpyrrolidones of medium molecular weight (PVP) and PVP hyaluronate [25]. Bovine serum aqueous solution and deionized water were used to dissolve pharmaceutical grade PVP K30 [29] with the molecular mass of PVP K30 being 40,000 g/mol [24]. PVP is an excellent lubricant with high biocompatibility. As the PVP concentration reached the highest level (i.e., 15% PVP deionized water solution and 10% PVP bovine serum solution), the friction coefficient was seen to substantially reduce but then taper off [24]. PVP appears to be a viable lubricating component for artificial joints in the future based on its superior lubricating qualities and biocompatibility in artificial joint pairs. PVP 40-50 G and PVP K30 were provided by Sigma-Aldrich, a provider of polymers for pharmaceutical uses, with a usual molecular weight of 40,000 g/mol (PVP K30 and PVP 40-50 G). Samples of 0%, 5%, 20% - PVP 40-50 G, and 10%, 20% - PVP K30 were prepared in sterile water. The samples were created by dissolving PVP K30 and PVP 40-50 G calculated volume fractions in sterile water. The samples were prepared at room temperature, and a stirrer set at 300 rpm was used to homogenize them for 15 min [26].

13.5 Conclusion

This chapter discussed the usefulness of PVP solution as a replacement SF and underlines the need of assessing the rheological characteristics of SF in a realistic setting. Additionally, different viscosupplements used in the treatment of osteoarthritis attempt to restore synovial fluid to its natural HA property features; therefore, utilizing viscosupplements that mirror healthy synovial fluid is a sensible strategy. It is observed that polyvinylpyrrolidones of medium molecular weight (PVP) and PVP hyaluronate showed the greatest resemblance to synovial fluid.

References

1. Zhang, Y. and Jordan, J.M., Epidemiology of osteoarthritis. *Clin. Geriatr. Med.*, 26, 3, 355–369, Aug. 2010.
2. More, S., Kotiya, A., Kotia, A., Ghosh, S.K., Spyrou, L.A., Sarris, I.E., Rheological properties of synovial fluid due to viscosupplements: A review for osteoarthritis remedy. *Comput. Methods Programs Biomed.*, 196, 105644, Dec. 2020.
3. Gale, L.R., *Biotribological Assessment for Artificial Synovial Joints: the Role of Boundary Lubrication*, PhD dissertation, Queensland University of Technology, Brisbane, Australia, 2007.
4. Anadere, I., Chmiel, H., Laschner, W., Viscoelasticity of 'normal' and pathological synovial fluid. *Biorheology*, 16, 3, 179–184, 1979.

5. Iwanaga, T., Shikichi, M., Kitamura, H., Yanase, H., Nozawa-Inoue, K., Morphology and functional roles of synoviocytes in the joint. *Arch. Histol. Cytol.*, 63, 1, 17–31, Feb. 2000.
6. Terčič, D. and Božič, B., The basis of the synovial fluid analysis, in: *13th Mediterranean Conference on Medical and Biological Engineering and Computing 2013*, pp. 461–464, Seville, Spain, Sep. 2013.
7. Eisenberg, J.M., Schumacher, H.R., Davidson, P.K., Kaufmann, L., Usefulness of synovial fluid analysis in the evaluation of joint effusions: Use of threshold analysis and likelihood ratios to assess a diagnostic test. *Arch. Intern. Med.*, 144, 4, 715–719, Apr. 1984.
8. Schumacher, H.R., Synovial fluid analysis and synovial biopsy, in: *Textbook of Rheumatology*, 7th ed, W.N. Kelley, E.D. Harris, S. Ruddy, C.B. Sledge (Eds.), pp. 612–619, WB Saunders, Philadelphia, PA, 1997.
9. Vicenti, G., Bizzoca, D., Carrozzo, M., Solarino, G., Moretti, B., Multi-omics analysis of synovial fluid: A promising approach in the study of osteoarthritis. *J. Biol. Regul. Homeost. Agents*, 32, 1, 9–13, Jan-Feb. 2019.
10. Brandt, K., *Diagnosis and Nonsurgical Management of Osteoarthritis*, Professional Communications Inc., Caddo, OK, 2000.
11. Wright, V. and Dowson, D., Lubrication and cartilage. *J. Anat.*, 121, 1, 107, 1976.
12. Dieppe, P., Osteoarthritis. *Acta Orthop. Scand.*, 69, 281, 2–5, 1998.
13. Hochberg, M.C. *et al.*, Guidelines for the medical management of osteoarthritis. Part II. Osteoarthritis of the knee. *Arthritis Rheumatol.*, 38, 11, 1541–1546, 1995.
14. Simon, L.S., The pharmacologic treatment of osteoarthritis, in: *Osteoarthritis: Diagnosis and Medical/Surgical Management*, pp. 267–286, 2007.
15. Hamerman, D., The biology of osteoarthritis. *N. Engl. J. Med.*, 320, 20, 1322–1330, 1989.
16. Balazs, E.A. *et al.*, Hyaluronic acid in synovial fluid. I. Molecular parameters of hyaluronic acid in normal and arthritic human fluids. *Arthritis Rheumatol.*, 10, 4, 357–376, 1967.
17. Moskowitz, R.W. (Ed.), *Osteoarthritis: Diagnosis and Medical/Surgical Management*, Lippincott Williams & Wilkins, USA, 2007.
18. Balazs, E.A. and Denlinger, J.L., Clinical uses of hyaluronan, in: *Ciba Foundation Symposium 143-The Biology of Hyaluronan: The Biology of Hyaluronan: Ciba Foundation Symposium 143*, pp. 265–285, John Wiley & Sons, Ltd, Chichester, UK, 2007.
19. Wobig, M. *et al.*, Viscosupplementation with hylan GF 20: A 26-week controlled trial of efficacy and safety in the osteoarthritic knee. *Clin. Ther.*, 20, 3, 410–423, 1998.
20. Gomis, A. *et al.*, Effects of different molecular weight elastoviscous hyaluronan solutions on articular nociceptive afferents. *Arthritis Rheumatol.*, 50, 1, 314–326, 2004.

21. Peyron, J.G., Intraarticular hyaluronan injections in the treatment of osteoarthritis: A state-of-the-art review. *J. Rheumatol. Suppl.*, 39, 10–15, 1993.
22. Bellamy, N. *et al.*, Viscosupplementation for the treatment of osteoarthritis of the knee. *Cochrane Database Syst. Rev.*, 2, 2006. https://doi.org/10.1002/14651858.CD005321.pub2Copy
23. Vadilenkaitis, V.V. and Matulis, A.A., Artificial synovial fluid for the intraarticular treatment of rheumatoid arthritis and osteoarthritis (chemical synthesis and clinico-experimental and biomechanical data). *Terapevticheskii Arkhiv*, 56, 11, 73–77, 1984.
24. Legre-Boyer, V., Viscosupplementation: Techniques, indications, results. *Orthop. Traumatol. Surg. Res.*, 101, 1, S101–S108, 2015.
25. Haaf, F., Sanner, A., Straub, F., Polymers of N-vinylpyrrolidone: Synthesis, characterization and uses. *Polym. J.*, 17, 1, 143–152, 1985.
26. Kotia, A., More, S., Yadav, A., Mohan, T.V.S.Y., Naidu, A.H., Rajesh, G., Sarris, I.E., Rheological properties and its effect on the lubrication mechanism of PVP K30 and PVP 40-50 G as artificial synovial fluids. *Inventions*, 6, 4, 61, 2021.
27. Wobig, M., Bach, G., Beks, P. *et al.*, The role of elastoviscosity in the efficacy of viscosupplementation for osteoarthritis of the knee: A comparison of hylan G-F 20 and a lowermolecular-weight hyaluronan. *Clin. Ther.*, 21, 9, 1549–1562, 1999.
28. Ruane, J.J., Barrett, J.R., Tennant, F., Treatment of osteoarthritis practical review of treatment options for spine, hand, hip, knee, and ankle osteoarthritis. *Pract. Pain. Manag.*, 15, 6, 2015.
29. Guo, Y., Hao, Z., Wan, C., Tribological characteristics of polyvinylpyrrolidone (PVP) as a lubrication additive for artificial knee joint. *Tribol. Int.*, 93, 214–219, 2016.

14

Artificial Human Heart Biofluid Simulation as a Boon to Humankind: A Review Study

Md Akhtar Khan

GITAM School of Technology, Hyderabad, India

Abstract

This chapter presents a review study of an artificial heart replicate as a boon for humankind, and it pumps blood through blood vessels to different organs. It is one of the most important organs in the human body. For a very long time, doctors and scientists have been working to develop something that is either similar or comparable to the heart. The current work exemplifies how the finite element method and computational fluid dynamics have been used to evaluate prosthetic heart valve fluid dynamics. The Reynolds number is also significant in developing an artificial heart for survivability. These interim implants can keep the patient alive until a donor heart becomes available. One of the key aspects of studying the development of an artificial heart is to ensure that the fluid dynamics inside the device is similar to that of a natural heart. This paper discusses the importance of biofluid simulation in the development of artificial human hearts and its potential to be a boon to humankind.

Keywords: Heart, artificial heart, evolution, finite element method, Reynolds number, computational fluid dynamics, Ansys, heart valve

14.1 Introduction

Artificial human heart biofluid simulation can significantly benefit humankind by improving our understanding of cardiovascular physiology and pathophysiology and facilitating the development of new diagnostic and

Email: khan.akhtar24@gmail.com

Jay Prakash Srivastava, Dražan Kozak, Vinayak Ranjan, Pankaj Kumar, Ranjan Kumar and Shubham Tayal (eds.) Mechanical Engineering in Biomedical Applications: Bio-3D Printing, Biofluid Mechanics, Implant Design, Biomaterials, Computational Biomechanics, Tissue Mechanics, (355–366) © 2024 Scrivener Publishing LLC

therapeutic tools for heart disease. The holy grail of open-heart surgery has been a completely artificial heart replacement for many years.

Increasing advancement of mechanical heart replacement prompted doctors to depart from the moral guidelines put forth by others. With over 30 million sufferers worldwide, heart failure accounts for a sizable portion of the illness burden [1]. The number of Americans diagnosed with heart disease in 2016 was expected to be around 28.1 million or 11.5% of the adult population [1]. Approximately one in every four fatalities in the United States is due to heart disease, a significant cause of death and morbidity in many developed and developing nations [2, 3].

The two types of artificial heart replacement are mechanical and bioprosthetic (made of tissue). Artificial replacement heart is continually being designed and improved via ongoing research. In some cases, mechanical artificial hearts are less biocompatible and more aggressively rejected by the body than bioprosthetic heart valves, which have a shorter life span [4].

As a result of being needed to control this continuous blood flow throughout the human body, the circulatory system plays an important role even-though it is a complex, crucial and intricate system. Transporting oxygen, nutrients, hormones, and other necessary materials to cells all over the body, as well as eliminating waste materials like carbon dioxide and metabolic wastes, is critical for maintaining homeostasis. The heart, blood vessels, and blood itself make up the circulatory system, which functions as a unit to guarantee healthy blood circulation [5]. Cardiovascular illness including heart failure is prominent cause of death worldwide. As a result, the number of donor hearts available for transplant is stagnant while the need for replacement heart is rising.

According to recent data, there were around 98,000 valve replacements in the USA and 195,300 overseas in 2008 [6]. Aortic valve replacement for aortic stenosis or aortic insufficiency is the most frequent valve procedure. Aortic insufficiency causes the aortic valve to lose its ability to hold blood as it should, enabling blood to spit back into the heart and potentially causing heart failure.

The future of the entire artificial heart seems a lot brighter than it did 10 years ago. Back then, no animal with an artificial heart implant had ever lasted more than 3 days. The Jarvik-7 heart (refer to Figure 14.1), one of the several artificial hearts in development, now holds the record, although neither it nor any of the other artificial hearts can even be used temporarily to replace a human heart due to the technology's rapid advancement [7].

Heart disease is predicted to claim the lives of about 17 million individuals per year. A heart transplant is one of the most effective therapies for serious heart disease. Unfortunately, the demand for donor hearts significantly outnumbers the supply, necessitating the development of an alternate option. The development of an artificial human heart is a potential solution

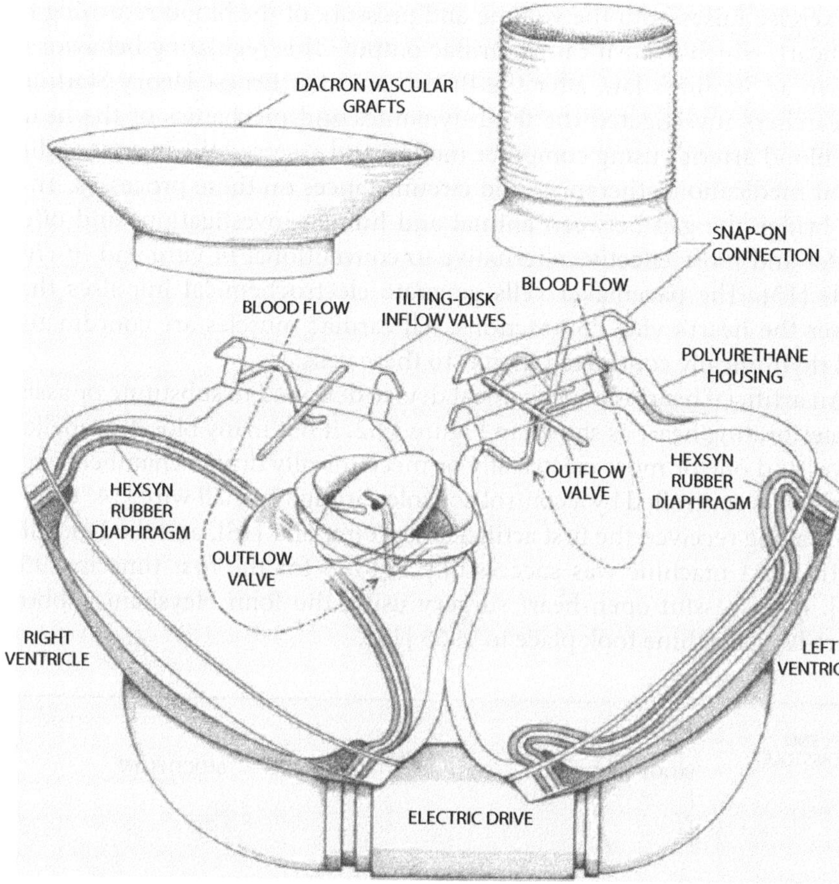

Figure 14.1 Artificial heart [7].

to this problem. However, creating an artificial heart that can replicate the functions of a natural heart is a challenging task. The development of such devices may greatly benefit from the use of biofluid simulation.

The two primary challenges are that very few donors are available for heart transplants and overcoming the immune system's propensity to reject a transplant in recipients is difficult. Around 60 million Americans today have some cardiovascular illness, according to the American Heart Association (AHA) [8, 9], and over one million people died as a result of cardiovascular disease each year [10, 11]. Congenital cardiac disease, coronary heart disease, cardiogenic shock, and infection are a few of them. According to the US National Heart, Lung, and Blood Institute, over 4.8 million Americans also suffer from congestive heart failure (CHF), a condition in which the heart is unable to pump enough blood [12].

Exercise raises both the volume and pressure of the blood returning to the heart, which in turn raises cardiac output. This regulating behavior is known as Starling's law, after the British scientist Ernest Henry Starling. Researchers investigated the fluid dynamics and mechanics of the heart and blood arteries using computer models and assessed the impact of different medications, therapies, and circumstances on these processes. This can bridge the gap between animal and human investigations and offer a safer and more effective alternative to conventional *in vitro* and *in vivo* trials [13]. The pacemaker cells generate electrochemical impulses that power the heart's vital contractions. All cardiac muscles are coordinated and rhythmically contracted thanks to these cells.

An artificial heart is a mechanical device designed to substitute or assist a deteriorating heart as shown in Figure 14.2. It has many blood-propelled valves and one or more electrically or mechanically driven chambers each of which is controlled by a control console through a small wire [14, 15]. In 1937, a dog received the first artificial heart implant [16], and the Dodrill–GMR heart machine was successfully utilized for the first time in 1952 [17]. A successful open-heart surgery using the John Heysham Gibbon heart-lung machine took place in 1953 [18].

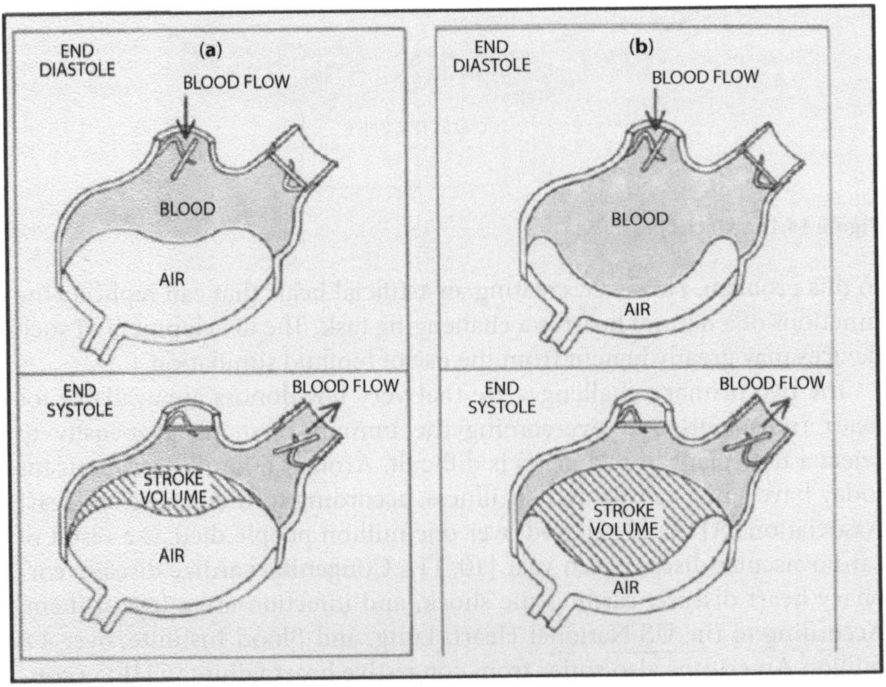

Figure 14.2 The regulatory mechanism of an artificial heart [7].

The output of an artificial heart increases when the venous return increases due to its regulatory mechanism. Here, two states of diastole (the filling stage) and systole (the pumping stage) of an air-driven mechanical heart are depicted. Figure 14.2 (a) shows the heart's output is around six litres per minute with venous blood flowing normally; however, as the animal in which the transplanted heart is located becomes more active. Figure 14.2 (b) represents the venous pressure rises and the output increases to approximately 11 litres per minute. The amount of venous return to the heart is mostly determined by the animal's vascular system, which is controlled by neurological and hormonal factors, so the body regulates cardiac output; the device reacts passively, without altering heart rate, by modulating the volume of the stroke. To meet the rising demand for heart replacements, the primary goal of the present review study is to use biofluid simulation to understand the fluid flow inside the artificial human hearts and its potential advantages to humankind. It is simple to use, full-size, and energy-efficient. Finite element modeling and computational fluid dynamics (CFD) simulation were used for the analysis of the flexible implantable cardioverter that was created and built.

14.2 Biofluid Simulation

Biofluid simulation is a technique used to study the flow of fluids inside biological systems. It involves using computational models to simulate the flow of fluids inside a system. Biofluid simulation has been used extensively in the field of cardiovascular research to study the flow of blood inside the human heart. This technique can also be used to study the fluid dynamics inside an artificial human heart. By simulating the flow of fluids inside an artificial heart, researchers can optimize the design of the device to ensure that it can replicate the functions of a natural heart.

One advantage of biofluid simulation is improved comprehension of blood flow patterns and wall shear stress in the cardiovascular system. This knowledge may be utilized to create novel diagnostic tools, such as models that forecast the development of atherosclerosis and aneurysms, as well as to enhance existing therapies, including optimizing stent placement and medication delivery.

Before being implanted in humans, novel cardiovascular devices like heart valves and ventricular assist devices can also be tested for effectiveness and safety using biofluid simulation. This can enhance patient outcomes and lower the risk of device-related problems. The model of the soft artificial heart geometry and the inlet and output valve using 3D SolidWorks was described by Saad Mahmood Ali *et al.*, as shown in Figure 14.3(a, b) [14]. Also, the authors observed that modeling and analysis of a new-generation

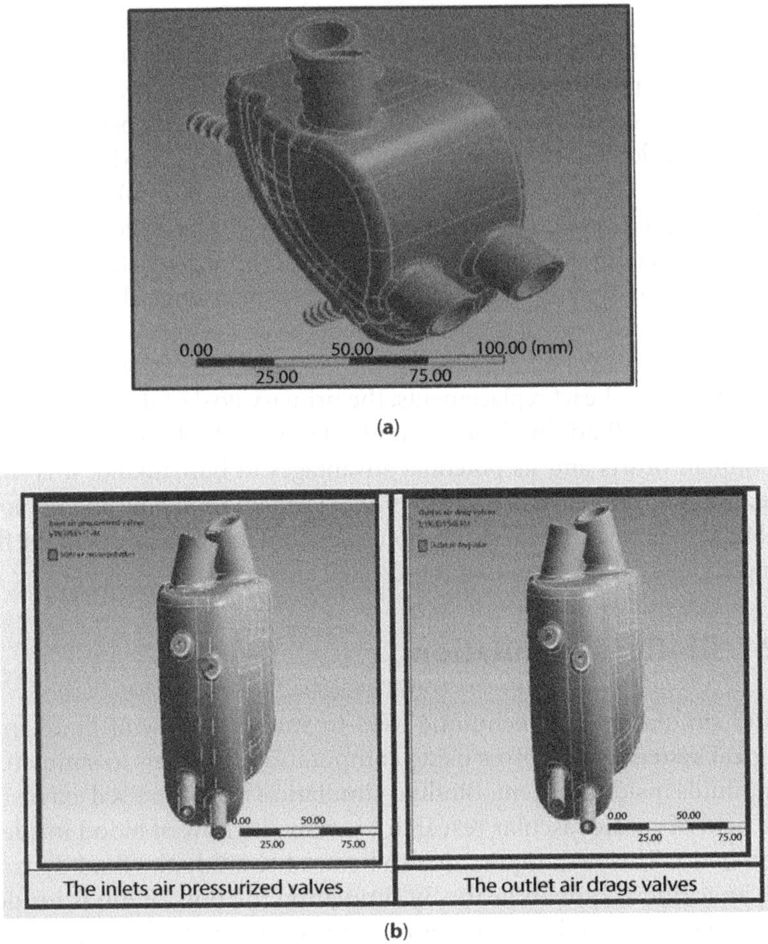

Figure 14.3 (a) The 3D model of artificial hear. (b) The inlet air-pressurized and the outlet air drag valves [14].

soft artificial heart that would completely replace the human heart may be supported by constructing 3D projections for the inner diaphragms, the right and left ventricles, and the drag and air pressure zone sections necessary for the soft artificial heart mechanics [14]. By using computer simulations, researchers can study the fluid dynamics and mechanics of the heart and blood vessels and test the effects of various drugs, interventions, and conditions on these processes. This can provide a safer and more efficient alternative to traditional *in vitro* and *in vivo* experiments and can also help to fill the gap between animal and human studies. To duplicate the modeled soft artificial heart, a number of numerical (Ansys) analytical systems were

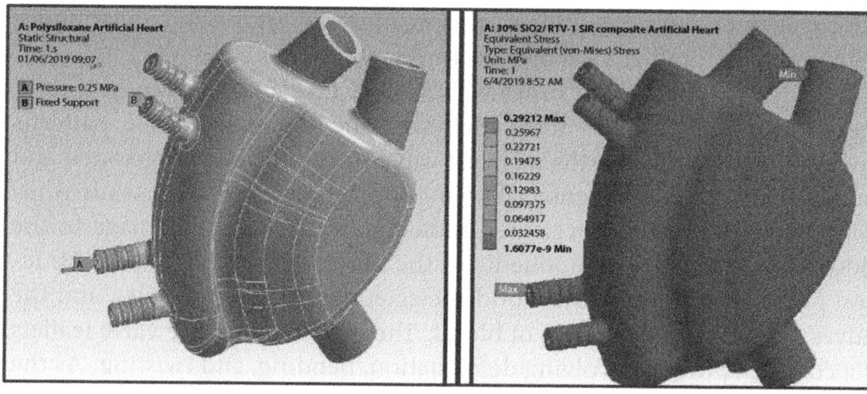

(a) (b)

Figure 14.4 (a) Boundary conditions and (b) equivalent (von Mises) stress of the designed soft artificial heart [7].

used by the researcher to determine the dynamic response throughout the activity of the blood flow cycles (refer to Figure 14.4(a, b)) [14].

14.3 Heart Valve Fluid Flow

The fluid mechanics of heart valves is important because abnormal flow patterns or forces can lead to valve dysfunction and other heart problems. For example, valve stenosis (narrowing) can occur when a valve becomes stiff and does not open properly, while valve regurgitation (leakage) can occur when a valve does not close properly, allowing blood to flow backward.

Researchers and clinicians use a variety of tools and techniques to study heart valve fluid mechanics, including CFD, magnetic resonance imaging (MRI), and echocardiography. These methods can provide detailed information about blood flow patterns, pressure gradients, and other factors that are important in understanding valve function.

Conversely, hemolysis, platelet activation, and thromboembolic events brought on by clot formation and subsequent detachment are associated with mechanical heart valves. These problems are believed to be caused by abnormal blood flow patterns near heart valves [15, 16]. The circumferential jet that breaks from the ball, hits the flow chamber wall, and travels along the wall is produced by the flow that exits the valve during the forward flow phase. A maximum forward flow velocity of 2.20 m/s was measured at the annulus in this forward flow jet under aortic conditions. Its velocity decreases to 1.80 m/s at 30 mm downstream of the valve. A region of low-velocity recirculating flow and a wake emerge just downstream of the cage's apex during the forward flow phase. A significant shear and a strong velocity gradient are seen at the forward flow's edge [16].

The design of artificial heart valves must take into account the fluid mechanics of the heart valve, including the forces and flow patterns involved. For example, artificial heart valves must be designed to open and close smoothly and efficiently, allowing for proper blood flow while minimizing turbulence and pressure gradients that can cause damage to the blood cells and other components of the blood. During systole (contraction phase of the heart), the blood is ejected out of the ventricles, and the valves open to allow the flow of blood. The movement of the valve leaflets is a complex process involving deformation, bending, and twisting. As the blood flows through the valve, it encounters various forces, including pressure, shear stress, and turbulence. These forces can influence the performance of the valve leaflets and the flow patterns of the blood.

The flow physics and movement in the heart valve are complex and interdependent, and small changes in either can have significant impacts on valve function. Researchers use CFD and other simulation tools to model and study the fluid dynamics of the heart valve and the movement of the valve leaflets. This knowledge can help to identify potential issues with valve function and guide the development of new treatments for heart valve disease.

In summary, the flow physics and movement in the heart valve are critical components of heart valve function. Understanding these processes is essential for identifying and treating valve disease and for developing new treatments and devices to improve outcomes for patients with heart valve problems.

14.4 Artificial Heart as a Boon to Humankind

The development of an artificial human heart has the potential to be a boon to humankind. An artificial heart can provide a viable alternative to heart transplants, which are limited by the availability of donor hearts. An artificial heart can also help save the lives of people who are not eligible for heart transplants due to various reasons, such as age or medical condition. Biofluid simulation can help researchers create an artificial heart that can replicate the functions of a natural heart, ensuring safety and reliability.

Devices that support the left ventricle's normal function and help the natural heart have been the subject of much research. More than 40 of these devices have been temporarily implanted in people with advanced cardiac disease as reported by Robert K. Jarvik [7]. A future configuration of an electrically powered artificial heart system is shown in Figure 14.5. A lead carrying a tiny cable through the skin would convey the electricity,

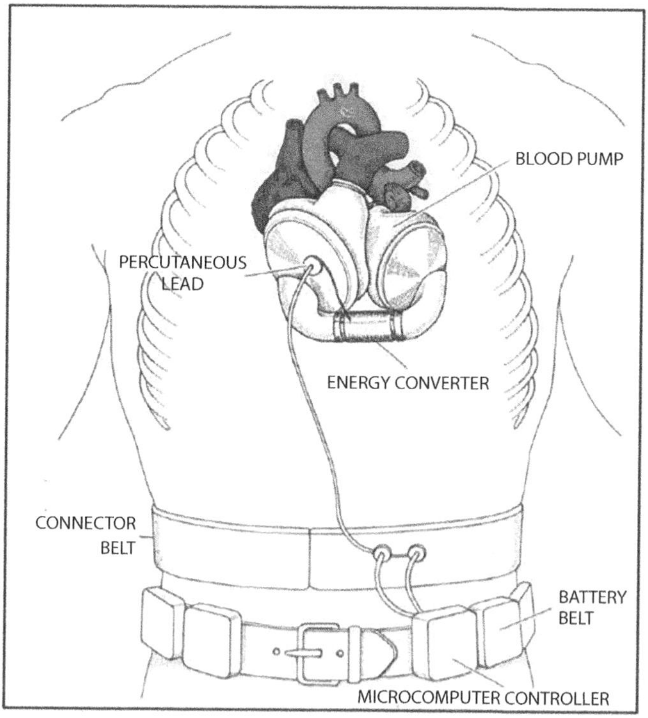

Figure 14.5 An electrically powered prosthetic heart system [7].

which would originate from batteries worn on a belt. Moreover, the control signals of a microcomputer controller would be part of the wire. The patient with an artificial heart would be totally mobile.

Left ventricular assist devices (LVADs) and complete artificial hearts are the two basic varieties of total artificial hearts (TAHs). The left ventricle, which is in charge of delivering oxygen-rich blood to the other parts of the body, is supported by LVADs. Although providing both left and right ventricular support, TAHs, on the other hand, are intended to replace the complete heart.

The current state of the art in artificial heart technology includes a range of devices that can provide temporary or long-term support to patients with heart failure. LVADs are the most commonly used devices, and several models are available, including the HeartMate II and the HeartWare HVAD. TAHs are less commonly used but have been successful in providing long-term support to patients who are not candidates for heart transplantation. Heart transplantation is a complex procedure, and patients who receive a total artificial heart (TAH) as a bridge to transplantation may face unique challenges during and after the surgery. Successful surgical intervention and prompt management of any postoperative complications are critical factors in improving the outcome of heart transplantation for these patients.

The TAH takes over the function of the failing heart, allowing the patient to stabilize and improve while waiting for a donor heart to become available. However, the TAH itself can cause complications, such as bleeding, infection, and blood clots. To ensure the best possible outcome for these patients, it is essential to have an experienced surgical team that can perform the heart transplantation and manage any postoperative complications promptly. In addition, close monitoring and preventive measures, such as prophylactic antibiotics and anticoagulation therapy, can help prevent complications from occurring in the first place. Overall, effective surgical intervention and rapid care and avoidance of postoperative complications are significant variables in enhancing the prognosis of heart transplantation in patients in the long term supported by a TAH.

Percutaneous procedures for the removal of diseased or calcified heart valves and their replacement with prosthetic valves have been developed and are known as transcatheter aortic valve replacement (TAVR) or transcatheter aortic valve implantation (TAVI). These procedures are less invasive than traditional open-heart surgery, and they involve placing a new valve within the diseased valve using a catheter inserted through the femoral artery or another access point. These techniques cannot be used to replace every valve though. Only aortic valve stenosis, a narrowing of the aortic valve that can lead to substantial symptoms and problems, is presently authorized for treatment with TAVR/TAVI. Some heart valve disorders, such as tricuspid valve disease or mitral valve regurgitation, still need open-heart surgery for valve replacement. It is also worth noting that while TAVR/TAVI is less invasive than open-heart surgery, it still carries risks and is not suitable for all patients. The decision to use TAVR/TAVI versus open-heart surgery depends on the individual patient's medical history, overall health, and the specific valve condition being treated.

Heart replacement, including heart transplantation and the use of artificial hearts, has been a life-saving procedure for many people with severe coronary failure. Heart failure is a gradual illness in which the heart weakens and becomes less effective in circulating blood across the body. It can cause tiredness, shortness of breath, and swelling in the legs and ankles, among other symptoms. In severe cases, heart failure can be life-threatening. Heart replacement procedures can provide a new, functioning heart or heart-like device that takes over the pumping function of the failing heart. This can significantly improve a patient's quality of life and extend their life span. However, heart replacement procedures are not without risks, and they require careful monitoring and management to ensure that the patient's body does not reject the new heart or device and to manage any postoperative complications.

14.5 Conclusion

Artificial human heart biofluid simulation is a rapidly developing field with significant potential for improving the diagnosis and treatment of heart disease. This chapter has highlighted the importance of understanding the fluid mechanics of the heart and its valves in developing new therapies for heart disease. Furthermore, the use of biofluid simulations has the potential to reduce the need for animal testing and clinical trials, making the development of new therapies faster and more cost-effective. By using simulations to model the behavior of artificial heart devices, researchers can test various design modifications and optimize device performance before conducting human trials. While there are still challenges to overcome in the development and use of artificial hearts, such as managing the risk of infection, improving durability, and optimizing the fluid mechanics of these devices, ongoing research and development hold great promise for the future. In conclusion, the development of artificial heart devices has been a significant boon to humankind. These devices have helped to improve the quality of life and extend the life span of many patients with severe heart disease. The development of artificial heart devices is a remarkable achievement in medical science and a significant boon to humankind. It has the potential to transform the lives of patients with heart disease and improve the outcomes for millions of people around the world.

References

1. Ziaeian, B. and Fonarow, G.C., Epidemiology and aetiology of heart failure. *Nat. Rev. Cardiol.*, 13, 6, 368–378.
2. Copeland, J.G., Smith, R.G., Arabia, F.A. *et al.*, Cardiac replacement with a total artificial heart as a bridge to transplantation. *N. Engl. J. Med.*, 351, 859–867, 2004.
3. Murphy, S.L., Xu, J., Kochanek, K.D., Curtin, S.C., Arias, E., Deaths: Final data for 2015. *Natl. Vital Stat. Rep.*, 66, 1–75, 2017.
4. Collin, S., *Preoperative planning and simulation for artificial heart implantation surgery signal and image processing*, Ph.D. Thesis, University of Rennes, 2018.
5. Robert, E. and Michler, M.D., Stem cell therapy for heart failure. *Methodist Debakey Cardiovasc. J.*, 9, 4, 1–8, 2013.

6. Friedewald, V.E., Bonow, R.O., Borer, J.S., Carabello, B.A., Kleine, P.P., Akins, C.W., Roberts, W.C., The editor's roundtable: Cardiac valve surgery. *Am. J. Cardiol.*, 99, 1269–1278, 2007.
7. Jarvik, R.K., The total artificial heart. *Sci. Am.*, 244, 74–81, 1981.
8. Yacoub, N. and Takkenberg, J., Will heart valve tissue engineering change the world? *Nat. Clin. Pract. Cardiovasc. Med.*, 2, 60–61, 2005.
9. American Heart Association, *The mechanical heart celebrates 50 lifesaving years*, New Research on Mechanical Pump, Heart Failure: AHA, America, 2002.
10. Lundberg, M.S., Johnson, T.B., Burkhart, D.B., Building a bioartificial heart, obstacles and opportunities. *J. Thorac. Cardiovasc. Surg.*, 748–750, 2017.
11. Akshay, P., Lianjun, W., Yonas, T., Artificial heart for a humanoid robot. *Electroactive Polymer Actuators And Devices*, vol. 1, pp. 1–16, 2014.
12. Mohammed E.S. and Ali, M.A., *Study of an Artificial Heart: Design and Simulation*, pp. Research Gate Publication, No. 297013927, pp. 1–9, 2016.
13. Stenberg, M., *Concept design and in vitro evaluation of a novel dynamic displacement ventricular assist device*, A thesis in Technology and Health, University of Florida, USA, 2006.
14. Ali, S.M., Ali, Z.J., Abd, M.M., Design and modelling of a soft artificial heart by using the solid works and ANSYS. *IOP Conf. Ser. Mater. Sci. Eng.*, 671, 012062, 2020.
15. Chandran, K.B., Khalighi, B., Chen, C.J., Experimental-study of physiological pulsatile flow past valve prostheses in a model of human aorta. *1. Caged ball valves. J. Biomech.*, 18, 763–765, 1985.
16. Figliola, R.S. and Mueller, T.J., Fluid stresses in vicinity of disk, ball, and tilting disk prosthetic heart-valves from *in vitro* measurements. *J. Biomech. Eng.*, 99, 173–177, 1977.
17. Gerosa, G., Gallo, M., Bottio, T., Tarzia, V., Successful heart transplant after 1374 days living with a total artificial heart. *Eur. J. Cardiothorac. Surg.*, 49, 4, e88–e89, 2016.
18. Dasi, L.P., Simon, H.A., Sucosky, P., Yoganathan, A.P., Fluid mechanics of artificial heart valves. *Clin. Exp. Pharmacol. Physiol.*, 36, 225–237, 2009.

Part IV
ROBOTICS

15

Robotics in Medical Science

Sourav Karmakar[1]*, Akanksha Mishra[2,3], Anand Kumar Mishra[4] and Jay Prakash Srivastava[5]

[1]International Institute of Information Technology, Hyderabad, India
[2]Department of Pediatrics, Albert Einstein College of Medicine, Bronx, NY, USA
[3]Dominick P. Purpura Department of Neuroscience, Albert Einstein College of Medicine, Bronx, NY, USA
[4]Department of Mechanical and Aerospace Engineering, Cornell University, Ithaca, USA
[5]Department of Mechanical Engineering, Gandhi Institute of Technology and Management, Hyderabad, Telangana, India

Abstract

This chapter presents the evolution, developments, and applications of biomedical robotics. Surgical robotics, soft robotics, sensor-based medical robotics applications, etc. are briefly discussed. The development of several new techniques due to robot-assisted minimally invasive surgery (RMIS) has decreased patient trauma, reduced hospital stays, and improved diagnostic and therapeutic outcomes. Through improvements in mechanical perspectives to sensing and intelligent robotic approaches, innovative assistive automated platforms are intended to enhance the physical capabilities of both patients and clinicians. This review summarizes the works of early investigations and recent advances in the development of various surgical robotic systems focusing on major scientific research in the age of artificial intelligence and machine learning. Further developments in manufacturing and materials, confirming the future potential for patient-specific soft robotics, are reviewed. This chapter clarifies the underlying concepts of minimally invasive surgery, human–robot interaction, robotic manipulation, human–computer interaction, soft robotics, biosensors, haptics, automation and control, robotics in dental application, and the role of CAD/CAM and modern manufacturing techniques in biomedical applications. This technical documentation envisions attracting more research in the area to improve the healthcare system by

**Corresponding author*: souravresearch16@gmail.com

Jay Prakash Srivastava, Dražan Kozak, Vinayak Ranjan, Pankaj Kumar, Ranjan Kumar and Shubham Tayal (eds.) Mechanical Engineering in Biomedical Applications: Bio-3D Printing, Biofluid Mechanics, Implant Design, Biomaterials, Computational Biomechanics, Tissue Mechanics, (369–396) © 2024 Scrivener Publishing LLC

adopting new technology and a supportive environment allowing a human–robot interaction to a new level.

Keywords: Robotics, surgery, artificial intelligence, sensors, machine learning

15.1 Introduction

Traditional medicine [1] comprises the knowledge, skills, and practices founded on local beliefs and experiences and used to promote well-being and treat sickness via prevention, detection, refining, and therapy. Some traditional medical techniques have been recorded in the literature, while others have been transmitted orally from one generation to the next. In many regions of the globe, traditional medicine continues to serve as the primary healthcare source for most people.

After the development of modern medicine and medications such as vaccinations and aspirin, it was discovered that traditional medicine was ineffective in treating more complicated diseases such as tuberculosis (TB) and cancer. As the global population expands, robotically assisted medical equipment has become essential. For instance, the Xenex Germ-Zapping Robot [2], created by the American corporation Xenex, employs UV radiation to disinfect hospitals and surgical settings, minimizing the risk of illness caused by microorganisms. Similarly, robotically assisted surgery has emerged as a viable alternative to conventional open surgery, such as laparoscopy, also known as minimal access surgery (MAS) [3]. Minimally invasive surgery (MIS) [4], which utilizes small incisions and specialized equipment, decreases tissue damage and blood loss during surgery, resulting in shorter recovery periods and reducing postoperative discomfort. The accuracy of total knee arthroplasty surgery is greatly enhanced by using robots. The advantages of robotic surgery include greater precision and accuracy.

The clinician's desire to "see the interior biological structure" has persisted throughout history. Hippocrates was the first to perform an illustrated endoscopy. He gave birth with a rectal speculum. The compliment for modern laparoscopy is also applicable to Bozzini. He created the "Lichleiter" illumination conductor [5, 6] to avoid the issue of inadequate lighting. Desormeaux introduced the "Lichtleiter" of Bozzini to a forbearing for the first time in 1853. Endoscopy and laparoscopy breakthroughs that served as the foundation for modern technologies resulted from several evolutions that occurred independently but almost concurrently. In 1901, Kelling used the term "coelioscopy" [7] to describe the procedure of inspecting the abdomen cavities of dogs using a cystoscope. In 1938, Veress developed spring-filled syringes to deflate the upper body and extract fluid and air. Its latest modification yields the "Veress" syringe, an appropriate tool for achieving pneumoperitoneum during laparoscopic surgery. Hasson developed a technique for doing laparoscopy using a small laparotomy incision in 1970. In 1982, the first rugged camera was introduced, and in 1981, at the commencement of "video-laparoscopy," Kurt Semm performed the first laparoscopic appendicectomy.

In 2008, Goodrich and Schultz [8] reviewed human–robot interaction (HRI) and opined that the usage of robotics for surgery or therapy [9] had increased exponentially in the recent past years. On March 3, 1997, in Belgium, Dr. Jaques Himpens was the first surgeon to perform the first robotic-assisted surgery on a living patient, a cholecystectomy [10], using a robot named Mona. Therefore, it has demonstrated an advantage over soft robotics [11], such as wearables [12], as wearables cannot aid in the patient's recovery.

The design specifications for industrial robots from the 1980s inspired Harmon's initial design standards for touch sensors, which were created in 1982. A piezoelectric rubber-based tactile array sensor was ultimately introduced by Drimus *et al.* [13]. They describe a method for identifying rigid and non-rigid objects, but the representation of the sensor is still the focus of their work. The Schunk PG70 parallel gripper [14] and the Schunk PowerCube robotic arm [15] comprise the robotic system. Similarly, Mira *et al.* [16] provided a technique for grabbing. A pair of Mitsubishi 7-degrees of freedom (DOF) manipulators were used in its construction [17]. One of the robots was outfitted with a Shadow five-finger hand [18] with a Tekscan tactile sensor that could measure the pressure in both the fingers and the palms. Figure 15.1 depicts the key challenges, techniques, and applications of microrobotics for biomedical research.

Delgado *et al.* [20] used two Kuka LWR arms connected with two Shadow hands which were embedded with a SynTouch biomimetic tactile sensor and Tekscan sensor [21]. Finger movements in the hands were controlled using both sensors. The arm can record the contact sites in addition to pressure, vibrations, and temperature. The same robotic setup was used by Zaidi *et al.* [22] with a new grasping approach based on a physical model that connects the touch and pressure feedback between the gripper and the object. In contrast, the robotic arm was designed with flexibility and simplicity. Moreover, to control the arm, the sliding mode control method could be an effective control model that reduces non-linear control problems into more straightforward solutions. In addition, pure tactile-based

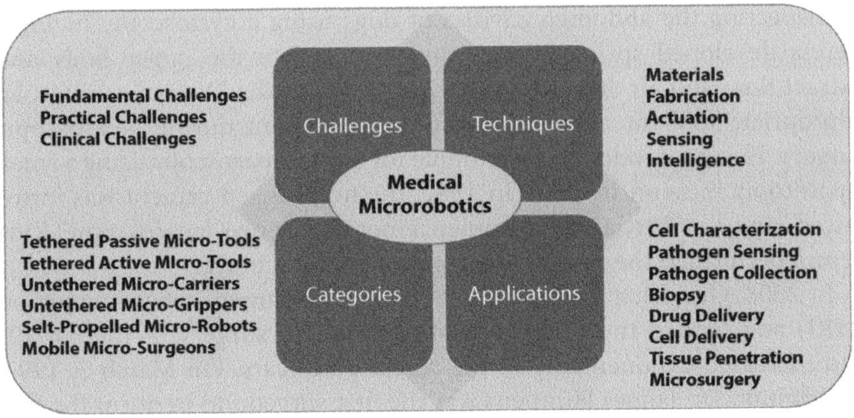

Figure 15.1 Schematic representation of medical microrobotics [19].

systems assess their performance against robotic grippers with many fingers to compare themselves to other objects. Contrary to earlier attempts, more sophisticated sensory systems—such as BioTac-type sensors—are increasingly used because of their multiple contact capability, high sensitivity, and spatial resolution [23, 24].

A group of scientists devised haptic/sensory tools [25] in 2003, a new medical technology that simplified open surgery. The term haptic is derived from the Greek word haptesthai, meaning "bodily feeling." A new type of technology imparts awareness and emotion to computers. Using a combination of force, vibration, and movement, haptic sensors replicate the user's cutaneous sense.

Using online-selectable patient replicas, virtual reality (VR) simulators [26] have been created over the past decade to overcome obstacles such as cost-effectiveness, hands-on training, medical expertise, and excess medical students. Using haptic training simulators, measuring the auxiliary force indicates an idealistic tool behavior, leading to systematic teaching for increased duty [26]. In addition, Talasaz *et al.* [27] have recently demonstrated the value of haptic feedback in robotic-assisted stitching. It is established that force feedback is necessary to enable the user to use the devices more efficiently and to decrease tissue damage, tissue strain, and unintended harm despite haptic technology being helpful in medicine [28] and surgery. However, a disadvantage, i.e., the surgeon cannot feel the region they are operating on, there is a significant chance of rupturing adjacent tissue. As a result, a sophisticated artificial tactile sensing system for MAS, also known as a robotic manipulator [29], was introduced to the market to compensate for this loss. This is a mechanical arm as it allows the surgeon to evaluate the region of concern from several perspectives, which is impossible with haptic/sensorized medical equipment, traditional medical tools, or soft robotics [30]. There are several individuals who utilize wearables to track their health and fitness.

Soft robotics holds an alternative approach to developing healthcare solutions [31], and it could play an integral part in surgery. The innovative and suitable design by engineers of soft robots with biomimetic features greatly assists surgery. These organ-mimicking and tissue-mimicking capabilities can aid biomechanics research [32]. With soft robotics, drug administration, and diagnostics, real-time monitoring is becoming a viable solution. Figure 15.2 signifies a schematic representation of state-of-the-art technology for the development of medical microrobots.

In the field of biomedical science, biocompatible materials are integral to contemporary medical treatments. Various materials in surgery or biomedicine have been evident in the millennia, such as soft and sterilized types of cotton, bandages, and stainless non-oxidized materials. As far back as 3,000 B.C., the first biomaterial was developed. Copper was the first metal humans used for weaponry and domestic purposes. Later, it was alloyed with zinc and utilized for medicinal purposes in conjunction with brass and bronze [34].

In addition, spring forces, which resembled sugar tongs and were made from bent strips of copper and bronze, were commonly employed

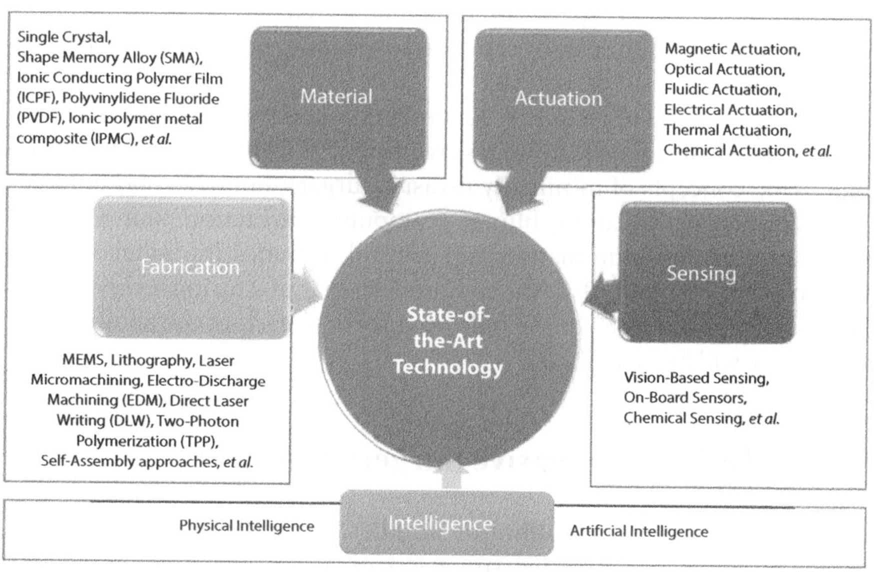

Figure 15.2 Schematic representation of state-of-the-art technology for the development of medical microrobots [33].

in surgery. Copper is an excellent material for constructing instruments since it is found in its natural state, is easily malleable, and can be sharpened and reshaped to provide a proper cutting edge. It was fashioned into bowls for medicine mixing, drainage tubes, and cups for dry cupping. Following this, several materials were utilized for medicinal reasons, including bone and ivory, which were widely used in medical areas for anything from needles to joint replacements. It was prevalent in the 1900s, even though its use for therapeutic purposes was banned. Other substances, such as rubbers, silvers, and nickel-plated stainless steel, dominated the medical instrument market. In 1986, additional novel or ungraded materials, such as liquid flexible soft material, and gases, such as helium–neon, krypton, and argon, were created for laser treatment. More research has been conducted and is underway to identify novel materials and elements for complicated treatment.

Another study shows that automation [35] has a promising future in the medical and surgical domains. In the 1970s, it was initially proposed for biomedical purposes [36] with the integration of artificial intelligence (AI) and machine learning (ML). The interest in AI significantly grew between the 1980s and the 1990s. AI techniques like Bayesian networks, hybrid intelligent systems, fuzzy expert systems, and artificial neural networks were used in district clinical programs. Compared with other fields, the healthcare sector did most of the AI research speculation in 2016 [37]. Because it is affordable, speeds up the patient's rehabilitation, and reduces pain and suffering, it has since expanded its leading position in the healthcare and medical professions [38].

In this summarized technical documentation, the application of robotics and related advanced techniques are presented. This chapter clarifies the underlying concepts of minimally invasive surgery, human–robot interaction, robotic manipulation, human–computer interaction, soft robotics, biosensors, haptics, automation and control, robotics in dental application, and the role of CAD/CAM and modern manufacturing techniques in biomedical applications. In the following section, each of the above-stated procedures will be described.

15.2 Minimally Invasive Surgery (MIS)

Surgery that requires a minimum number of incisions and stitches is known as minimally invasive surgery. Unlike an open operation, MIS physicians use several methods to perform surgery while putting the least amount of stress on the body. Because of this, MIS has outperformed traditional surgery. Nevertheless, this area of surgery is constantly evolving. Doctors continue to search for creative ways to improve patient outcomes because, among other things, safer and more effective therapies are being developed. While this procedure has been used for many generations, the first notable development occurred in the 1860s due to Lister's groundbreaking antiseptic operation. The capacity of surgeons to review or re-evaluate their specialty has been fostered and greatly enhanced by technological advances in addition to these efforts. Some of the major technological milestones related to surgery are outlined in Figure 15.3. Anesthesia, radiology, microbiology, histology, immunology, oncology, and other allied departments are needed for recent surgeries.

The rod-lens endoscope, created by Hopkins in the 1960s, simplified the MIS. MIS had a period of clinical evaluation and technological advancement in the 1970s and 1980s before overtaking and becoming accepted as the standard of care for most surgical operations. In keeping with its well-established hypernym, MIS aims to minimize the number and size of visible skin incisions or holes, minimizing postoperative discomfort, speeding up recovery, improving cosmesis, and assuring overall cost-effectiveness. Even though the advantages of MIS are readily apparent to the patient population, modern operational technologies necessitate that

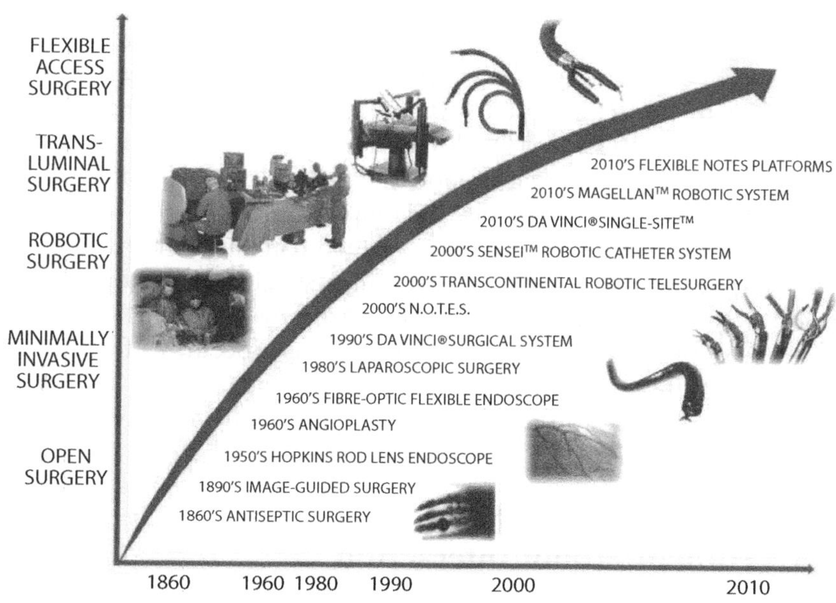

Figure 15.3 Outline of some of the major technological milestones related to surgery [39].

surgeons remain current and knowledgeable about the newest tools or platforms, adjust to changes in the functionality of standard surgical procedures, and make up for occasionally unreasonable ergonomic principles. Recently, robotic operating platforms have been developed in the medical industry to solve these constraints. Robotic manipulation, or increased dexterity and manipulability, and better stability and motion accuracy are all features of the solutions that robotics provides. By utilizing single-slit and natural opening access techniques and MIS recommendations for more complex surgical approaches, contemporary robotic technology [40] has made it possible to reduce further the number of skin cuts and openings during clinical operations. The main prerequisite for such a task is the capacity to enter a specific target anatomy through locations that are not the most obvious or convenient. These groundbreaking methods for the subsequent stage of MIS advancement are described as accurately as possible in flexible access surgery [41]. The technology's current challenge is adapting to the demands of flexible access surgery [42].

The most common operational robots in use right now are surgical extenders and remote manipulators. The da Vinci robot, initially intended for cardiac surgery, is one of the most popular and lucrative robots in its category. The surgeon performs on the patient while sitting at a master panel next to them in this master–slave telemanipulator [43] system through the slave's arms. The operator uses an endoscope to observe the patient's internal organs, and by moving the master manipulator, they may change where the slave robot is placed. The operator closes the servo-control loop by correcting soft tissue movement using visual feedback. The German Aerospace Center (DLR) has upgraded the MiroSurge system [44], a cutting-edge telesurgery robot. The methods include three robotic manipulators, two operating tools, and one endoscope.

Nevertheless, surgeons often face a steep learning curve when stiff or flexible yet unstable tools are used in a surgical environment that is indirectly monitored with MIS [45]. Moreover, any MIS platform should be built with AI in mind to clarify the procedures for the surgeon. Safety, effectiveness, and an easy-to-use, dependable system are the critical clinical requirements for all minimally invasive surgical platforms.

15.3 Human–Robot Interaction

Human–robot interaction (HRI) is human and robot cooperation research and evaluation. As a result, it is often referred to as HRI by researchers.

The interactions between humans and machines, artificial intelligence, robotics, natural language processing, design, and psychology are all included in the multidisciplinary field of human–robot interaction. Regarding HRI, the robots' senses allow them to interact with humans and their environments. They can provide human-like traits, making interaction with technology more exciting and acceptable. The master–slave control system [46], an example of robotic manipulation and haptic feedback, is predominantly used in operational robots, such as the most successful and profitable minimally invasive surgical robot da Vinci [47, 48] (produced by Intuitive Surgical Inc.), a type of surgical robot used in MIS that primarily consists of surgeon comfort, stereoscopic illustrating system, and patient care [49–52]. As the human–robot interface of RMIS, the primary user transfers the surgeon's tasks to the slave robot and provides the operator with touch feedback. Da Vinci's primary manipulator is an 8-degrees-of-freedom (DOF) sequential robotic arm that can be manipulated by the user's hand during the operation procedure. Da Vinci's prime operator applies cable-driven technology and gear transmission process. In order to promote surgical compliance, a control technique for terminating the strength and torque provided to the surgeon is used. Awareness of effort-free situations is the key to improving the efficiency of surgeons and the operation success rate.

One of the medical applications of human–robot interaction is developing a potential strategy for enhancing the sociocommunicative skills of children and adolescents with autism spectrum disorder (ASD) [53]. According to the World Health Organization, about 160 children worldwide have an autism spectrum disorder, and most children are not diagnosed with this disorder until after the age of 4 years [3, 4]. There have been several robotic platforms developed with specific and compliant designs that can interact physically with children. A few of these robots include Teddy Bear Huggable [26], Keepon [54], Paro [27, 28], KASPAR [14, 22, 29–31, 55], and Probo [32, 34] which encourage physical contact in therapies [35]. Furthermore, robots have helped children develop social skills by engaging them in playful interactions.

Innovative commercial products based on user-centric, practical operations of analytic neuroscience and perception research are also being promoted by modern technology in neuroprosthetics and robotics. The brain–computer interface (BCI) [55] coupled with haptic systems, interactive robotics, and autonomous systems [56] was researched as modern applications of neuroprosthetics and human–robot interaction in this unique method. These modern technologies demonstrate the advantages of using a multidisciplinary approach based on cognitive neuroscience, person and situational perception, neural network, and deep learning. These technologies offer a natural and highly efficient interaction between people, neuroprostheses, and robotic self-extensions since the human teacher plays a significant part in this learning process.

15.4 Robotic Manipulation

Robots aided jobs that are repetitive, risky, and monotonous. Most existing robotic manipulators are designed and built to increase stiffness to decrease end-effector vibration and achieve excellent positional precision. The amount of rigidity is reached using heavy materials and a massive design. Therefore, the conventional cumbersome stiff manipulators are unsatisfactory in terms of energy usage and shipment speed. In addition, the performance of highly valued robots is severely hindered by their continued energetic deflection for some time following a move [57, 58]. Nature is replete with examples of manipulation, including animals, insects, and plants. The key benefits of biology or nature over robots include intelligence, the ability to spread, the ability to reproduce, and adaptability at a level beyond the capabilities of robotics or AI. From the use of primitive tools for hunting or farming to the creation of implements, it is reasonable to assume that manipulation has played a central part in human or animal evolution. Thus, human performance differs from robotics in that people can hold or grab objects with dexterity and vary their grasp with precision, but robots are more dependent on grasping. The primary benefit of robots over humans/biology is intelligence in the designing approach. Engineers regularly create [59] gadgets that can outpace or defeat some animal chores. Kinematics of the robotic manipulator is shown in Figure 15.4.

The design of the manipulator is determined by four important factors: accuracy, structural compliance, degrees of freedom (DOF), and force application, as shown in Figure 15.5. The axes only display changes in quantity. The diagonal line of the entire design space, which combines soft and strong materials, has human hands lying on it. Robotic spraying or painting enables the manipulation of significant objects, exceptional precision, operation in extreme settings, and reproducibility, to name a few examples. Therefore, we discovered via experience that manipulation in robots is far more complex than in people who are adept manipulators. However, robotic manipulation has recently produced innovations in manufacturing, healthcare, logistics, and other industries. A comprehensive examination may possess difficulties [61]. In recent years, there has

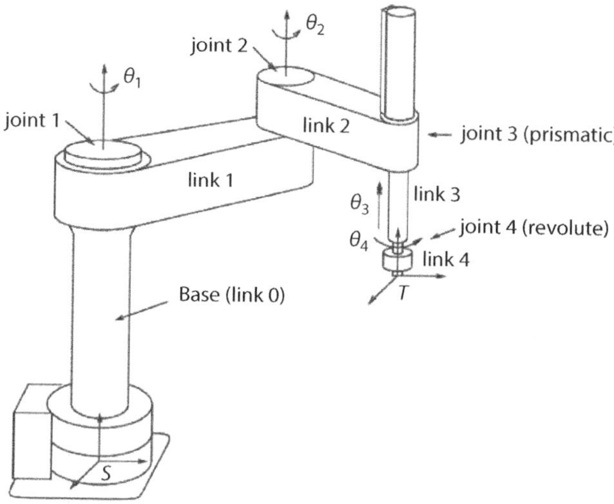

Figure 15.4 Kinematics of the robotic manipulator.

been an increasing desire for robots that can adapt to their surroundings in real time and determine their behavior and actions depending on their link to the surrounding environment. Human-arm sign recognition is a significant research challenge in the natural human–computer interface (HCI) or artificial intelligence and machine learning (AI and ML).

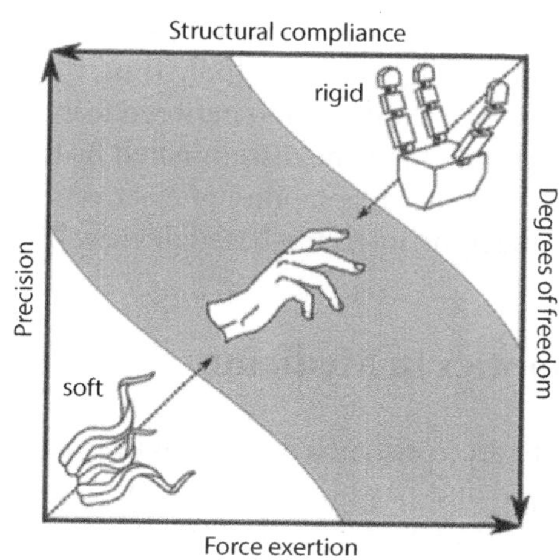

Figure 15.5 Schematic representation of soft and rigid manipulators in the 2D design space [60].

15.5 The Role of Human–Computer Interaction (HCI)

In designing, implementing, and evaluating immersive computing systems for human usage and conducting research on its critical phenomena, HCI [62] is a regulatory concern. Information and communication technology (ICT) for development is a unique field (ICT4D: ICT for growth in developing countries) [63] owing to the user's interaction with the local region. Digital technologies have shown to be incredibly successful in spreading knowledge and information for preventive healthcare in the present pandemic crisis. One example is using digital technology in health campaigns to improve awareness and preventative healthcare. Digital technology may also assist people who suspect they may have a disease or are self-diagnosing to learn more about their symptoms and condition using online knowledge and dictionary resources. Additionally, they can get information from patient-recommended social networks on how other people have handled similar situations. A well-known illustration of one of these services is PatientsLikeMe [64, 65].

Human–machine interaction has played a positive and influential role in wearable intelligent systems [66] with broad applications in bionics, mobility, and robotics. Particularly beneficial for intelligent robots and prosthetics [67] is the invention of artificial skin or prosthetic hands based on human-like sensory systems with multifunctionality and high stretchability. Despite advancements in prosthetic limbs, they cannot entirely replace genuine human limbs due to incompatibilities between artificial touch sensors and human digital signals [68]. Thus, the initial stage of data mining functions as a typical interaction between features since it provides a considerable stimulus that may be disseminated by tactile sense [69, 70]. In addition, the autonomous adaptability of electronic systems is vital for enhancing human–machine interface-based devices [71–73].

15.6 Soft Robotics in Medicine

Conventional robots are constructed from rigid materials that are resistant to elasticity. Rigid and flexible robots can be accurate and robust but lack the versatility and safety of living beings. As the discipline of robotics extends beyond manufacturing and industrial automation into healthcare, field exploration, and human–robot interaction or cooperation, the agents involved are becoming more adaptable and receptive to trained inherent compliance. Soft robotics focuses on building, controlling, and designing robots with viscoelastic material instead of rigid linkages, which is suitable for the field of medicine and biomedicine [74].

Since 1964, the creation or concept of soft robotics has been inspired by the "robot dance" [75, 76] since it imitates the motions and restrictions of traditional rigid robots; also, there was a demand for a ubiquitous sort of robot in automated manufacturing and material handling. It also demonstrates that industrial robots have relied on traditional "hard" robots for almost four decades and still do. The past several years has witnessed a rapid expansion of soft robotics, technologies that are significantly distinct from their rigid counterparts. Soft robotics will influence substantially many aspects of our society and businesses, from manufacturing and consumer electronics to medical applications and wearable technologies [77].

The development of soft robots necessitates the use of materials, such as gels and elastomers, which are often studied as organic materials and are expected to be biocompatible [78–81]. Thus, the inherent qualities of soft robotics make them useful for various biomedical applications, including wearable robots for human–robot interactions, surgical robotic manipulators, and interactive robots for safety interactions, among others.

Rapid advancement in artificial life over the past few decades has produced a significant need for scientists to incorporate research ideas into creating the next generation of biomaterials. Biomaterials research is essential for identifying novel materials for flexible, soft robots required for medical treatment. Numerous biological components, including sugars, lipids, and proteins, are readily accessible and can be utilized as biodegradable structural elements—implementations of soft robotics that are biocompatible and environmentally friendly.

Soft robotics hold border potential to develop more adaptable, bioinspired, safe, and dexterous surgical robots. With soft robotics, the researcher can achieve minimally invasive surgical robotic systems to provide remote teleoperated diagnosis and treatment through small incisions using their skillful, adaptive, and thin robotic manipulators to overcome the long-standing constraints of modern surgical instruments. It is also apparent that soft robotic MIS has a well-established demand for surgical instruments that can pass through small gaps, connect with delicate organs, and move, disfigure, or modify stiffness [82]. For example, STIFF-FLOP, a curved and modular design arm using soft material, was presented [82]. The creation of flexible surgical devices can speed up a systematic strategy that one research team devised and experimentally tested to offer more durable and ideal solutions for the STIFF-FLOP techniques. The STIFF-FLOP arm's breakthrough comes from bioinspiration and soft robotics.

It is made from soft materials that provide increased manipulation for surgical applications [83], have several bending degrees, are independent, and may be stretched and compressed. However, due to their inherent softness, soft surgical robots pose new difficulties in sensing and controlling their shape and motion. Their interaction region can be quite large and is not restricted to the effector tip as it is for rigid robots [84–87].

Surgeons need an interface that can provide information about the manipulator's shape inside the body because the motion is frequently counterintuitive to human operators and causes confusion. Medical robots are also expected to work more autonomously, perform intelligently in unstructured and changing environments, and be more responsive and safer in the event of accidental contact.

The compliance and construction of soft robots prohibit the use of several conventional sensors, such as encoders, haptic sensors [88], metal or semiconductor strain gauges, and inertial measurement units (IMUs). Although piezoelectric polymer-inspired bendable sensors are commercially available, they may not be ideal since all system components must be bendable and extended [89]. Figure 15.6 represents earthworm-inspired soft robots classified based on actuation methods and number of body segments.

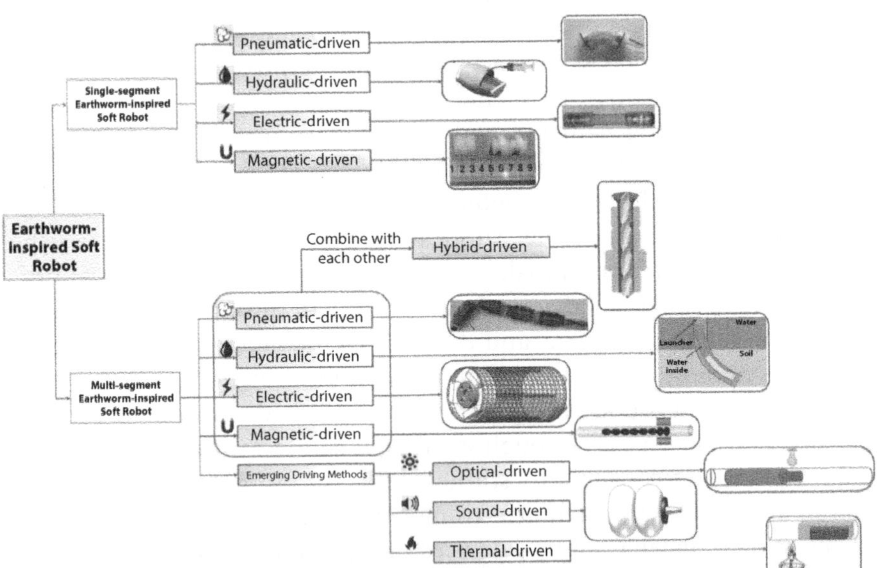

Figure 15.6 Earthworm-inspired soft robots classified based on actuation methods and number of body segments [90].

Soft, flexible electronics might make novel sensory modalities conceivable [91, 92]. Proprioceptive sensors [93] for a soft robot [94] are often composed of contactless or low-coefficient elastomers combined with liquid-phase materials. Since soft robots are controlled by producing curvatures and kinesthetics, which rely on convexity sensors, they are controlled by convexity sensors. The low integral of suggested elastomer sensors [95] (with a modulus in the range of 105–106 Pa) conveys the tiniest change in the essential structures' resistance [96].

These sensors often have lined topologies in which several thin elastomer layers have microfluidic channels cut out using soft lithography [97, 98]. The liquid conductor material (such as gallium-containing alloys like EGaIn or eutectic gallium–indium) was poured into these tubes. Custom sensors may be made to calculate or estimate various stresses using layered channel forms, such as shear, curvature, or tensile. In order to explore the fabrication challenges related to inserting complex channel networks alongside liquid conductors, recent studies have analyzed or assessed mask building of the conductor or direct 3D printing of conductive material. However, the curvature of a soft robot's body parts may be evaluated in real time via exteroceptive sensing. Soft robotics applications may be expanded by using biological sensors and the appropriate chemicals [99]. These sensors, as opposed to the standard audio and optical recorders used with traditional robots, may be more adaptable for usage with soft robots [100].

It would be more versatile, flexible, and pleasant, yielding when necessary, such as when the owner or holder could immediately move and transition to active muscular support. With the aid of material science, a surgeon might employ ultrathin, soft robotic gloves to handle the manufacturing problems of using complex channel networks and liquid conductors. Research has examined or studied mask buildup of the conductor and direct 3D printing of conductive material—enhanced tactile sensory input and recording capabilities during operation [101].

The gripper arms of soft robotics [102] are typically constructed utilizing two rubber carrier belts and material science. As a result of the rubber's flexibility, it may readily wrap around the object to be held [103–105].

In addition to robot sensors that simulate the functionality of a human point of view (speech recognition, computer vision, etc.) [106], robot-specific ideal capabilities-serving detectors are being developed. These sensors arranged a machine to study individuals and their surroundings in ways inaccessible to humans [107]. Biological signals, such as heart rate, blood pressure, and galvanic skin reaction (GSR, the measurement of skin conductance using a galvanometer), provide information about the user's emotional state that cannot be detected in any other way. Mower *et al.*

used GSR in an HRI system to depict and anticipate when a user will likely abandon a rehabilitation-type task. Pose and gesture are crucial sources of information for social interaction. Social and expressive motions, for instance, are vital to human–human and human–robot interaction. Such information can be provided for computer illustration in specific scenarios. Wearable sensors, an example of soft robotics, may capture accurate real-time data on human behavior in certain situations. Such wearable approaches have been utilized in HRI tasks for stroke therapy and social interaction [108, 109]. In addition, researchers are creating algorithms for combining the multisensor, multimodal data intrinsic to HRI domains to design new and improved sensors for HRI's specific demands. For instance, Kapoor and Picard created a system for effective detection that fuses several sensors using Gaussian models [110]. Incorporating person monitoring, facial recognition, sound source localization, and leg detection, multimodal sensing [111] has also been used by a robot to sense the attention of human users and assess if a user is addressing the robot.

For years, robots have assisted humans in accomplishing repetitive, high-stress, and quantifiable activities, therefore reducing the burden on humanity. Moreover, small and medium-sized businesses are rapidly modernizing due to the availability of inexpensive solutions and user-friendly collaborative robots. Many cooperative robots move modestly in human-populated, chaotic, and dynamic situations [112]. Consequently, human–robot collaboration solutions have become cutting-edge in various medical services and industrial applications, in which people and robots share information and knowledge in a networked environment.

15.7 Haptics in Medicine

Haptics has numerous applications in medicine, particularly in medical simulation and training. It allows medical professionals to practice complex procedures and hone their skills in a safe, controlled environment. For example, haptic simulators can be used to train surgeons in laparoscopic and robotic surgery, allowing them to practice manipulating surgical instruments [113] in a virtual environment that closely resembles the experience of operating on a patient [114]. Haptic technology is also used to train healthcare professionals in ultrasound scanning and other diagnostic imaging techniques, where the ability to feel the texture and resistance of internal organs is critical to making an accurate diagnosis. Additionally, haptics is used in the development of prosthetic limbs and other assistive devices to help amputees regain their sense of touch and control [115].

A remote machine is managed by an operator using a machine interface, shown in Figure 15.7. Distributed pressure sensors on a remote machine collect haptics data, which is then merged with vision data by changing the pressure data to a visual expression and sending the composite image to the operator. The operator instinctively divides the combined vision and haptic information into vision and haptic information. This proposal is based on the observation that the majority of surgeons believed that haptic presentation was not actually necessary because they could properly infer the sense of haptic based just on visual information. They also believed that recognition of haptics information is due to both the sense of vision and the sense of haptics.

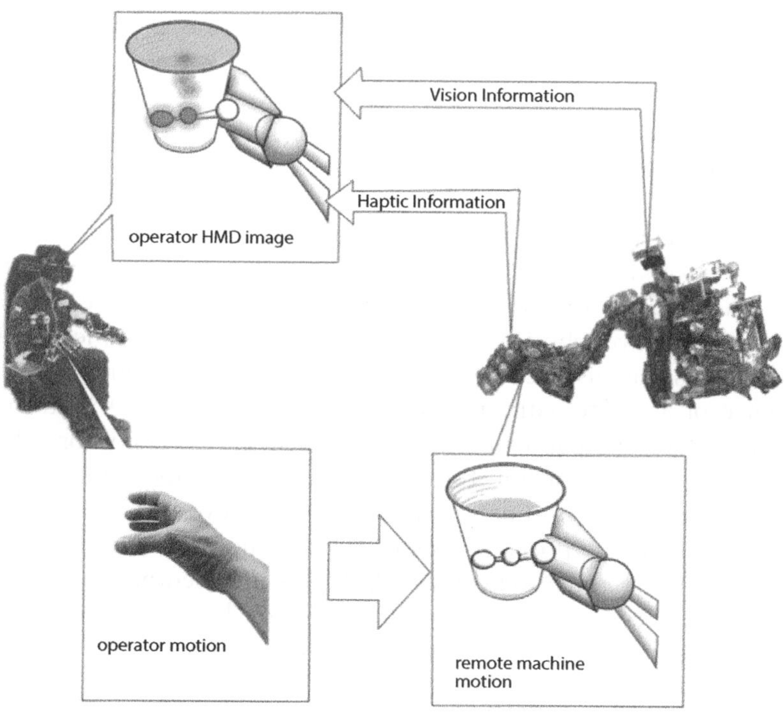

Figure 15.7 Schematic representation of visual haptics image for manipulation [116].

15.8 Automation and Control

AI, ML, and automation are creating significant interest in the conventional examination of medical data. Commendable is the rapid commercialization of artificial intelligence [117]. The expanding corpus of information on AI techniques has recently had a profoundly positive effect on healthcare [118] due to its influence on the global landscape of biomedical and healthcare research. Deep learning (DL) is a subset of ML that employs network-based replicas to replicate the human brain's ability to analyze vast volumes of complex data, such as language processing, drug discovery, and image identification, as a decision support system for humans [119]. Researchers have looked imaginatively at AI and ML to provide an evidence-based strategy in their respective fields in their investigations of therapies for illnesses such as pancreatic, ovarian, breast, and lung cancer. In many studies evaluating the use of AI for breast screening based on social and ethical challenges of acceptance and radiologist implementation, there is evidence of novel results such as DL establishing the existence of lymph node metastases from breast cancer. In colorectal screening, the processing of endoscopic images, stills, and videos was accelerated using CADe ML models [120, 121].

15.9 Dental

The goal of dental and orthopedic regulation [122] has been to improve the efficiency and coordination of all patient treatments since it was established. New dental orthopedics treatments may develop due to many enabling innovative robotics technologies. This is accurate for the COVID-19 crisis-era social distance and distance research criteria. Due to the high incidence of asymptomatic individuals contracting COVID-19 through aerosol contact, the pandemic has highlighted the necessity for robotic applications in dental orthopedics [123]. Another method is material science as a nanomaterial in nanotechnology, which over the past 10 years, has made tremendous progress in dental orthopedics, delivering systematic and successful treatment outcomes [124]. This included nanobiotechnology, nanorobotics, and nanomaterials, also referred to as the prerequisites for designing and building nanorobots made of materials at or very close to the nanoscale [125]. Nanorobots have been used to stabilize dental movement in animals using nanoelectromechanical systems (NEMS) [126–128] and nano low-intensity pulsed ultrasound (LIPUS) devices [129, 130].

15.10 CAD/CAM

3D-printed soft robots [131], by combining liquid metal sensors with fluidic actuation, have the ability to sense [132]. Given the increasing importance of automated design procedures, this perspective would let robots have many more sensors than would be possible with human manufacturing methods [133]. Taylor and Stoianovici [134] proposed a primary, high-level classification of surgical robots by splitting them into two major categories: surgical assistants and surgical computer-aided design/manufacturing (CAD/CAM) techniques. Surgical CAD/CAM techniques are demonstrated to aid in planning and intraoperative operation via the reformatting of preoperative images and construction of three-dimensional (3D) representations, indexation of these data to the patient in the operating room, and application of robots and image layer cover displays to facilitate the precise execution of the planned assessment action. Other subcategories of surgical help systems include surgical extenders and additional surgical supports. With a focus on intraoperative problem-solving support and skill development, surgical extenders work directly with the operator to enhance human abilities in performing several operational duties.

It is fair to expect that advancements in modern technology and the creation of considerably new enhancements will determine the future of robotic surgical systems [135, 136]. These advancements include shrinking the size of robotic arms, haptic feedback and proprioception, contemporary tissue estimation and hemostasis systems, bending the shafts of mechanical implements, instrumentation of the natural orifice transluminal endoscopic surgery (NOTES) concept, combining navigation techniques with augmented reality usages [137], and finally, autonomous rotorcraft.

15.11 Conclusion

In conclusion, the field of robotics has proven to be a game changer in the medical industry. With the development of soft robotics, material science, artificial intelligence, haptic sensors, and human–computer interaction, we are seeing advancements in robotic technology that are revolutionizing surgical procedures. Soft robotic surgery, in particular, has emerged as a promising approach to minimally invasive procedures, enabling greater precision, safety, and control in surgeries. With the ability to mimic the natural movements of the human body, these robots are able to provide surgeons with better dexterity and more flexibility in performing complex procedures. As a result, patients can experience faster recovery times, reduced pain, and fewer complications. As research in this field continues to expand, we can expect to see even more innovative applications of robotics in medicine, ultimately leading to better patient outcomes and a more effective healthcare system.

References

1. Iwu, M.M. and Gbodossou, E., The role of traditional medicine. *Lancet*, 356, S3, 2000.
2. Phan, N.D.M. *et al.*, An ultraviolet C light-emitting robot design for disinfection in the operating room, in: *Recent Trends in Mechatronics Towards Industry 4.0*, Springer, Singapore, 2022.
3. Rogers, D.A., Elstein, A.S., Bordage, G., Improving continuing medical education for surgical techniques: Applying the lessons learned in the first decade of minimal access surgery. *Ann. Surg.*, 233, 2, 370, 377, 2001.
4. Fuchs, K.H., Minimally invasive surgery. *Endoscopy*, 34, 2, 370, 377, 2002.
5. Khoo, L.T., Ozer, A.F., Cosar, M., History of minimally invasive spine surgery, in: *Minimally Invasive Procedures in Spine Surgery (e-book)*, 2016.
6. Quraishi, M.S. and Bradley, P.J., *British Society for the History of ENT*, Abstracts from meeting, 1996.
7. St. Peter, S.D. and Holcomb III, G.W., *History of Minimally Invasive Surgery*, Elsevier Health Sciences, Philadelphia, 2008.
8. Goodrich, M.A. and Schultz, A.C., *Human-Robot Interaction: A Survey*, vol. 1, pp. 203–275, Foundations and Trends® in Human–Computer Interaction, 2008.
9. Cespedes, N., Meza-Kubo, L., Sepulveda, F., Social human-robot interaction for gait rehabilitation. *IEEE Trans. Neural Syst. Rehabil. Eng.*, 28, 6, 1446–1455, 2020.
10. Dubois, F., Berthelot, G., Levard, H., Cholecystectomy by coelioscopy. *Presse Med.*, 18, 19, 980–982, 1989.
11. Cianchetti, M. *et al.*, Biomedical applications of soft robotics. *Nat. Rev. Mater.*, 3, 6, 143–153, 2018.
12. Bonato, P., Wearable sensors and systems. *IEEE Eng. Med. Biol. Mag.*, 29, 3, 25–36, 2010.
13. Qi, Y. *et al.*, Piezoelectric ribbons printed onto rubber for flexible energy conversion. *Nano Lett.*, 10, 2, 524–528, 2010.
14. Bogh, S. *et al.*, Integration and assessment of multiple mobile manipulators in a real-world industrial production facility. *ISR/Robotik 2014; 41st International Symposium on Robotics*, VDE, 2014.

15. Duguleana, M., Barbuceanu, F.G., Mogan, G., Evaluating human-robot interaction during a manipulation experiment conducted in immersive virtual reality. *International Conference on Virtual and Mixed Reality*, Springer, Berlin, Heidelberg, pp. 135–144, 2011.
16. Patel, R.V., Talebi, H.A., Jayender, J., Shadpey, F., Robot-assisted minimally invasive surgery, in: *Springer Handbook of Robotics*, B. Siciliano and O. Khatib (Eds.), pp. 1781–1814, Springer, Berlin Heidelberg, 2009.
17. Siciliano, B., Khatib, O., Kröger, T. (Eds.), *Springer Handbook of Robotics*, vol. 200, Springer, Berlin, 2008.
18. Liu, H. *et al.*, Multisensory five-finger dexterous hand: The DLR/HIT hand II. *2008 IEEE/RSJ International Conference on Intelligent Robots and Systems*, IEEE, 2008.
19. Zhang, D., Gorochowski, T.E., Marucci, L., Lee, H.-T., Gil, B., Li, B., Hauert, S., Yeatman, E., Advanced medical micro-robotics for early diagnosis and therapeutic interventions. *Front. Robot. AI*, 9, 372, 2022.
20. Bischoff, R. *et al.*, The KUKA-DLR lightweight robot arm-A new reference platform for robotics research and manufacturing. *ISR 2010 (41st International Symposium on Robotics) and ROBOTIK 2010 (6th German Conference on Robotics)*, VDE, 2010.
21. Ruppel, P. *et al.*, Simulation of the SynTouch BioTac sensor, in: *International Conference on Intelligent Autonomous Systems*, Springer, Cham, 2018.
22. Karmakar, S. *et al.*, Helical SMA actuator based artificial muscle and arm with sliding mode control, in: *Advances in Robotics-5th International Conference of The Robotics Society*, 2021.
23. Nadon, F., Valencia, A.J., Payeur, P., Multi-modal sensing and robotic manipulation of non-rigid objects: A survey. *Robotics*, 7, 74, 5–10, 3–4, 2018.
24. Stenslie, S., *Virtual Touch: A Study of the Use and Experience of Touch in Artistic, Multimodal and Computer-based Environments*, Oslo School of Architecture and Design, 2011.
25. Panait, L., Akkary, E., Bell, R., Roberts, K., Dudrick, S., Duffy, A., The role of haptic feedback in laparoscopic simulation training. *J. Surg. Res.*, 156, 2, 312–316, 2009.
26. Adamovich, S.V. *et al.*, Design of a complex virtual reality simulation to train finger motion for persons with hemiparesis: A proof of concept study. *J. Neuroeng. Rehabil.*, 6, 1, 372, 2009.
27. Licona, A.R. *et al.*, Applications of haptics in medicine, in: *Haptic Interfaces for Accessibility, Health, and Enhanced Quality of Life*, 2020.
28. Shackelford, J.F., *Materials Science for Engineers*, Pearson; 8th edition (10 July 2014), 2000.
29. Yu, K.T. *et al.*, More than a million ways to be pushed. a high-fidelity experimental dataset of planar pushing, in: *2016 IEEE/RSJ International Conference on Intelligent Robots and Systems (IROS)*, IEEE, 2016.

30. Lunni, D. *et al.*, A closed loop shape control for bio-inspired soft arms, in: *Conference on Biomimetic and Biohybrid Systems*, Springer, Cham, 2017.
31. Tan, Z. *et al.*, Composite hydrogel: A high fidelity soft tissue mimic for surgery. *Mater. Des.*, 160, 373, 2018.
32. Lou, Z. *et al.*, Reviews of wearable healthcare systems: Materials, devices and system integration. *Mater. Sci. Eng. R Rep.*, 140, 373, 2020.
33. Zhang, D., Gorochowski, T.E., Marucci, L., Lee, H.-T., Gil, B., Li, B., Hauert, S., Yeatman, E., Advanced medical micro-robotics for early diagnosis and therapeutic interventions. *Front. Robot. AI*, 9, 373, 2022.
34. Shackelford, J.F., *Materials Science for Engineers*, Pearson; 8th edition (10 July 2014), 2000.
35. Kinyua, J. and Awuah, L., AI/ML in security orchestration, automation and response: Future research directions. *Intell. Autom. Soft Comput.*, 28, 2, 374, 377, 2021.
36. Thakur, A. *et al.*, Application of artificial intelligence in pharmaceutical and biomedical studies. *Curr. Pharm. Des.*, 26, 29, 374, 2020.
37. Casino, F., Dasaklis, T.K., Patsakis, C., A systematic literature review of blockchain-based applications: Current status, classification and open issues. *Telemat. Inform.*, 36, 374, 2019.
38. Malik, P., Pathania, M., Rathaur, V.K., Overview of artificial intelligence in medicine. *J. Family Med. Prim. Care*, 8, 7, 374, 2019.
39. Vitiello, V. *et al.*, Emerging robotic platforms for minimally invasive surgery. *IEEE Rev. Biomed. Eng.*, 6, 111–126, 2012.
40. Mack, M.J., Minimally invasive and robotic surgery. *J. Am. Med. Assoc.*, 285, 568–572, Feb. 2001.
41. Li, C. *et al.*, A robotic system with multichannel flexible parallel manipulators for single port access surgery. *IEEE Trans. Industr. Inform.*, 15, 3, 376, 2018.
42. Mintz, Y. and Brodie, R., Introduction to artificial intelligence in medicine. *Minim. Invasive Ther. Allied Technol.*, 28, 2, 376, 2019.
43. Battina, D.S., Artificial intelligence in software test automation: A systematic literature review. *JETIR*, 2019.
44. Hagn, U. *et al.*, DLR MiroSurge: A versatile system for research in endoscopic telesurgery. *Int. J. Comput. Assist. Radiol. Surg.*, 5, 2, 183–193, Mar. 2010.
45. Su, H. *et al.*, Toward teaching by demonstration for robot-assisted minimally invasive surgery. *IEEE Trans. Autom. Sci. Eng.*, 18, 2, 376, 2021.
46. Krahnstoever, N. *et al.*, Collaborative real-time control of active cameras in large scale surveillance systems, in: *Workshop on Multi-Camera and Multi-Modal Sensor Fusion Algorithms and Applications-M2SFA2*, 2008.
47. Rosen, J., Hannaford, B., Satava, R.M. (Eds.), *Surgical Robotics: Systems Applications and Visions*, Springer Science & Business Media, 2011.
48. Moran, M.E., Out of the darkness: Two monks, two mythical humanoid machines. *J. Endourol.* Mary Ann. Liebert Inc., 20, 140, 2006.

49. Ghalamzan, E.A.M. *et al.*, Human-in-the-loop optimisation: Mixed initiative grasping for optimally facilitating post-grasp manipulative actions, in: *2017 IEEE/RSJ International Conference on Intelligent Robots and Systems (IROS)*, IEEE, 2017.
50. Pratheep, V.G. *et al.*, Design and fabrication of multitasking transit robot. *IOP Conf. Ser. Mater. Sci. Eng.*, 1057, 1, 377, 2021.
51. Hanly, E.J. *et al.*, Multiservice laparoscopic surgical training using the daVinci surgical system. *Am. J. Surg.*, 187, 2, 377, 2004.
52. Eslamian, S., Reisner, L.A., Pandya, A.K., Development and evaluation of an autonomous camera control algorithm on the da Vinci surgical system. *Int. J. Med. Robot. Comput. Assist. Surg.*, 16, 2, 377, 2020.
53. Raptopoulou, A. *et al.*, Human-robot interaction for social skill development in children with ASD: A literature review. *Healthc. Technol. Lett.*, 8, 4, 377, 2021.
54. Talasaz, A., Trejos, A.L., Patel, R.V., The role of direct and visual force feedback in suturing using a 7-DOF dual-arm teleoperated system. *IEEE Trans. Haptics*, 10, 2, 377, 2016.
55. Oliveira, B. *et al.*, Characterization of the workspace and limits of operation of laser treatments for vascular lesions of the lower limbs. *Sensors*, 22, 19, 377, 2022.
56. Wolpaw, J.R. *et al.*, Brain-computer interface technology: A review of the first international meeting. *IEEE Trans. Rehabil. Eng.*, 8, 2, 377, 2000.
57. Rosell, J., PMK: A knowledge processing framework for autonomous robotics perception and manipulation. *Sensors*, 19, 5, 1–23, 2019.
58. Roweis, S. and Ghahramani, Z., A unifying review of linear Gaussian models. *Neural Comput.*, 11, 2, 378, 1999.
59. Khan, H.M., Jianhua, Y., Tareen, S.A.K., Locomotion patterns of a vermicular robot, in: *2016 2nd International Conference on Robotics and Artificial Intelligence (ICRAI)*, IEEE, 2016.
60. Hughes, J. *et al.*, Soft manipulators and grippers: A review. *Front. Robot. AI*, 3, 69, 2016.
61. DiCarlo, J.J., Zoccolan, D., Rust, N.C., How does the brain solve visual object recognition? *Neuron*, 73, 3, 378, 2012.
62. Stuart, T.T., Baecker, R., Carey, T., Gasen, J., Mantei, M., Perlman, G., Strong, G., Verplank, W., *ACM SIGCHI Curricula for Human-Computer Interaction*, 1992.
63. Bon, A., Akkermans, H., Gordijn, J., Developing ICT services in a low-resource development context. *Complex Syst. Inf. Modeling Q.*, 9, 380, 2016.
64. Wicks, P. *et al.*, Sharing health data for better outcomes on PatientsLikeMe. *J. Med. Internet Res.*, 12, 2, 380, 2010.
65. Blandford, A., HCI for health and wellbeing: Challenges and opportunities. *Int. J. Hum.-Comput. Stud.*, 131, 380, 2019.

66. Yin, R., *et al.*, Wearable sensors-enabled human–machine interaction systems: From design to application. *Adv. Funct. Mater.*, 31.11, 2008936, 12–14, 2021. https://onlinelibrary.wiley.com/doi/10.1002/adfm.202008936
67. Weiner, P. *et al.*, Designing prosthetic hands with embodied intelligence: The KIT prosthetic hands. *Front. Neurorobot.*, 16, 380, 2022. https://www.frontiersin.org/articles/10.3389/fnbot.2022.815716/full
68. Dargahi, J. *et al.*, *Tactile Sensing and Displays: Haptic Feedback for Minimally Invasive Surgery and Robotics*, John Wiley & Sons, 2012.
69. Flores De Jesus, K., *A Versatile Method of Resolution Enhancement for Tactile Sensor Array Used as Synthetic Skin: Modeling and Implementation*, 2013.
70. Dahiya, R.S. *et al.*, Tactile sensing—From humans to humanoids. *IEEE Trans. Robot.*, 26, 1, 1–20, Feb. 2009.
71. Motlagh, Y., Göhner, P., Adaptive human-machine-interface of automation systems, in: *Technological Innovation for Collective Awareness Systems. DoCEIS 2014. IFIP Advances in Information and Communication Technology*, Camarinha-Matos, L.M., Barrento, N.S., Mendonça, R. (eds), vol. 423, Springer, Berlin, Heidelberg, 2014. https://doi.org/10.1007/978-3-642-54734-8_20
72. Yetisen, A.K., Martinez-Hurtado, J.L., Ünal, B., Khademhosseini, A., Butt, H., Wearables in machine. *Adv. Mater.* 2018, 30, 1706910. https://doi.org/10.1002/adma.201706910.
73. Przekora, A., A concise review on tissue engineered artificial skin grafts for chronic wound treatment: Can we reconstruct functional skin tissue *in vitro*? *Cells*, 9, 6, 1383, 2020.
74. Robert Jr., L.P. *et al.*, A review of personality in human–robot interactions. *Found. Trends® Inf. Syst.*, 4, 2, 85–199, 2020.
75. Shinozaki, K., Iwatani, A., Nakatsu, R., Concept and construction of a robot dance system, in: *ICMIT 2007: Mechatronics, MEMS, and Smart Materials*, vol. 6794, SPIE, 2008.
76. Rossiter, J. and Hauser, H., Soft robotics—The next industrial revolution. *IEEE Robot. Autom. Mag.*, 23, 3, 20–30, Sep. 2016.
77. Lou, Z. *et al.*, Reviews of wearable healthcare systems: Materials, devices and system integration. *Mater. Sci. Eng. R Rep.*, 140, 100523, Sep. 2020.
78. Dietrich, J. *et al.*, Development of a peristaltically actuated device for the minimal invasive surgery with a haptic sensor array, in: *Micro- and Nanostructuers of Biological Systems*, 2004.
79. Glozman, D. *et al.*, A self-propelled inflatable earthworm-like endoscope actuated by single supply line. *IEEE Trans. Biomed. Eng.*, 57, 6, 1476–1485, Jun. 2010.
80. Goetzke, H.H., Pattrick, J.G., Federle, W., Froghoppers jump from smooth plant surfaces by piercing them with sharp spines. *Proc. Natl. Acad. Sci.*, 116, 8, 2836–2841, 2019.
81. Gomes, P., Surgical robotics: Reviewing the past, analysing the present, imagining the future. *Robot. Comput.-Integr. Manuf.*, 27, 2, 261–275, 2011.

82. Fras, J., Macias, M., Czarnowski, J., Brancadoro, M., Menciassi, A., Glowka, J., Soft manipulator actuation module—With reinforced chambers, in: *Soft and Stiffness-Controllable Robotics Solutions for Minimally Invasive Surgery: The STIFF-FLOP Approach*, River Publishers, Gistrup, Denmark, 2018.
83. Lunni, D., Fruscalzo, L., Bonomo, M., Dimino, M., Epasto, A.S., A closed loop shape control for bio-inspired soft arms, in: *Conference on Biomimetic and Biohybrid Systems*, pp. 157–169, Springer, Cham, 2017.
84. Visentin, F., Mishra, A.K., Naselli, G.A., Mazzolai, B., Simplified sensing and control of a plant-inspired cable driven manipulator, in: *2019 2nd IEEE International Conference on Soft Robotics (RoboSoft)*, pp. 422–427, IEEE, 2019.
85. Noh, Y., Wang, H., Sitti, M., Optical force and torque sensor for flexible robotic manipulators, in: *Soft and Stiffness-Controllable Robotics Solutions for Minimally Invasive Surgery: The Stiff-Flop Approach*, 2018.
86. Dawood, A.B., Kasim, M., Yang, G., Fusing dexterity and perception for soft robot-assisted minimally invasive surgery: What we learnt from STIFF-FLOP. *Appl. Sci.*, 11, 14, 6436, 2021.
87. Konstantinova, J., Wurdemann, H., Shafti, A. (Eds.), *Soft and Stiffness-Controllable Robotics Solutions for Minimally Invasive Surgery: The STIFF-FLOP Approach*, CRC Press, 2022.
88. Okamura, A.M., Haptic feedback in robot-assisted minimally invasive surgery. *Curr. Opin. Urol.*, 19, 1, 87–92, 2009.
89. Fay, S., *The Characterisation of Blast Loading for Shallow Buried Explosives*, PhD Dissertation, University of Sheffield, Sheffield, UK, 2020.
90. Liu, J., Li, P., Zuo, S., Actuation and design innovations in earthworm-inspired soft robots: A review. *Front. Bioeng. Biotechnol.*, 11, 382, 2023.
91. Teixeira, T., Dublon, G., Savvides, A., *A Survey of Human-Sensing: Methods for Detecting Presence, Count, Location, Track, and Identity*, Technical Report, ENALAB, Yale University, 2010.
92. Kortenkamp, D., Simmons, R., Brugali, D., *Springer Handbook of Robotics*, B. Siciliano and O. Khatib (Eds.), Springer, Berlin, Heidelberg, 2008.
93. Salter, T. *et al.*, Using proprioceptive sensors for categorizing human-robot interactions, in: *2007 2nd ACM/IEEE International Conference on Human-Robot Interaction (HRI)*, pp. 11–18, 2007.
94. Zhou, X. *et al.*, A master-slave system to acquire biometric imagery of humans at distance, in: *First ACM SIGMM International Workshop on Video Surveillance*, pp. 87–94, 2003.
95. Ni, N. and Zhang, L., Dielectric elastomer sensors, in: *Elastomers*, pp. 25–48, InTech, Rijeka, 2017.
96. Hellebrekers, T., Kroemer, O., Majidi, C., Soft magnetic skin for continuous deformation sensing. *Adv. Intell. Syst.*, 1, 4, 1900033, 2019.
97. Ionov, L. *et al.*, Smart microfluidic channels. *Adv. Funct. Mater.*, 16, 9, 1323–1328, 2006.

98. Burgin, T., Choong, V.-E., Maracas, G., Large area submicrometer contact printing using a contact aligner. *Langmuir*, 16, 12, 5261–5266, 2000.
99. Noguchi, N. *et al.*, Development of a master–slave robot system for farm operations. *Comput. Electron. Agric.*, 44, 1, 17–27, 2004.
100. Chen, S. *et al.*, Generalized way to make temperature tunable conductor–insulator transition liquid metal composites in a diverse range. *Mater. Horiz.*, 6, 9, 1925–1931, 2019.
101. Palmann, C., Mavromatis, S., Sequeira, J., A new approach for registering remote sensing images from various modalities, in: *Image and Signal Processing for Remote Sensing XV*, vol. 7477, SPIE, pp. 74770E–74770E-12, 2009.
102. Camarillo, D.B. *et al.*, Mechanics modeling of tendon-driven continuum manipulators. *IEEE Trans. Robot.*, 24, 6, 1262–1273, 2008.
103. Pusalkar, N., Karmakar, S., Aggrawal, R., Mali, P., Singh, A., Sarkar, A., Krishna, K.M., SMA actuated dual arm flexible gripper, in: *Proceedings of the Advances in Robotics 2019*, pp. 1–6, 2019.
104. Thieffry, M., *Model-Based Dynamic Control of Soft Robots*, PhD Dissertation, Université Polytechnique des Hauts-de-France, 2019.
105. Baek, S.-F., Lee, S.-H., Chang, J.H., Design and control of a robotic finger for prosthetic hands, in: *Proceedings 1999 IEEE/RSJ International Conference on Intelligent Robots and Systems. Human and Environment Friendly Robots with High Intelligence and Emotional Quotients (Cat. No. 99CH36289)*, vol. 1, IEEE, 1999.
106. Krichmar, J.L., Neurorobotics—A thriving community and a promising pathway toward intelligent cognitive robots. *Front. Neurorobot.*, 12, 383, 2018.
107. Valadão, C.T. *et al.*, Analysis of the use of a robot to improve social skills in children with autism spectrum disorder. *Res. Biomed. Eng.*, 32, 2, 161–175, 2016.
108. Bonomi, A.G. *et al.*, Atrial fibrillation detection using a novel cardiac ambulatory monitor based on photo-plethysmography at the wrist. *J. Am. Heart Assoc.*, 7, 15, 384, 2018.
109. Poon, C. *et al.*, Monitoring of postoperative patients using a wearable wireless sensor platform, in: *2011 Annual International Conference of the IEEE Engineering in Medicine and Biology Society (EMBC)*, Boston, MA, USA, Aug. 30-Sept. 3, 2011.
110. Kapoor, A. and Picard, R.W., Multimodal affect recognition in learning environments, in: *Proceedings of the 13th Annual ACM International Conference on Multimedia*, Singapore, Nov. 6-11, 2005.
111. Diab, M. *et al.*, A low-cost, modular, open-source robotic prosthetic hand. *IEEE Access*, 7, 42367–42376, 2019.
112. Askren, K., *Survey of Deep Neural Networks Handling Plan Development using Simulations of Real-World Environments*, Thesis, University Honors College Middle Tennessee State University, 2019.

113. Albrecht, T. *et al.*, Three-dimensional endoscopic visualization in functional endoscopic sinus surgery. *Eur. Arch. Otorhinolaryngol.*, 273, 11, 3655–3660, Nov. 2016.
114. Moran, M.E., The da Vinci robot. *J. Endourol.*, 20, 12, 1059–1062, Dec. 2006.
115. Meccariello, G. *et al.*, An experimental study about haptic feedback in robotic surgery: May visual feedback substitute tactile feedback? *J. Robot. Surg.*, 10, 1, 21–26, Mar. 2016.
116. Haruna, M., Ogino, M., Koike-Akino, T., Proposal and evaluation of visual haptics for manipulation of remote machine system. *Front. Robot. AI*, 7, 529040, 2020.
117. Sridharan, M. and Stone, P., Color learning on a mobile robot: Towards full autonomy under changing illumination. *IJCAI*, 2007.
118. ATM Theunissen, K. *et al.*, Barriers to and facilitators of partner notification for Chlamydia trachomatis among health care professionals. *BMC Health Serv. Res.*, 14, 1, 386, 2014.
119. Musa, I.H. *et al.*, Artificial intelligence and machine learning in cancer research: A systematic and thematic analysis of the top 100 cited articles indexed in Scopus database. *Cancer Control*, 29, 386, 2022.
120. Yang, L. *et al.*, CADE: Detecting and explaining concept drift samples for security applications, in: *30th USENIX Security Symposium (USENIX Security 21)*, 2021.
121. Mishra, A.K. *et al.*, SIMBA: Tendon-driven modular continuum arm with soft reconfigurable gripper. *Front. Robot. AI*, 4, 386, 2017.
122. Webster, T.J., Nanophase ceramics: The future orthopedic and dental implant material, in: *Advances in Chemical Engineering*, vol. 27, pp. 125–166, Academic Press, 2001.
123. Adel, S. *et al.*, Robotic applications in orthodontics: Changing the face of contemporary clinical care. *BioMed. Res. Int.*, 2021, 386, 2021.
124. Khatria, H. *et al.*, Nano-orthodontics: Small is the new big. *EC Dent. Sci.*, 18, 386, 2019.
125. Sierra, D.P. *et al.*, A review of research in the field of nanorobotics, United States, 2005.
126. Gambhir, R.S. *et al.*, Nanotechnology in dentistry: Current achievements and prospects. *J. Orofac. Sci.*, 5, 1, 386, 2013.
127. Krogh, A., What are artificial neural networks? *Nat. Biotechnol.*, 26, 2, 386, 2008.
128. Govindankutty, D., Applications of nanotechnology in orthodontics and its future implications: A review. *Int. J. Appl. Dent. Sci.*, 1, 4, 386, 2015.
129. Kuroda, S. *et al.*, Clinical use of miniscrew implants as orthodontic anchorage: success rates and postoperative discomfort. *Am. J. Orthod. Dentofacial Orthop.*, 131, 1, 386, 2007.
130. Harrison, A. *et al.*, Mode & mechanism of low intensity pulsed ultrasound (LIPUS) in fracture repair. *Ultrasonics*, 70, 386, 2016.

131. Wehner, M., Truby, R., Fitzgerald, D. *et al.*, An integrated design and fabrication strategy for entirely soft, autonomous robots. *Nature*, 536, 451–455, 2016. https://doi.org/10.1038/nature19100
132. Truby, R.L. *et al.*, Soft somatosensitive actuators via embedded 3D printing. *Adv. Mater.*, 30, 15, 387, 2018.
133. Tapia, J. *et al.*, Makesense: Automated sensor design for proprioceptive soft robots. *Soft Robot.*, 7, 3, 387, 2020.
134. Taylor, R.H. *et al.*, Medical robotics and computer-integrated surgery, in: *Springer Handbook of Robotics*, Springer, Cham, 2016.
135. Hagn, U. *et al.*, The DLR MIRO: A versatile lightweight robot for surgical applications. *Ind. Robot. Int. J.*, 35, 4, 324–336, 2008.
136. Moustris, G.P. *et al.*, Evolution of autonomous and semi-autonomous robotic surgical systems: A review of the literature. *Int. J. Med. Robot. Comput. Assist. Surg.*, 7, 4, 375–392, Dec. 2011.
137. Grobbelaar, W., Verma, A., Shukla, V.K., Analyzing human robotic interaction in the food industry. *J. Phys. Conf. Ser.*, 1817, 1, 012021, Feb. 2021.

16

A Research Perspective on Ankle–Foot Prosthetics Designs for Transtibial Amputees

Vidyapati Kumar, Pushpendra Gupta and Dilip Kumar Pratihar*

Department of Mechanical Engineering, Indian Institute of Technology, Kharagpur, West Bengal, India

Abstract

Amputees may regain their lost limbs and continue daily activities using prosthetic ankle–foot devices, and researchers have presented various prosthetic foot designs that could improve functionality during the last several decades. However, a vast number of individuals are denied access to this technology, since most of the developed prosthetic devices are exorbitantly bulky in design and expensive. Thus, engaging amputees in different activities, such as walking, jogging, dancing, cycling, golfing, swimming, and other sports with a single design, becomes impossible. The current study aims to synthesize existing information on ankle–foot prosthetics based on demographics, engineering designs, and biomechanics underlying the biological foot to identify significant challenges and limitations of existing work to lead the future development of low-cost ankle prostheses. Based on the findings of the previous studies, it can be concluded that future research should focus on a variety of factors, including but not limited to low cost, domestic accessibility, robustness, moisture and corrosion resistance, susceptibility to easy fabrication, lightweight, suitable aesthetic, and psychosocial acceptability.

Keywords: Prosthesis, foot biomechanics, ankle–foot prosthesis, below-knee amputee

**Corresponding author*: dkpra@mech.iitkgp.ac.in

Jay Prakash Srivastava, Dražan Kozak, Vinayak Ranjan, Pankaj Kumar, Ranjan Kumar and Shubham Tayal (eds.) Mechanical Engineering in Biomedical Applications: Bio-3D Printing, Biofluid Mechanics, Implant Design, Biomaterials, Computational Biomechanics, Tissue Mechanics, (397–412) © 2024 Scrivener Publishing LLC

16.1 Introduction

The Rig Veda, an Indian religious book, documented the development of the first ankle prosthetics or lower limb prostheses between 3,500 and 1,800 BC [1]. It describes how the queen Vishpla was amputated in the battlefield and how an iron leg was attached to her to return to the battle. Currently, 38 million individuals globally have had a significant amputation, with roughly 85% of them having had their lower limbs amputated [2]. Amputation is caused through surgical procedures for various causes, including cancer, blood vessel abnormalities, injuries, trauma, diabetes, and peripheral vascular disease [3]. In this context, prosthetics are artificial equipment that substitutes lost biological feet and, therefore, are meticulously engineered to imitate the biomechanical anatomy of the biological feet.

Consequently, in these amputation cases, the primary objective is to substitute the lost limb with a prosthetic foot, which enables amputees to regain mobility and drive to lead a new healthy life and participate in all routine activities. The development of prosthetic component uses the ability of storing energy in elastic components and then utilizing it during the next phase. The deformed elastic components assist in storing the energy during loading, and this energy is used in the late stance which in turn pushes the ankle to propel the body forward. The ankle prosthesis should be built with plantar flexion, dorsiflexion, neutral posture, and toe hyperextension in mind during the early, mid, and late stances, respectively [4].

The performance limit is heavily influenced by the capabilities provided by different components of traditional ankle–foot prostheses. The contribution of these components to improved performance is often validated throughout the design phase. There are several methods for developing prosthetic foot designs; nevertheless, the iterative design approach has gained prominence in recent decades. Its design was created by analyzing prosthetic foot users. Even though the design is a very significant component, it got relatively little attention; therefore, it covered a very small percentage of work in this sector. Modern prostheses are unable to demonstrate comparable measurable performance benefits over past designs using conventional procedures [5–7]. However, device design impacts performance, since systematic differences in device behavior have shown substantial effects on results [8–10].

The next-generation design for ankle–foot prosthetics has improved walking performance [11–13]. These recent findings to improve performance depend on alternative designs that vary from conventional designs in a number of ways but are unable to monitor the critical design features that have been found to have a strong impact on the ability to develop more effective prostheses. Furthermore, given individual variances in physiology and experience, it is unknown if all users will benefit from these innovative technologies. These innovative technologies also raise the cost by 10 to 100 times when compared with existing equipment, affecting user data collection since it is uncertain which categories would benefit the most. Based only on intuition and observation, determining what prosthetic behavior would increase certain performance results for a given user is deceptively difficult. Simulation-based design techniques, which are used in other fields, are not currently relevant for designing these systems since they have not proven the capacity to reliably foresee human behavior in response to prospective device designs for average or individual participants. In light of these constraints, it is required to establish new approaches for the optimum design and selection of prosthesis behavior, which are driven by rigorous testing and explicitly characterize correlations between device behavior and walking performance. Furthermore, these processes are critical for improving our existing knowledge about the science behind changes in human responses to specific design characteristics, reducing in design in preparation, accelerating revolutionary prosthetic design, and ultimately providing better results to people who have had lower limb amputations.

It has been discovered that below-knee amputations in the Indian context are caused by railway accidents, road traffic accidents, vascular reasons, diabetes gangrene/foot, combat operation, malignancy, crush injury from machinery, and other variables [14], as seen in Figure 16.1.

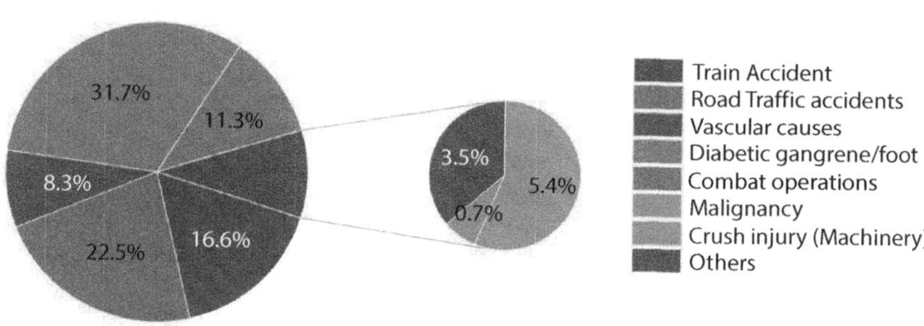

Figure 16.1 Leading causes of below-knee amputation in the Indian context.

16.2 Biomechanics of Biological Ankle and Foot

The biomechanics of the human body has been studied using three planes, namely, the sagittal plane, the coronal plane, and the transverse plane, as depicted in Figure 16.2 [15]. The rotational axis of the foot joint is located in different planes.

To develop a powered ankle prosthesis, it is critical to understand the properties of a healthy person's ankle joint. According to the inverted pendulum model [16], the human body's one leg controls the landing, while the other leg propels the body slightly higher, providing an ideal parabolic trajectory. A correct synchronization between two legs occurs when walking to create the parabolic trajectory, and the patients fail to maintain the normal parabolic trajectory owing to lower limb impairment. As a result, the function of a lower limb prosthetic or orthotic device is to bridge this gap and aid in recovery. The ankle joint is important in maintaining the parabolic trajectory when walking.

Walking has five primary objectives that must be accomplished. The primary goal of walking for a healthy individual is to move in a sagittal direction at a set speed and direction. The second purpose of walking is to spend as little energy as possible to accomplish the first goal. The body does this by pushing forward in an attempt to maintain a straight line. Walking requires the least amount of energy since the body moves very little up and down. The ultimate goal for persons with painful foot ailments is to cause as little discomfort as possible. To do this, our brains use a variety of tactics, such

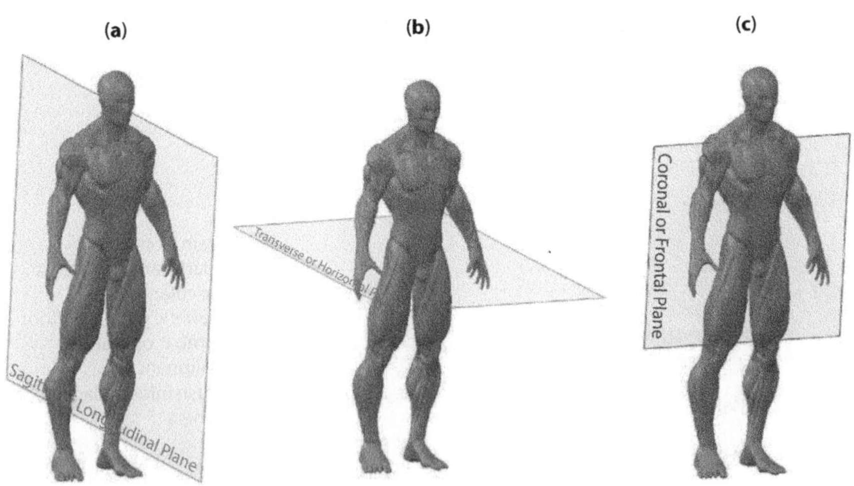

Figure 16.2 Various planes of motion such as (a) sagittal, (b) transverse, and (c) coronal [15].

as alleviating pressure on a hurting foot or modifying the foot position while walking to reduce discomfort. The fourth objective focuses on foot adjustment over uneven terrain in order to lessen the impact created by the body. These little adjustments act as a shock absorber. The fifth objective is to impart forward momentum to the body, which is accomplished by the support foot acting as a lever at the end of the gait cycle. As a result, understanding how prostheses work requires understanding foot motions such as dorsiflexion, plantar flexion, eversion–inversion, and internal–external axial rotation, as depicted in Figure 16.3.

Human walking consists of two main phases, viz., the stance phase and the swing phase. The stance phase, in which both legs are on the ground, accounts for around 60% of the walking duration. Running clearly shortens the stance phase, which becomes the float phase when both legs are in the air. During the swing phase, only one leg is on the ground; the leg in the air is referred to as the swing leg. The ankle biomechanics while walking is split into many stages, such as heel strike (HS) to foot flat (FF) or controlled plantar flexion, FF to heel off (HO) or controlled dorsiflexion, HO to toe off (TO) or powered plantar flexion, and the swing phase. The ankle characteristic was given greater attention throughout the controlled dorsiflexion and powered plantar flexion phases since this research aims to lower the motor's power needs. The ankle torque steadily rises throughout the controlled dorsiflexion phase to support the body, while the contralateral leg stays in the air. The stance leg supports the body weight. At the

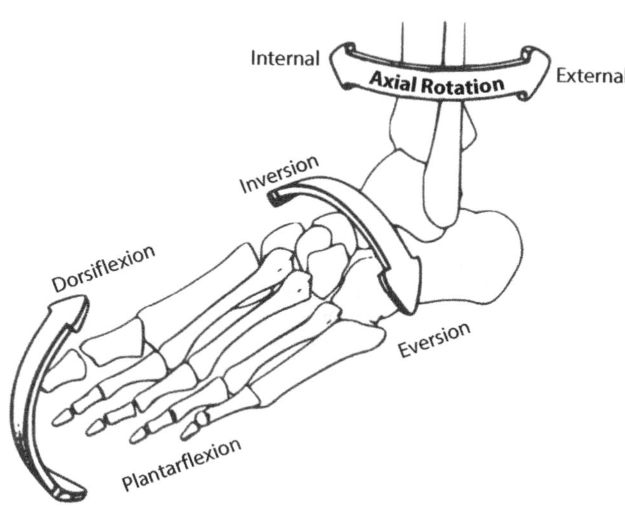

Figure 16.3 Various movements of the foot [17].

conclusion of the controlled dorsiflexion phase and the start of the powered plantar flexion phase, the maximum ankle torque is needed. During the powered plantar flexion phase, the ankle joint pushes the body center of mass against gravity to follow a parabolic trajectory. The ankle joint returns to its original position during the swing phase. The preceding discussion is visualized in Figure 16.4 [18], which depicts the gait cycle of a healthy biological ankle with numerous stages.

Additionally, while at the stance phase of walking, it is important to take into account the five subphases that one foot passes through. These stages are heel strike, early flatfoot, late flatfoot, heel rise, and toe off and are shown in Figure 16.5.

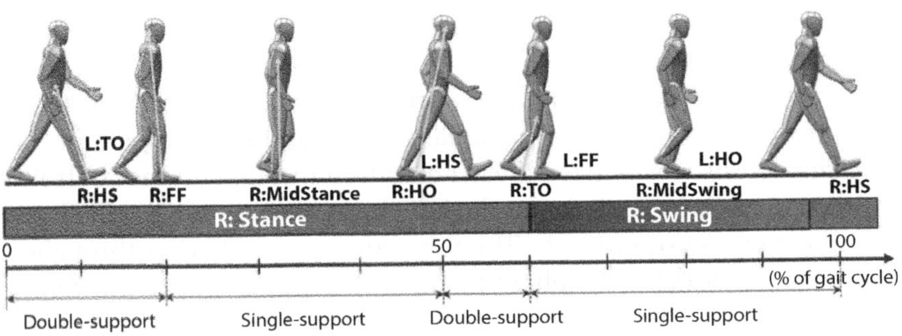

Figure 16.4 Gait cycle of the biological right foot with different phases [18].

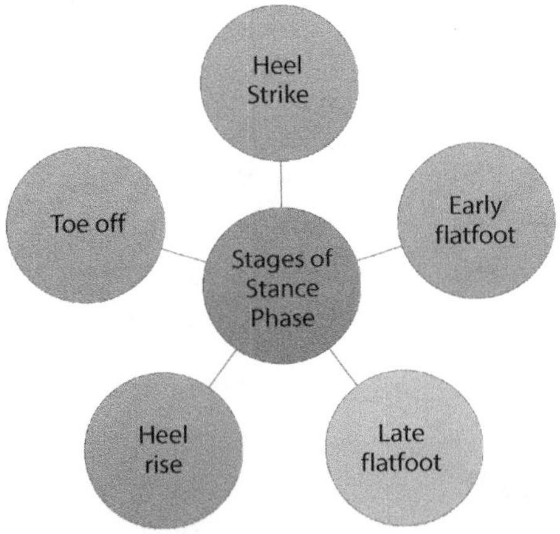

Figure 16.5 Various stages of the stance phase.

The heel-striking event begins when the heel initially strikes the surface and continues till the entire leg is on the surface. The "early flatfoot" stage starts when the whole heel makes contact with the surface. Early flatfooting ceases whenever the body's center of gravity (CG) passes beneath the top of the leg. Whenever a person stands or walks, the CG of the person is around the pelvic region in the anterior of the spinal column; therefore, the central objective of the "early flatfoot" phase is to endow the feet with the potential to operate as a damper, aiding to minimize the impact of the weight falling on the feet. When the body's CG has migrated ahead of the balanced posture, a person is considered to be in the late flatfoot condition. The "late flatfoot" phase of gait comes to an end when the heel clears the ground. During the "late flatfoot" phase of gait, the foot must transition from a flexible shock absorber to a stiff lever capable of propelling the body forward. As the name implies, the heel rise phase occurs when the heel commences to lift off the surface. At this stage, the feet function as a veritable tool to propel the body forward. At this stage of walking, the foot is exposed to substantial strains, which are frequently two to three times a person's body weight. This is because the foot may form a lever arm with its focus on the ankle, increasing the forces generated by body weight. As mentioned above, high forces occur, and taking into account the fact that a normal person actually takes 3,000–5,000 steps per day, the normal biological foot can easily develop chronic stress-like conditions like metatarsalgia, bunions, posterior tibial tendon dysfunction, peroneal tendonitis, and sesamoiditis. Therefore, the toe off phase starts as soon as the toes rise off the surface, which causes the biological foot's swing phase to begin.

16.3 Prosthetic Foot

This section summarizes the literature review of previous work on designing and developing ankle prosthetic devices. Transtibial amputees often employ ankle prosthetics to restore the amputated leg portion. The principles of several passive and powered ankle prosthesis designs, as well as their benefits and drawbacks, are covered in this section.

16.3.1 Design of Passive Prosthetic Ankle–Foot

Passive prostheses do not need an external power source to actuate the leg joints, making them look lightweight and cost-effective. Popular passive ankle prosthesis designs include the solid ankle, cushioned heel (SACH) foot [6], flex-foot [19], and controlled energy storage and release (CESR) [20].

The SACH foot, which stands for solid ankle and cushioned heel, is an example of a conventional passive prosthetic design. The heel is cushioned to absorb the strong impulse created by the heel hit. Aside from this, no additional mechanism was engaged that may help with walking. Flex-foot is made out of a carbon composite leaf spring that stores and releases energy when walking. CESR is a more sophisticated kind of passive ankle prosthesis in which energy, i.e., collision energy, is stored during heel strike and released during push-off [21]. The energy was held via a gripping mechanism, and the precise moment of release was controlled by a computer and sensor assembly. As a result, the push-off force was raised while metabolic energy consumption was lowered when compared with traditional passive ankle prosthetics. An energy storage and release mechanism, on the other hand, might release the same or less energy accumulated in the early stance phase. As a result, CESR could not conduct net positive work, limiting its use to slow-speed walking and limiting its usage to terrains requiring net positive effort, such as stair and ramp ascent.

16.3.2 Design of Powered Ankle–Foot Prosthetics

The drawbacks of the passive ankle prosthesis might be solved by a powered one. It generates net positive work by inserting an external power source that might deliver the amount of power that the muscles of the missing limb are meant to create. Aside from that, fine control was possible with the motorized ankle prosthesis, which helped to restore the amputee's regular walking pattern to some degree. The powered ankle prosthetics were designed using either a pneumatic [22] or an electric actuator as an external power source. Electric actuators were performed to pneumatic actuators owing to the latter's colossal installation space requirements, high maintenance cost, control complexity, and so on. Ossur's Proprio Foot™, the first commercial active ankle prosthesis, uses an electric motor to alter the ankle angle during the swing phase [23].

Alimusaj *et al.* [24] validated the aforementioned concept during stair ambulation using Proprio Foot. Increased dorsiflexion was anticipated to improve the biomechanics of the ankle joint during the stance phase of walking [24]. Although it performed somewhat better than the passive-elastic foot, it was unable to fully leverage the advantages of utilizing an electric motor to enhance the ankle joint's behavior during the stance phase.

A totally powered ankle prosthesis would need a lot of power to regulate all of the intermediate stages of a gait cycle. As a result, compliance mechanisms that integrated active and passive processes were investigated for the creation of powered ankle prosthetics. Among the existing compliance mechanisms, series elastic actuators [25, 26], robotic tendon actuators [26], and mechanically adjustable compliance and controllable equilibrium position actuators (MACCEPA) [27, 28] have the potential to acquire a lot of traction. There are two types of compliance actuators: screw transmission-based compliance actuators and link-based compliance actuators. The mechanism that actively adjusts the stiffness of the elastic element is used to divide the elastic element [29, 30]. The principles of powered ankle prosthesis designs that made use of compliance mechanisms include screw transmission-based series elastic actuators and link-based actuators.

16.3.3 Classification of the Ankle Prosthesis

Based on the foregoing research, it has been discovered that ankle prostheses are commonly categorized as ESAR, CERS, and active variants according to their constituents and functioning as shown in Figure 16.6.

Furthermore, as illustrated in Figure 16.7 [17], ESAR prosthesis may be conceptualized into five distinct kinds of designs. Figure 16.7(a) represents the overall configuration of an ESAR prosthesis, Figure 16.7(b) portrays a customized ESAR prosthesis, Figure 16.7(c) exhibits an ESAR with split plates, Figure 16.7(d) describes an ESAR with cushions, and Figure 16.7(e) demonstrates an ESAR with a damping system. According to the previous assessments, the general structure of an ESAR prosthesis is similar to the one shown in Figure 16.7(a) and generally varies in form; occasionally, a single talon plate is aggregated, or the arrangement of the plates

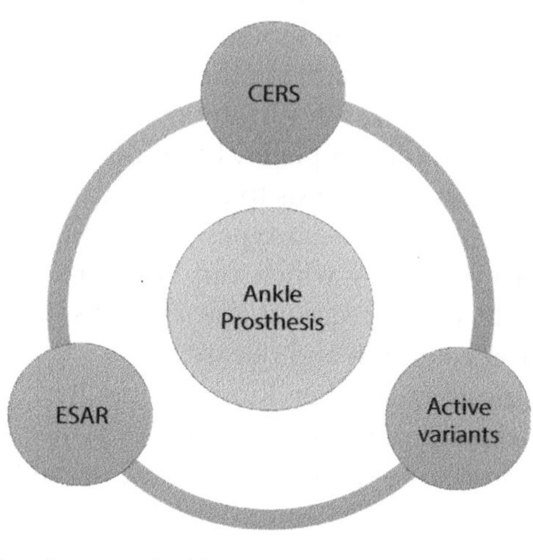

Figure 16.6 Broad classification of ankle prosthetics.

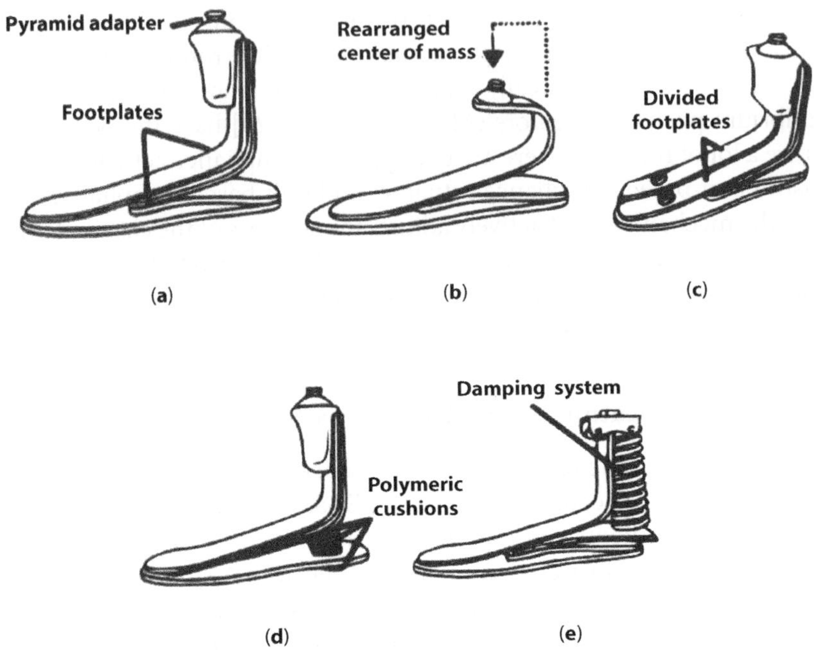

Figure 16.7 ESAR prosthesis. (a) Generalized framework of ESAR. (b) Customized variant of ESAR. (c) Split plate-based ESAR. (d) Cushion-based ESAR. (e) Damper-based ESAR [17].

may change. In other circumstances, as seen in Figure 16.7(b), the center of mass is shifted and the plates are reorganized. Figure 16.7(c) depicts a variant in which the foot plates are separated, allowing the prosthesis to simulate eversion and inversion motions. Some polymeric cushions are aggregated in Figure 16.7(d), removing the necessity for additional plates. Figure 16.7(e) depicts the use of various dampening systems (springs, actuators, etc.) to replace certain plates. Pyramid adapters are used in all of these designs to link the prosthesis to the transtibial components [31].

There are various ESAR prosthesis types that employ a basic plate design to control the return of energy which can be referred to Figure 16.8(a). Figure 16.8(b) depicts another variant that employs just a singular spring arm to regulate power retention and release.

The CESR prosthesis is another kind of prosthesis that is based on the concept of a lockable framework that stores power in the spring and then releases such stored power during the leg mobility throughout the

(a) (b)

Figure 16.8 ESAR prosthesis. (a) Multiple plates of ESAR. (b) Single spring prosthesis design of ESAR [17].

terminating stride. This precise instinctive reaction, in conjunction with the ESAR foot, generates the necessary power throughout the walking phase. Endo Ken devised the conventional CERS prosthesis shown in Figure 16.9 [17].

According to the usefulness of the different components, active prostheses are often divided into three kinds, namely, multi-array prostheses (MAP), low-powered prostheses (LPP), and controlled adaptive stiffness (CAS). The ESAR composite foot (E), a DC motor (A), and a Maxon motor connected to a ball-screw transmission (C) are the components of a typical multi-array prosthesis. These components are responsible for moving the linkage system (D), which translates the motoric rotating motion into linear motion. According to functionality, the timing belt is used to link the motor to (C) and is used in place of the spring (G). According to the distinct types of prostheses, (F) and (G) illustrate springs or actuators like pneumatic, electric, or hydraulic ones. Finally, the transtibial component

Figure 16.9 CERS prosthesis [17].

and prosthesis are connected by the pyramidal structure (H) seen in Figure 16.10 [17].

The LPP model, which is demonstrated in Figure 16.11 [17], aims to reduce the amount of power required by the actuators. The distinct foot-plates (G and C) are included, as shown in Figure 16.11, even though in some designs they are integrated into a single plate. Figure 16.12 [17] shows the CAS prosthesis that is premised on the ESAR foot (D), with the

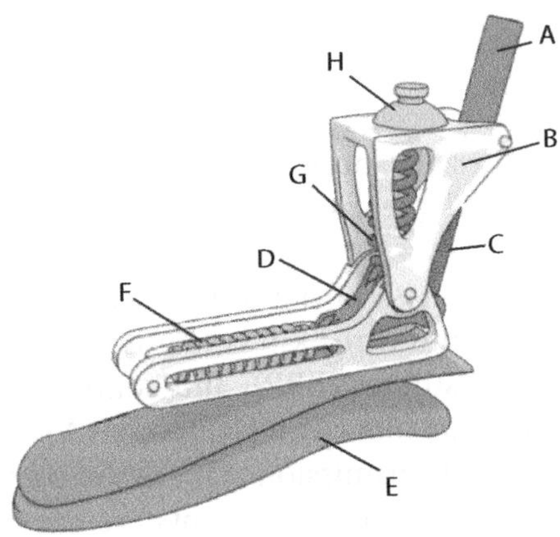

Figure 16.10 Multi-array active prostheses [17].

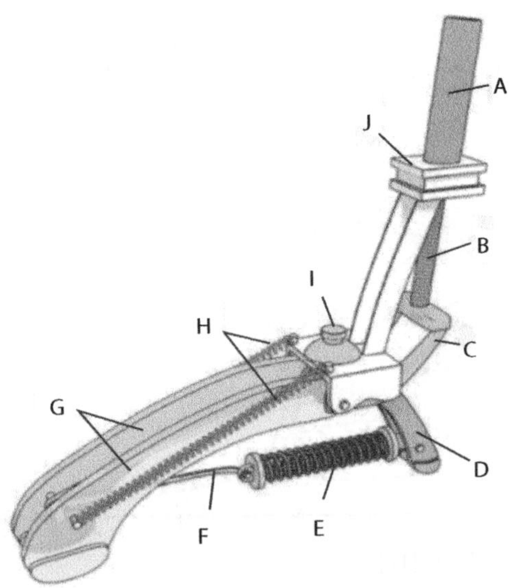

Figure 16.11 LPP active prosthesis [17].

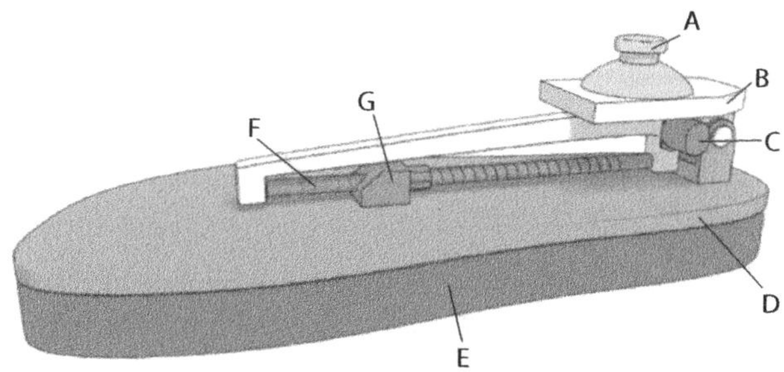

Figure 16.12 CAS prosthesis [17].

addition of a cushion in certain cases (E). This prosthetics' sole function is to modulate stiffness at different parts of the gait cycle.

16.4 Conclusion

Several investigations in this domain have already been conducted, and it can be contended that this area has already piqued the interest of researchers. However, in order to mimic the human ankle–foot, researchers must have a strong understanding of foot biomechanics and anatomy. When designing a prosthetic foot, it is critical to have a detailed understanding of power output production at the lower limb throughout the stance phase of gait, so that it can execute realistic biological gait, minimize motion asymmetries, and decrease the power needed for ambulation. Previously, wood prosthetics were used, but now ESER variants of prosthetics are very popular due to significant pushing off. As a result of ESER's increased pushing off effectiveness, which mimics the biological foot more accurately, less energy is ultimately required. Mobility is also improved, the body is more balanced throughout the gait cycle, and high mechanical energy is supplied. Furthermore, it has the advantage of minimal stress on the limb, which minimizes the risk of the leading limb colliding with the surface. Various studies have shown that amputees are more prone to develop knee osteoarthritis in the limb because all prosthetic feet do not have very excellent foot push off, resulting in an increase in stress on the healthy limb. As a result, designers should work on improving ankle push off to totally remove this issue. Based on previous research findings, it is possible to conclude that future research should focus on a variety of factors, including but not limited to low cost, domestic accessibility, robustness, moisture and corrosion resistance, susceptibility to easy fabrication, lightweight, suitable aesthetic, and psychosocial acceptability.

References

1. Grimmer, M. and Seyfarth, A., Mimicking human-like leg function in prosthetic limbs, in: *Neuro-Robotics: From Brain Machine Interfaces to Rehabilitation Robotics*, pp. 105–155, 2014.
2. Kishner, S., Laborde, J.M., Early, J.S., Gait analysis after amputation: Overview, gait cycle, adaptive strategies of those who have undergone amputations. *Medscape*, 2022. Available: https://emedicine.medscape.com/article/1237638-overview [Online; Accessed 20-Sept-2022]
3. Biddiss, E.A. and Chau, T.T., Upper limb prosthesis use and abandonment. *Prosthet. Orthot. Int.*, 31, 3, 236–257, 2007.
4. Versluys, R., Beyl, P., Van Damme, M., Desomer, A., Van Ham, R., Lefeber, D., Prosthetic feet: State-of-the-art review and the importance of mimicking human anklefoot biomechanics. *Disabil. Rehabil. Assist. Technol.*, 4, 2, 65–75, 2009.
5. Barr, A.E. *et al.*, Biomechanical comparison of the energy-storing capabilities of SACH and carbon copy II prosthetic feet during the stance phase of gait in a person with below-knee amputation. *Phys. Ther.*, 72, 5, 344–354, 1992.
6. Casillas, J.M., Dulieu, V., Cohen, M., Marcer, I., Didier, J.P., Bioenergetic comparison of a new energy-storing foot and SACH foot in traumatic below-knee vascular amputations. *Arch. Phys. Med. Rehabil.*, 76, 1, 39–44, 1995.
7. Lehmann, J.F., Price, R., Boswell-Bessette, S., Dralle, A., Questad, K., DeLateur, B.J., Comprehensive analysis of energy storing prosthetic feet: Flex foot and seattle foot versus standard SACH foot. *Arch. Phys. Med. Rehabil.*, 74, 11, 1225–1231, 1993.
8. Adamczyk, P.G., Roland, M., Hahn, M.E., Sensitivity of biomechanical outcomes to independent variations of hindfoot and forefoot stiffness in foot prostheses. *Hum. Mov. Sci.*, 54, 154–171, 2017.
9. Klodd, E., Hansen, A., Fatone, S., Edwards, M., Effects of prosthetic foot forefoot flexibility on oxygen cost and subjective preference rankings of unilateral transtibial prosthesis users. *J. Rehabil. Res. Dev.*, 47, 6, 543–552, 2010.
10. Major, M.J., Twiste, M., Kenney, L.P.J., Howard, D., The effects of prosthetic ankle stiffness on ankle and knee kinematics, prosthetic limb loading, and net metabolic cost of trans-tibial amputee gait. *Clin. Biomech.*, 29, 1, 98–104, 2014.
11. Herr, H.M. and Grabowski, A.M., Bionic ankle-foot prosthesis normalizes walking gait for persons with leg amputation. *Proc. R. Soc. B Biol. Sci.*, 279, 1728, 457–464, 2012.

12. Zelik, K.E. *et al.*, Systematic variation of prosthetic foot spring affects center-of-mass mechanics and metabolic cost during walking. *IEEE Trans. Neural Syst. Rehabil. Eng.*, 19, 4, 411–419, 2011.
13. Grabowski, A.M., Rifkin, J., Kram, R., K3 promoter™ prosthetic foot reduces the metabolic cost of walking for unilateral transtibial amputees. *J. Prosthet. Orthot.*, 22, 2, 113–120, 2010.
14. Shankar, P., Grewal, V.S., Agrawal, S., Nair, S.V., A study on quality of life among lower limb amputees at a tertiary prosthetic rehabilitation center. *Med. J. Armed Forces India*, 76, 1, 89–94, 2020.
15. Chauhan, P., Singh, A.K., Raghuwanshi, N.K., The state of art review on prosthetic feet and its significance to imitate the biomechanics of human ankle-foot. *Mater. Today Proc.*, 62, P12, 6364–6370, 2022.
16. Kuo, A.D., Donelan, J.M., Ruina, A., Energetic consequences of walking like an inverted pendulum: Step-to-step transitions. *Exerc. Sport Sci. Rev.*, 33, 2, 88–97, 2005.
17. Zagoya-López, J., Zúñiga-Avilés, L.A., Vilchis-González, A.H., Ávila-Vilchis, J.C., Foot/ankle prostheses design approach based on scientometric and patentometric analyses. *Appl. Sci.*, 11, 12, 5591, 2021.
18. Alamdari, A. and Krovi, V.N., A review of computational musculoskeletal analysis of human lower extremities, in: *Human Modeling for Bio-Inspired Robotics: Mechanical Engineering in Assistive Technologies*, pp. 37–73, 2017.
19. Leonard, J.A., Esquenazi, A., Fisher, S.V., Hicks, J.E., Meier, R.H., Nelson, V.S., Prosthetics, orthotics, and assistive devices. 1. General concepts. *Arch. Phys. Med. Rehabil.*, 70, 5–S, S195–201, 1989.
20. Segal, A.D. *et al.*, The effects of a controlled energy storage and return prototype prosthetic foot on transtibial amputee ambulation. *Hum. Mov. Sci.*, 31, 4, 918–931, 2012.
21. Collins, S.H. and Kuo, A.D., Recycling energy to restore impaired ankle function during human walking. *PloS One*, 5, 2, 1–6, 2010.
22. Sup, F., Bohara, A., Goldfarb, M., Design and control of a powered transfemoral prosthesis. *Int. J. Robot. Res.*, 27, 2, 263–273, 2008.
23. W. Koniuk, Self-adjusting prosthetic ankle apparatus. US Pat. App. 09/816,902, 2001.
24. Alimusaj, M., Fradet, L., Braatz, F., Gerner, H.J., Wolf, S., II, Kinematics and kinetics with an adaptive ankle foot system during stair ambulation of transtibial amputees. *Gait Posture*, 30, 3, 356–363, 2009.
25. Dong, D., Ge, W., Convens, B., Sun, Y., Verstraten, T., Vanderborght, B., Design, optimization and energetic evaluation of an efficient fully powered ankle-foot prosthesis with a series elastic actuator. *IEEE Access*, 8, 61491–61503, 2020.
26. Hollander, K.W., Ilg, R., Sugar, T.G., Herring, D., An efficient robotic tendon for gait assistance. *J. Biomech. Eng.*, 128, 5, 788–791, 2006.
27. Kumar, V., Babbar, A., Sharma, A., Kumar, R., Tyagi, A., Polymer 3D bioprinting for bionics and tissue engineering applications, in: *Additive*

Manufacturing of Polymers for Tissue Engineering, pp. 17–39, CRC Press, Boca Raton, 2022.

28. Van Ham, R., Vanderborght, B., Van Damme, M., Verrelst, B., Lefeber, D., MACCEPA, the mechanically adjustable compliance and controllable equilibrium position actuator: Design and implementation in a biped robot. *Rob. Auton. Syst.*, 55, 10, 761–768, 2007.
29. Kumar, V., Prakash, C., Babbar, A., Choudhary, S., Sharma, A., Uppal, A.S., Additive manufacturing in biomedical engineering, in: *Additive Manufacturing Processes in Biomedical Engineering*, pp. 143–164, 2022.
30. Babbar, A., Tian, Y., Kumar, V., Sharma, A., 3D bioprinting in biomedical applications, in: *Additive Manufacturing of Polymers for Tissue Engineering*, pp. 1–16, CRC Press, Boca Raton, 2022.
31. Babbar, A., Jain, V., Gupta, D., Sharma, A., Prakash, C., Kumar, V., Goyal, K.K., Additive manufacturing for the development of biological implants, scaffolds, and prosthetics, in: *Additive Manufacturing Processes in Biomedical Engineering*, pp. 27–46, 2022.

Index

Printed and bound by CPI Group (UK) Ltd, Croydon, CR0 4YY
19/09/2026
14965102-0002